全栈开发技术丛书

Spring Boot+Vue. js
全栈开发从入门到实战 IntelliJ IDEA版·微课视频版

陈恒 主编

关菁华 张立杰 孙国元 李敏 副主编

清华大学出版社
北京

内 容 简 介

本书全面介绍 Spring Boot 3 + Vue.js 3 全栈开发的基本思想、方法和技术。全书共 13 章，内容涵盖 Spring 基础、Spring MVC 基础、Spring Boot 核心知识、Web 开发、数据访问、MyBatis 与 MyBatis-Plus 框架基础、安全控制、异步消息、单元测试、Vue.js 基础与进阶、电子商务平台的设计与实现（Spring Boot + Vue.js 3 + MyBatis-Plus）等。书中实例侧重实用性、通俗易懂，通过本书的学习，读者能够快速掌握 Spring Boot 3 + Vue.js 3 全栈开发的基础知识、编程技巧以及完整的开发体系，为大型项目开发打下坚实的基础。

本书可作为大学计算机及相关专业的教材或教学参考书，也可作为 Java 技术的培训教材，适合具有 Java 和 Java Web 编程基础的读者，尤其适合广大 Java EE 应用开发人员阅读与使用。

本书封面贴有清华大学出版社防伪标签，无标签者不得销售。
版权所有，侵权必究。举报：010-62782989，**beiqinquan@tup.tsinghua.edu.cn**。

图书在版编目（CIP）数据

Spring Boot + Vue.js 全栈开发从入门到实战：IntelliJ IDEA 版：微课视频版/陈恒主编. —北京：清华大学出版社，2024.7（2024.8重印）
（全栈开发技术丛书）
ISBN 978-7-302-65955-6

Ⅰ. ①S… Ⅱ. ①陈… Ⅲ. ①网页制作工具－JAVA 语言－程序设计 Ⅳ. ①TP312.8 ②TP393.092.2

中国国家版本馆 CIP 数据核字（2024）第 065027 号

策划编辑：魏江江
责任编辑：王冰飞
封面设计：刘　键
责任校对：时翠兰
责任印制：沈　露

出版发行：清华大学出版社
　　　　网　　址：https://www.tup.com.cn，https://www.wqxuetang.com
　　　　地　　址：北京清华大学学研大厦 A 座　　邮　　编：100084
　　　　社 总 机：010-83470000　　邮　　购：010-62786544
　　　　投稿与读者服务：010-62776969，c-service@tup.tsinghua.edu.cn
　　　　质 量 反 馈：010-62772015，zhiliang@tup.tsinghua.edu.cn
　　　　课 件 下 载：https://www.tup.com.cn，010-83470236
印 装 者：三河市铭诚印务有限公司
经　　销：全国新华书店
开　　本：185mm×260mm　　印　张：22　　字　数：581 千字
版　　次：2024 年 7 月第 1 版　　　　印　次：2024 年 8 月第 2 次印刷
印　　数：1501～3000
定　　价：79.80 元

产品编号：101614-01

前言

党的二十大报告指出，教育、科技、人才是全面建设社会主义现代化国家的基础性、战略性支撑。必须坚持科技是第一生产力、人才是第一资源、创新是第一动力，深入实施科教兴国战略、人才强国战略、创新驱动发展战略，开辟发展新领域新赛道，不断塑造发展新动能新优势。

时至今日，脚本语言和敏捷开发大行其道，基于 Spring 框架的 Java EE 开发显得烦琐许多，开发者经常遇到两个非常头疼的问题：（1）大量的配置文件；（2）与第三方框架整合。Spring Boot 的出现颠覆了 Java EE 开发，可以说具有划时代的意义。Spring Boot 的目标是帮助开发者编写更少的代码实现所需功能，遵循"约定优于配置"原则，从而使开发者只需要很少的配置，或者使用默认配置，就可以快速搭建项目。虽然 Spring Boot 给开发者带来了开发效率，但 Spring Boot 并不是什么新技术，完全是一个基于 Spring 的应用。例如 Spring Boot 的最大优点"自动配置"是通过 Spring 的@Conditional 注解实现的，所以读者在学习 Spring Boot 之前最好快速学习 Spring 与 Spring MVC 的基础知识。

本书系统地介绍了 Spring Boot 3 + Vue.js 3 全栈开发的主要技术，主要包括 3 个方面的内容。

（1）快速开发一个 Web 应用系统（Spring 与 Spring MVC 基础、Thymeleaf 与 Vue.js 3 视图技术、MyBatis 与 MyBatis-Plus 数据访问技术）。

（2）Spring Boot 的高级特性（自动配置、部署、单元测试以及安全机制）。

（3）分布式架构技术（REST、MongoDB、Redis、Cache 以及异步消息）。

本书的重点不是简单地介绍基础知识，而是精心设计了大量实例。读者通过本书可以快速地掌握 Spring Boot 3 + Vue.js 3 全栈开发的实践应用，提高 Java EE 应用的开发能力。

全书共 13 章，各章的具体内容如下。

第 1 章：介绍 Spring 的基础知识，包括 Spring 开发环境的构建、Spring IoC、Spring AOP 以及 Spring Bean 等内容。

第 2 章：介绍 Spring MVC 的基础知识，包括 Spring MVC 的工作原理、Spring MVC 的工作环境、基于注解的控制器、JSON 数据交互以及 Spring MVC 的基本配置等内容。

第 3 章：主要介绍如何快速构建第一个 Spring Boot 应用，包括 Spring Boot 的优点、主要特性以及如何使用 IntelliJ IDEA 快速构建第一个 Spring Boot 应用。

第 4 章：介绍 Spring Boot 的核心，包括核心注解、基本配置、自动配置原理、条件注解以及自定义 Starters 等内容。

第 5 章：介绍 Spring Boot 的 Web 开发相关技术，包括 Spring Boot 的 Web 开发支持、Thymeleaf 模板引擎技术、JSON 数据交互、文件上传与下载、异常统一处理以及对 JSP 的支持。

第 6 章：主要讲解 Spring Boot 访问数据库的解决方案，包括 Spring Data JPA、Spring Boot 整合 REST、Spring Boot 整合 MongoDB、Spring Boot 整合 Redis、数据缓存技术等内容。

第 7 章：重点介绍 MyBatis 与 MyBatis-Plus 的基础知识，并详细介绍 Spring Boot 如何整合 MyBatis 与 MyBatis-Plus。

第 8 章：介绍 Spring Security，包括 Spring Security 快速入门、基于 Spring Data JPA 的 Spring Security 操作实例等内容。

第 9 章：介绍企业级系统间的异步消息通信，包括消息模型、JMS 与 AMQP 企业级消息代理、Spring Boot 对异步消息的支持以及异步消息通信实例等内容。

第 10 章：主要介绍 Spring Boot 单元测试的相关内容，包括 JUnit 5 的注解、断言以及单元测试用例。

第 11 章：主要介绍 Vue.js 的基础知识，包括 Vue.js 的安装、Vue.js 的生命周期、插值与表达式、计算属性与监听器属性、内置指令与自定义指令、组件、响应性以及 setup 组件选项等内容。

第 12 章：主要介绍 Vue.js 的进阶知识，包括单文件组件与 webpack、Vue CLI、路由 Vue Router、setup 语法糖、Element Plus UI 组件库等内容。

第 13 章：以电子商务平台的设计与实现为综合案例，讲述如何使用 Spring Boot + Vue.js 3 + MyBatis-Plus 开发一个前、后端分离的应用程序。

附录 A 给出 11 个与本书对应的实验，可以提高学生的实际动手能力。

为了方便各类高等院校选用教材和读者自学，本书配有教学大纲、教学课件、思政教案、程序源码、教学进度表、实验大纲、在线题库、习题答案和 1300 分钟的微课视频等配套资源。

资源下载提示

课件等资源：扫描封底的"课件下载"二维码，在公众号"书圈"下载。

素材（源码）等资源：扫描目录上方的二维码下载。

在线自测题：扫描封底的作业系统二维码，再扫描自测题二维码在线做题。

微课视频：扫描封底的文泉云盘防盗码，再扫描书中相应章节的视频讲解二维码，可以在线学习。

本书的出版得到清华大学出版社相关人员的大力支持，在此表示衷心的感谢。同时，本书编者参阅了相关书籍、博客以及其他官网资源，在此对这些资源的贡献者与分享者深表感谢。由于 Spring Boot+Vue.js 全栈开发框架技术发展迅速，并持续改进与优化，加上编者水平有限，书中难免会有不足之处，敬请各位专家和读者批评指正。

本书是辽宁省一流本科课程"工程项目实训"以及辽宁省普通高等学校一流本科教育示范专业"大连外国语大学计算机科学与技术专业"的建设成果。

编 者
2024 年 7 月

源码下载

目录

第1章　Spring 基础

1.1　Spring 概述 ··· 1
  1.1.1　Spring 的由来 ····························· 1
  1.1.2　Spring 的体系结构 ························ 2
1.2　Spring 开发环境的构建 ······················· 4
  1.2.1　配置 IntelliJ IDEA 的 Web 服务器 ···· 4
  1.2.2　Spring 的下载 ····························· 6
  1.2.3　第一个 Spring 入门程序 ················ 6
1.3　Spring IoC ·· 10
  1.3.1　Spring IoC 的基本概念 ················ 10
  1.3.2　Spring 的常用注解 ······················ 10
  1.3.3　基于注解的依赖注入 ···················· 11
1.4　Spring AOP ····································· 14
  1.4.1　Spring AOP 的基本概念 ·············· 14
  1.4.2　基于注解开发 AspectJ ················· 16
1.5　Spring Bean ····································· 20
  1.5.1　Bean 的实例化 ··························· 20
  1.5.2　Bean 的作用域 ··························· 22
  1.5.3　Bean 的初始化和销毁 ·················· 24
本章小结 ··· 25
习题 1 ·· 26

第2章　Spring MVC 基础

2.1　Spring MVC 的工作原理 ···················· 27
2.2　Spring MVC 的工作环境 ···················· 28
  2.2.1　Spring MVC 所需要的 JAR 包 ······· 28
  2.2.2　使用 IntelliJ IDEA 开发 Spring MVC 应用 ······· 29
  2.2.3　基于 Java 配置的 Spring MVC 应用 ······· 35

2.3 基于注解的控制器 37
　　2.3.1 Controller 注解类型 37
　　2.3.2 RequestMapping 注解类型 37
　　2.3.3 编写请求处理方法 39
　　2.3.4 Controller 接收请求参数的常见方式 40
　　2.3.5 重定向与转发 44
　　2.3.6 应用@Autowired 进行依赖注入 45
　　2.3.7 @ModelAttribute 47
2.4 JSON 数据交互 48
　　2.4.1 JSON 数据结构 48
　　2.4.2 JSON 数据转换 49
2.5 Spring MVC 的基本配置 52
　　2.5.1 静态资源配置 52
　　2.5.2 拦截器配置 53
　　2.5.3 文件上传配置 54
本章小结 59
习题 2 59

第 3 章　Spring Boot 入门

3.1 Spring Boot 概述 60
　　3.1.1 什么是 Spring Boot 60
　　3.1.2 Spring Boot 的优点 60
　　3.1.3 Spring Boot 的主要特性 61
3.2 第一个 Spring Boot 应用 61
　　3.2.1 Maven 简介 61
　　3.2.2 Maven 的 pom.xml 61
　　3.2.3 使用 IntelliJ IDEA 快速构建 Spring Boot 应用 62
本章小结 65
习题 3 65

第 4 章　Spring Boot 核心

4.1 Spring Boot 的基本配置 66
　　4.1.1 启动类和核心注解@SpringBootApplication 66
　　4.1.2 Spring Boot 的全局配置文件 67
　　4.1.3 Spring Boot 的 Starters 67
4.2 读取应用配置 68

        4.2.1 使用 Environment 类 68
        4.2.2 使用@Value 注解 69
        4.2.3 使用@ConfigurationProperties 注解 70
        4.2.4 使用@PropertySource 注解 71
    4.3 日志配置 72
    4.4 Spring Boot 的自动配置原理 73
    4.5 Spring Boot 的条件注解 75
        4.5.1 条件注解 75
        4.5.2 自定义条件 76
        4.5.3 自定义 Starters 78
    本章小结 84
    习题 4 84

第 5 章 Spring Boot 的 Web 开发

   5.1 Spring Boot 的 Web 开发支持 85
   5.2 Thymeleaf 模板引擎 85
       5.2.1 Spring Boot 的 Thymeleaf 支持 86
       5.2.2 Thymeleaf 的基础语法 87
       5.2.3 Thymeleaf 的常用属性 90
       5.2.4 通过 Spring Boot 与 Thymeleaf 实现页面信息的国际化 93
       5.2.5 Spring Boot 与 Thymeleaf 的表单验证 97
       5.2.6 基于 Thymeleaf 和 BootStrap 的 Web 开发实例 101
   5.3 使用 Spring Boot 处理 JSON 数据 104
   5.4 Spring Boot 的文件上传与下载 108
   5.5 Spring Boot 的异常统一处理 112
       5.5.1 自定义 error 页面 113
       5.5.2 @ExceptionHandler 注解 116
       5.5.3 @ControllerAdvice 注解 117
   5.6 Spring Boot 对 JSP 的支持 118
   本章小结 120
   习题 5 120

第 6 章 Spring Boot 的数据访问

   6.1 Spring Data JPA 122
       6.1.1 Spring Boot 的支持 123

	6.1.2	简单条件查询	124
	6.1.3	关联查询	131
	6.1.4	@Query 和@Modifying 注解	148
	6.1.5	排序与分页查询	148
6.2	REST		152
	6.2.1	REST 简介	152
	6.2.2	Spring Boot 整合 REST	154
	6.2.3	Spring Data REST	155
	6.2.4	REST 服务测试	156
6.3	MongoDB		161
	6.3.1	安装 MongoDB	161
	6.3.2	Spring Boot 整合 MongoDB	161
	6.3.3	增、删、改、查	162
6.4	Redis		165
	6.4.1	安装 Redis	166
	6.4.2	Spring Boot 整合 Redis	169
	6.4.3	使用 StringRedisTemplate 和 RedisTemplate	170
6.5	数据缓存 Cache		174
	6.5.1	Spring 缓存支持	174
	6.5.2	Spring Boot 缓存支持	176
	6.5.3	使用 Redis Cache	180
本章小结			181
习题 6			181

第 7 章 MyBatis 与 MyBatis-Plus

7.1	Spring Boot 整合 MyBatis		182
7.2	MyBatis 基础		185
	7.2.1	\<select>元素	185
	7.2.2	\<insert>、\<update>以及\<delete>元素	189
	7.2.3	动态 SQL	189
7.3	MyBatis-Plus 快速入门		191
	7.3.1	MyBatis-Plus 简介	191
	7.3.2	Spring Boot 整合 MyBatis-Plus	192
7.4	MyBatis-Plus 基础		194
	7.4.1	MyBatis-Plus 注解	194
	7.4.2	CRUD 接口	197
	7.4.3	条件构造器	204

本章小结 ·· 208
习题 7 ··· 208

第 8 章 Spring Boot 的安全控制

8.1 Spring Security 快速入门 ··· 209
    8.1.1 什么是 Spring Security ·· 209
    8.1.2 Spring Security 的用户认证 ·· 209
    8.1.3 Spring Security 的请求授权 ·· 211
    8.1.4 Spring Security 的核心类 ··· 211
    8.1.5 Spring Security 的验证机制 ·· 213
8.2 Spring Boot 的支持 ·· 213
8.3 实际开发中的 Spring Security 操作实例 ··································· 213
本章小结 ·· 223
习题 8 ··· 223

第 9 章 异步消息

9.1 消息模型 ··· 224
    9.1.1 点对点式 ·· 224
    9.1.2 发布/订阅式 ·· 224
9.2 企业级消息代理 ··· 225
    9.2.1 JMS ·· 225
    9.2.2 AMQP ··· 226
9.3 Spring Boot 的支持 ·· 228
    9.3.1 JMS 的自动配置 ··· 228
    9.3.2 AMQP 的自动配置 ··· 228
9.4 异步消息通信实例 ·· 228
    9.4.1 JMS 实例 ·· 228
    9.4.2 AMQP 实例 ··· 233
本章小结 ·· 238
习题 9 ··· 238

第 10 章 Spring Boot 单元测试

10.1 JUnit 5 ·· 239
    10.1.1 JUnit 5 简介 ·· 239

10.1.2 JUnit 5 注解 ··· 240
10.1.3 JUnit 5 断言 ··· 241
10.2 单元测试用例 ··· 242
10.2.1 测试环境的构建 ··· 242
10.2.2 测试 Mapper 接口 ··· 244
10.2.3 测试 Service 层 ··· 245
10.3 使用 Postman 测试 Controller 层 ··· 246
本章小结 ··· 249
习题 10 ··· 249

第 11 章 Vue.js 基础

11.1 网站交互方式 ··· 250
    11.1.1 多页应用程序 ··· 250
    11.1.2 单页应用程序 ··· 251
11.2 MVVM 模式 ··· 252
11.3 Vue.js 是什么 ··· 253
11.4 安装 Vue.js ··· 253
11.5 第一个 Vue.js 程序 ··· 253
    11.5.1 安装 Visual Studio Code 及其插件 ··· 254
    11.5.2 创建第一个 Vue.js 应用 ··· 255
    11.5.3 声明式渲染 ··· 256
    11.5.4 Vue.js 的生命周期 ··· 257
11.6 插值与表达式 ··· 259
    11.6.1 文本插值 ··· 259
    11.6.2 原始 HTML 插值 ··· 259
    11.6.3 JavaScript 表达式 ··· 259
11.7 计算属性和监听器属性 ··· 259
    11.7.1 计算属性 computed ··· 260
    11.7.2 监听器属性 watch ··· 263
11.8 内置指令 ··· 264
    11.8.1 v-bind 指令 ··· 265
    11.8.2 条件渲染指令 v-if 和 v-show ··· 266
    11.8.3 列表渲染指令 v-for ··· 267
    11.8.4 事件处理 ··· 268
    11.8.5 表单与 v-model ··· 270
    11.8.6 实战：购物车实例 ··· 272

11.9 组件 ·············· 274
11.9.1 组件的注册 ·············· 275
11.9.2 使用 props 传递数据 ·············· 277
11.9.3 组件的通信 ·············· 278
11.9.4 动态组件与异步组件 ·············· 283
11.9.5 实战：正整数数字输入框组件 ·············· 285

11.10 自定义指令 ·············· 286
11.10.1 自定义指令的注册 ·············· 286
11.10.2 实战：实时时间转换指令 ·············· 288

11.11 响应性 ·············· 290
11.11.1 什么是响应性 ·············· 290
11.11.2 响应性的原理 ·············· 290

11.12 setup 组件选项 ·············· 291
11.12.1 setup 函数的参数 ·············· 291
11.12.2 setup 函数的返回值 ·············· 293
11.12.3 使用 ref 创建响应式引用 ·············· 294
11.12.4 在 setup 内部调用生命周期钩子函数 ·············· 295

本章小结 ·············· 295

习题 11 ·············· 295

第 12 章 Vue.js 进阶

12.1 单文件组件与 webpack ·············· 298

12.2 安装 Node.js 和 NPM ·············· 299
12.2.1 安装 Node.js ·············· 299
12.2.2 NPM 常用的命令 ·············· 300

12.3 Vue Router ·············· 302
12.3.1 Vue Router 的安装 ·············· 302
12.3.2 Vue Router 的基本用法 ·············· 306
12.3.3 Vue Router 的高级应用 ·············· 311
12.3.4 路由钩子函数 ·············· 321
12.3.5 路由元信息 ·············· 324
12.3.6 登录权限验证示例 ·············· 325

12.4 setup 语法糖 ·············· 328
12.4.1 属性与方法的绑定 ·············· 328
12.4.2 路由 ·············· 330
12.4.3 组件传值 ·············· 331

12.5 Element Plus UI 组件库 333
- 12.5.1 Element Plus 的安装 334
- 12.5.2 Element Plus 组件的介绍 334
- 12.5.3 Element Plus 组件的应用 335
- 12.5.4 按需引入 Element Plus 336

本章小结 338
习题 12 338

第 13 章 电子商务平台的设计与实现（Spring Boot + Vue.js 3 + MyBatis-Plus）

.... 339
- 13.1 系统设计 339
- 13.2 数据库设计 339
- 13.3 使用 IntelliJ IDEA 构建后端系统 339
- 13.4 使用 Vue CLI 构建前端系统 339
- 13.5 测试运行 339
- 13.6 本章小结 339

习题 13 339

附录 A　实验指导

.... 340

第 1 章　Spring 基础

学习目的与要求

本章重点讲解 Spring 的基础知识。通过本章的学习，要求读者了解 Spring 的体系结构，理解 Spring IoC 与 AOP 的基本原理，了解 Spring Bean 的生命周期、实例化以及作用域。

本章主要内容

- Spring 开发环境的构建
- Spring IoC
- Spring AOP
- Spring Bean

Spring 是当前主流的 Java 开发框架，为企业级应用开发提供了丰富的功能，掌握 Spring 框架的使用已经是 Java 开发者必须具备的技能之一。本章将学习如何使用 IntelliJ IDEA 开发 Spring 程序，注意在此之前需要构建 Spring 的开发环境。

1.1　Spring 概述

▶ 1.1.1　Spring 的由来

Spring 是一个轻量级 Java 企业级应用程序开发框架，最早由 Rod Johnson 创建，目的是解决企业级应用开发的业务逻辑层和其他各层的耦合问题。它是一个分层的 Java SE/EE full-stack（一站式）轻量级开源框架，为开发 Java 应用程序提供全面的基础架构支持。Spring 负责基础架构，因此 Java 开发者可以专注于应用程序的开发。

Spring Framework 6.0 于 2022 年 11 月正式发布，这是 2023 年及以后新一代框架的开始，包含 OpenJDK 和 Java 生态系统中当前和即将到来的创新。Spring Framework 6.0 作为重大更新，要求使用 Java 17 或更高版本，并且已迁移到 Jakarta EE 9+（在 jakarta 命名空间中取代了以前基于 javax 的 API），以及对其他基础设施的修改。基于这些变化，Spring Framework 6.0 支持最新的 Web 容器，例如 Tomcat 10，以及最新的持久性框架 Hibernate ORM 6.1。这些特性仅可用于 Servlet API 和 JPA 的 jakarta 命名空间变体。

在基础架构方面，Spring Framework 6.0 引入了 Ahead-Of-Time（AOT）转换的基础以及对 Spring 应用程序上下文的 AOT 转换和相应的 AOT 处理支持的基础，能够事先将应用程序中或 JDK 中的字节码编译成机器码。在 Spring Framework 6.0 中还有许多新功能和改进可用，例如 HTTP 接口客户端、对 RFC 7807 问题详细信息的支持以及 HTTP 客户端的基于 Micrometer 的可观察性。

1.1.2 Spring 的体系结构

Spring 的功能模块被有组织地分散到约 20 个模块中，这些模块分布在核心容器（Core Container）层、数据访问/集成（Data Access/Integration）层、Web 层以及面向切面编程（Aspect-Oriented Programming，AOP）模块、植入（Instrumentation）模块、消息传输（Messaging）和测试（Test）模块中，如图 1.1 所示。

图 1.1　Spring 的体系结构

❶ Core Container

Spring 的 Core Container 是其他模块建立的基础，由 Beans（spring-beans）、Core（spring-core）、Context（spring-context）和 Expression（spring-expression，Spring 表达式语言）等模块组成。

spring-beans 模块：提供了 BeanFactory，是工厂模式的一个经典实现，Spring 将管理对象称为 Bean。

spring-core 模块：提供了框架的基本组成部分，包括控制反转（Inversion of Control，IoC）和依赖注入（Dependency Injection，DI）功能。

spring-context 模块：建立在 spring-beans 和 spring-core 模块的基础之上，提供一个框架式的对象访问方式，是访问定义和配置的任何对象媒介。

spring-expression 模块：提供了强大的表达式语言去支持运行时查询和操作对象图。这是对 JSP 2.1 规范中规定的统一表达式语言（Unified EL）的扩展。该语言支持设置和获取属性值、属性分配、方法调用、访问数组、集合和索引器的内容、逻辑和算术运算、变量命名以及从 Spring 的 IoC 容器中以名称检索对象。它还支持列表投影、选择以及常见的列表聚合。

❷ AOP 和 Instrumentation

Spring 框架中与 AOP 和 Instrumentation 相关的模块有 AOP（spring-aop）模块、Aspects（spring-aspects）模块以及 Instrumentation（spring-instrument）模块。

spring-aop 模块：提供了一个符合 AOP 要求的面向切面的编程实现，允许定义方法拦截器和切入点，将代码按照功能进行分离，以便干净地解耦。

spring-aspects 模块：提供了与 AspectJ 的集成功能，AspectJ 是一个功能强大且成熟的 AOP 框架。

spring-instrument 模块：提供了类植入（Instrumentation）支持和类加载器的实现，可以在特定的应用服务器中使用。Instrumentation 提供了一种虚拟机级别支持的 AOP 实现方式，使得开发者无须对 JDK 做任何升级和改动就可以实现某些 AOP 的功能。

❸ Messaging

Spring 4.0 以后新增了 Messaging（spring-messaging）模块，该模块提供了对消息传递体系结构和协议的支持。

❹ Data Access/Integration

数据访问/集成层由 JDBC（spring-jdbc）、ORM（spring-orm）、OXM（spring-oxm）、JMS（spring-jms）和 Transactions（spring-tx）模块组成。

spring-jdbc 模块：提供了一个 JDBC 的抽象层，消除了烦琐的 JDBC 编码和数据库厂商特有的错误代码解析。

spring-orm 模块：为流行的对象-关系映射（Object-Relational Mapping）API 提供集成层，包括 JPA 和 Hibernate。使用 spring-orm 模块可以将这些 O/R 映射框架与 Spring 提供的所有其他功能结合使用，例如声明式事务管理功能。

spring-oxm 模块：提供了一个支持对象/XML 映射的抽象层实现，例如 JAXB、Castor、JiBX 和 XStream。

spring-jms 模块（Java Messaging Service）：指 Java 消息传递服务，包含用于生产和使用消息的功能。自 Spring 4.1 以后，提供了与 spring-messaging 模块的集成。

spring-tx 模块（事务模块）：支持用于实现特殊接口和所有 POJO（普通 Java 对象）类的编程和声明式事务管理。

❺ Web

Web 层由 Web（spring-web）、WebMVC（spring-webmvc）、WebSocket（spring-websocket）和 WebFlux（spring-webflux）模块组成。

spring-web 模块：提供了基本的 Web 开发集成功能。例如多文件上传功能、使用 Servlet 监听器初始化一个 IoC 容器以及 Web 应用上下文。

spring-webmvc 模块：也称为 Web-Servlet 模块，包含用于 Web 应用程序的 Spring MVC 和 REST Web Services 实现。Spring MVC 框架提供了领域模型代码和 Web 表单之间的清晰分离，并与 Spring Framework 的所有其他功能集成。本书后续章节将会详细讲解 Spring MVC 框架。

spring-websocket 模块：Spring 4.0 以后新增的模块，它提供了 WebSocket 和 SockJS 的实现，主要是与 Web 前端的全双工通信的协议。

spring-webflux 模块：一个新的非堵塞函数式 Reactive Web 框架，可以用来建立异步的、非阻塞、事件驱动的服务，并且扩展性非常好（该模块是 Spring 5 新增的模块）。

❻ Test

Test（spring-test）模块支持使用 JUnit 或 TestNG 对 Spring 组件进行单元测试和集成测试。

扫一扫

视频讲解

1.2 Spring 开发环境的构建

在使用 Spring 框架开发 Spring 应用之前，应该先搭建其开发环境。

▶ 1.2.1 配置 IntelliJ IDEA 的 Web 服务器

如果用户的计算机上已经安装了 IntelliJ IDEA（本书使用的是 ideaIU-2022.2.1），那么可以使用 IntelliJ IDEA 便捷地构建 Java Web 应用。虽然 IntelliJ IDEA 自带了 OpenJDK，但是建议在使用之前先安装并配置 JDK 和 Web 服务器。其具体步骤如下：

❶ 安装 JDK

安装并配置 JDK（本书采用的 JDK 是 jdk-18_windows-x64_bin.exe），在按照提示安装完成 JDK 以后，需要配置"环境变量"的"系统变量"Java_Home 和 Path。在 Windows 10 系统下，系统变量示例如图 1.2 和图 1.3 所示。

图 1.2 新建系统变量 Java_Home

图 1.3 编辑系统变量的环境变量 Path

❷ Web 服务器

目前，比较常用的 Web 服务器有 Tomcat、JRun、Resin、WebSphere、WebLogic 等，本书采用的是 Tomcat 10.0。

登录 Apache 软件基金会的官方网站（http://jakarta.Apache.org/tomcat），下载 Tomcat 10.0 的免安装版（本书采用 apache-tomcat-10.0.23-windows-x64.zip）。登录网站之后，首先在 Download 中选择 Tomcat 10，然后在 Binary Distributions 的 Core 中选择相应版本即可。

在安装 Tomcat 之前需要先安装 JDK 并配置系统环境变量 Java_Home。将下载的 apache-tomcat-10.0.23-windows-x64.zip 解压到某个目录下，解压缩后将出现如图 1.4 所示的目录结构。

图 1.4 Tomcat 目录结构

通过执行 Tomcat 根目录下 bin 文件夹中的 startup.bat 来启动 Tomcat 服务器。执行 startup.bat 启动 Tomcat 服务器会占用一个 MS-DOS 窗口，如果关闭当前 MS-DOS 窗口将关闭 Tomcat 服务器。

Tomcat 服务器启动后，在浏览器的地址栏中输入"http://localhost:8080"，将出现如图 1.5 所示的 Tomcat 测试页面。

图 1.5 Tomcat 测试页面

❸ 集成 Tomcat

启动 IntelliJ IDEA，选择 File→Settings 命令，在弹出的对话框中选择 Build, Execution, Deployment 下的 Application Servers 选项，然后在右侧单击+按钮，在弹出的菜单中选择 Tomcat Server 命令，打开如图 1.6 所示的 Tomcat Server 界面，在其中选择 Tomcat 目录。

图 1.6　Tomcat 配置界面

在图 1.6 中单击 OK 按钮即可完成 Tomcat 配置。

至此可以使用 IntelliJ IDEA 创建 Java Web 应用，并在 Tomcat 下运行。

▶ 1.2.2　Spring 的下载

在使用 Spring 框架开发应用程序时需要引用 Spring 框架自身的 JAR 包。Spring Framework 6.0.0 的 JAR 包可以从 Maven 中央库获得。

在 Spring 的 JAR 包中有 4 个基础包 spring-core-6.0.0.jar、spring-beans-6.0.0.jar、spring-context-6.0.0.jar 和 spring-expression-6.0.0.jar，分别对应 Spring 核心容器的 4 个模块 spring-core 模块、spring-beans 模块、spring-context 模块和 spring-expression 模块。

对于 Spring 框架的初学者，在开发 Spring 应用时只需要将 Spring 的 4 个基础包和 Spring Commons Logging Bridge 对应的 JAR 包 spring-jcl-6.0.0.jar 复制到 Web 应用的 WEB-INF/lib 目录下即可。

▶ 1.2.3　第一个 Spring 入门程序

本小节通过一个简单的入门程序向读者演示 Spring 框架的使用过程，具体见例 1-1。

【例 1-1】一个简单的入门程序。

其具体实现步骤如下：

❶ 使用 IntelliJ IDEA 创建模块并导入 JAR 包

首先使用 IntelliJ IDEA 创建一个名为 ch1 的项目，如图 1.7 所示。

然后在项目 ch1 中创建一个名为 ch1_1 的模块，如图 1.8 所示。

第 1 章　Spring 基础

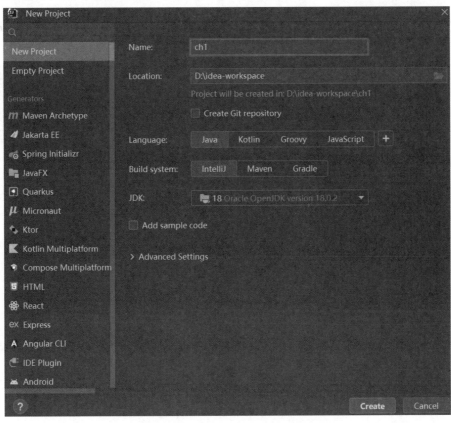

图 1.7　使用 IntelliJ IDEA 创建项目

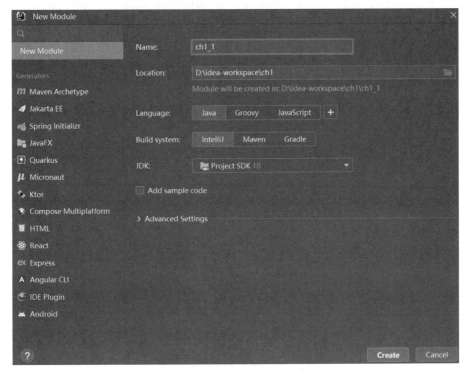

图 1.8　使用 IntelliJ IDEA 创建模块

接着右击模块名 ch1_1，在弹出的快捷菜单中选择 Add Framework Support 命令，给模块 ch1_1 添加 Web Application，如图 1.9 所示。

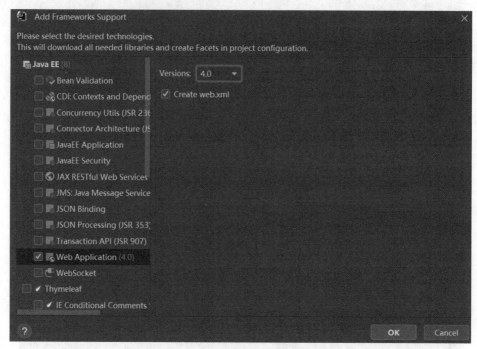

图 1.9　给模块添加 Web Application

最后将 Spring 的 4 个基础包和 Spring Commons Logging Bridge 对应的 JAR 包 spring-jcl-6.0.0.jar 复制到 ch1_1 的 WEB-INF/lib 目录中。具体做法是在 WEB-INF 目录下创建 lib 目录，将 JAR 包复制到 lib 中，然后右击选择 Add as Library 命令，添加为模块依赖，如图 1.10 所示。

图 1.10　给模块添加依赖

> **注意**：在讲解 Spring MVC 框架之前，本书的实例并没有真正运行 Java Web 应用。创建 Java Web 应用是为了方便添加相关 JAR 包。

❷ 创建接口 TestDao

Spring 解决的是业务逻辑层和其他各层的耦合问题，因此它将面向接口的编程思想贯穿于整个系统应用。

在模块 ch1_1 的 src 目录下创建一个名为 dao 的包，并在 dao 包中创建接口 TestDao，在该接口中定义一个 sayHello()方法，代码如下：

```
package dao;
public interface TestDao {
    public void sayHello();
}
```

❸ 创建接口 TestDao 的实现类 TestDaoImpl

在 dao 包下创建接口 TestDao 的实现类 TestDaoImpl，代码如下：

```
package dao;
public class TestDaoImpl implements TestDao{
    @Override
    public void sayHello() {
        System.out.println("Hello, Study hard!");
    }
}
```

❹ 创建配置文件 applicationContext.xml

在模块 ch1_1 的 src 目录下创建 Spring 的配置文件 applicationContext.xml，并在该文件中使用实现类 TestDaoImpl 创建一个 id 为 test 的 Bean，代码如下：

```xml
<?xml version="1.0" encoding="UTF-8"?>
<beans xmlns="http://www.springframework.org/schema/beans"
    xmlns:xsi="http://www.w3.org/2001/XMLSchema-instance"
    xsi:schemaLocation="http://www.springframework.org/schema/beans
        http://www.springframework.org/schema/beans/spring-beans.xsd">
    <!--将指定类 TestDaoImpl 配置给 Spring，让 Spring 创建其实例-->
    <bean id="test" class="dao.TestDaoImpl" />
</beans>
```

注意：配置文件的名称可以自定义，但习惯上命名为 applicationContext.xml，有时候也命名为 beans.xml。有关 Bean 的创建将在本书后续章节详细讲解，在这里读者只需了解即可。另外，配置文件信息不需要读者手写，可以从 Spring 的帮助文档中复制（首先使用浏览器打开 "https://docs.spring.io/spring-framework/docs/current/reference/html/core.html#spring-core"，在 1.2.1Configuration Metadata 小节下即可找到配置文件的约束信息）。

❺ 创建测试类

在模块 ch1_1 的 src 目录下创建一个名为 test 的包，并在 test 包中创建 Test 类，代码如下：

```java
package test;
import org.springframework.context.ApplicationContext;
import org.springframework.context.support.ClassPathXmlApplicationContext;
import dao.TestDao;
public class Test {
    private static ApplicationContext appCon;
    public static void main(String[] args) {
        appCon = new ClassPathXmlApplicationContext("applicationContext.xml");
        //从容器中获取 test 实例
        TestDao tt = appCon.getBean("test", TestDao.class);
        //test 为配置文件中的 id
        tt.sayHello();
    }
}
```

执行上述 main 方法以后，将在控制台中输出"Hello, Study hard!"。在上述 main 方法中并没有使用 new 运算符创建 TestDaoImpl 类的对象，而是通过 Spring IoC 容器来获取实现类的对象，这就是 Spring IoC 的工作机制。

1.3　Spring IoC

1.3.1　Spring IoC 的基本概念

控制反转（Inversion of Control，IoC）是一个比较抽象的概念，是 Spring 框架的核心，用来消减计算机程序的耦合问题。依赖注入（Dependency Injection，DI）是 IoC 的另外一种说法，只是从不同的角度描述相同的概念。下面通过实际生活中的一个例子解释 IoC 和 DI。

当人们需要一件东西时，第一反应就是找东西，比如想吃面包，在没有面包店时，最直接的做法可能是按照自己的口味制作面包，也就是需要主动制作面包。然而，现在各种网店、实体店盛行，已经没有必要自己制作面包，想吃面包了，去网店或实体店购买即可。注意这里自己并没有制作面包，而是由店家制作，但是完全符合自己的口味。

上面只是列举了一个非常简单的例子，但包含了控制反转的思想，即把制作面包的主动权交给店家。下面通过面向对象编程思想继续探讨这两个概念。

当某个 Java 对象（调用者，比如自己）需要调用另一个 Java 对象（被调用者，即被依赖对象，比如面包）时，在传统编程模式下，调用者通常会采用"new 被调用者"的代码方式来创建对象（比如自己制作面包）。这种方式会增加调用者与被调用者之间的耦合性，不利于后期代码的升级与维护。

当 Spring 框架出现以后，对象的实例不再由调用者来创建，而是由 Spring 容器（比如面包店）来创建。Spring 容器会负责控制程序之间的关系（比如面包店负责控制顾客与面包的关系），而不是由调用者的程序代码直接控制，这样控制权由调用者转移到 Spring 容器，控制权发生了反转，这就是 Spring 的控制反转。

从 Spring 容器角度来看，Spring 容器负责将被依赖对象赋值给调用者的成员变量，相当于为调用者注入它所依赖的实例，这就是 Spring 的依赖注入，主要目的是解耦，体现一种"组合"的理念。

综上所述，控制反转是一种通过描述（在 Spring 中可以是 XML 或注解）并通过第三方去产生或获取特定对象的方式。在 Spring 中实现控制反转的是 IoC 容器，其实现方法是依赖注入。

1.3.2　Spring 的常用注解

在 Spring 框架中，尽管使用 XML 配置文件可以很简单地装配 Bean，但是如果应用中有大量的 Bean 需要装配，会导致 XML 配置文件过于庞大，不方便以后的升级与维护，因此更多的时候推荐开发者使用注解（annotation）的方式去装配 Bean。

需要注意的是，基于注解的装配需要使用<context:component-scan>元素或@ComponentScan 注解定义包（注解所在的包）扫描的规则，然后根据定义的规则找出哪些类（Bean）需要自动装配到 Spring 容器中，再交由 Spring 进行统一管理。

Spring 框架基于 AOP 编程（面向切面编程）实现注解解析，因此在使用注解编程时需要导入 spring-aop-6.0.0.jar 包。

在 Spring 框架中定义了一系列的注解，常用注解如下。

❶ 声明 Bean 的注解

1）@Component

该注解是一个泛化的概念，仅表示一个组件对象（Bean），可以作用在任何层次上，没有明确的角色。

2）@Repository

该注解用于将数据访问层（DAO）的类标识为 Bean，即注解数据访问层 Bean，其功能与@Component 相同。

3）@Service

该注解用于标注一个业务逻辑组件类（Service 层），其功能与@Component 相同。

4）@Controller

该注解用于标注一个控制器组件类（Spring MVC 的 Controller），其功能与@Component 相同。

❷ 注入 Bean 的注解

1）@Autowired

该注解可以对类成员变量、方法及构造方法进行标注，完成自动装配的工作。通过使用@Autowired 来消除 setter 和 getter 方法。默认按照 Bean 的类型进行装配。

2）@Resource

该注解与@Autowired 的功能一样。它们的区别在于，该注解默认是按照名称来装配注入的，只有当找不到与名称匹配的 Bean 时才会按照类型来装配注入；而@Autowired 默认是按照 Bean 的类型进行装配，如果想按照名称来装配注入，则需要结合@Qualifier 注解一起使用。

@Resource 注解有 name 和 type 两个属性。name 属性指定 Bean 实例名称，即按照名称来装配注入；type 属性指定 Bean 类型，即按照 Bean 的类型进行装配。

3）@Qualifier

该注解与@Autowired 注解配合使用。当@Autowired 注解需要按照名称来装配注入时，则需要结合该注解一起使用，Bean 的实例名称由@Qualifier 注解的参数指定。

在上面几个注解中，虽然@Repository、@Service 和@Controller 等注解的功能与@Component 相同，但为了使标注类的用途更加清晰（层次化），在实际开发中推荐使用@Repository 标注数据访问层（DAO 层）、使用@Service 标注业务逻辑层（Service 层）以及使用@Controller 标注控制器层（控制层）。

▶ 1.3.3 基于注解的依赖注入

Spring IoC 容器（ApplicationContext）负责创建和注入 Bean。Spring 提供了使用 XML 配置、注解、Java 配置以及 Groovy 配置实现 Bean 的创建和注入。本书尽量使用注解（@Component、@Repository、@Service 以及@Controller 等业务 Bean 的配置）和 Java 配置（全局配置，例如数据库、MVC 等相关配置）完全代替 XML 配置，这也是 Spring Boot 推荐的配置方式。

下面通过一个简单实例演示基于注解的依赖注入的使用过程。

【例 1-2】基于注解的依赖注入的使用过程。

其具体实现步骤如下：

❶ 使用 IntelliJ IDEA 创建模块 ch1_2 并导入 JAR 包

参考 1.2.3 节，在项目 ch1 中创建一个名为 ch1_2 的模块，同时给模块 ch1_2 添加 Web Application，并将 Spring 的 4 个基础包、Spring Commons Logging Bridge 对应的 JAR 包 spring-jcl-6.0.0.jar、jakarta.annotation.Resource 注解类对应的 JAR 包 annotations-api.jar 以及 spring-aop-6.0.0.jar（本节扫描注解，需要事先导入 Spring AOP 的 JAR 包）复制到 ch1_2 的 WEB-INF/lib 目录中，添加为模块依赖。

❷ 创建 DAO 层

在模块 ch1_2 的 src 中创建 annotation.dao 包，在该包下创建 TestDao 接口和 TestDaoImpl 实现类，并将实现类 TestDaoImpl 使用@Repository 注解标注为数据访问层。

TestDao 的代码如下：

```java
package annotation.dao;
public interface TestDao {
    public void save();
}
```

TestDaoImpl 的代码如下：

```java
package annotation.dao;
import org.springframework.stereotype.Repository;
@Repository("testDaoImpl")
/**相当于@Repository，但如果在 Service 层使用@Resource(name="testDaoImpl")，
   testDaoImpl 不能省略**/
public class TestDaoImpl implements TestDao{
    @Override
    public void save() {
        System.out.println("testDao save");
    }
}
```

❸ 创建 Service 层

在模块 ch1_2 的 src 中创建 annotation.service 包，在该包下创建 TestService 接口和 TestServiceImpl 实现类，并将实现类 TestServiceImpl 使用@Service 注解标注为业务逻辑层。

TestService 的代码如下：

```java
package annotation.service;
public interface TestService {
    public void save();
}
```

TestServiceImpl 的代码如下：

```java
package annotation.service;
import jakarta.annotation.Resource;
import org.springframework.stereotype.Service;
import annotation.dao.TestDao;
@Service("testServiceImpl")          //相当于@Service
public class TestServiceImpl implements TestService{
    @Resource(name="testDaoImpl")
    /**相当于@Autowired，@Autowired 默认按照 Bean 类型注入**/
    private TestDao testDao;
    @Override
    public void save() {
        testDao.save();
        System.out.println("testService save");
    }
}
```

❹ 创建 Controller 层

在模块 ch1_2 的 src 中创建 annotation.controller 包，在该包下创建 TestController 类，并将 TestController 类使用@Controller 注解标注为控制器层。

TestController 的代码如下：

```java
package annotation.controller;
import org.springframework.beans.factory.annotation.Autowired;
```

```
import org.springframework.stereotype.Controller;
import annotation.service.TestService;
@Controller
public class TestController {
    @Autowired
    private TestService testService;
    public void save() {
        testService.save();
        System.out.println("testController save");
    }
}
```

❺ 创建配置类

本书尽量不使用 Spring 的 XML 配置文件，而使用注解和 Java 配置，因此这里需要使用@Configuration 创建一个 Java 配置类（相当于一个 Spring 的 XML 配置文件），并通过@ComponentScan 扫描使用注解的包（相当于在 Spring 的 XML 配置文件中使用<context:component-scan base-package="Bean 所在的包路径"/>语句）。

在模块 ch1_2 的 annotation 包下创建名为 ConfigAnnotation 的配置类。

ConfigAnnotation 的代码如下：

```
package annotation;
import org.springframework.context.annotation.ComponentScan;
import org.springframework.context.annotation.Configuration;
@Configuration          //声明当前类是一个配置类，相当于一个Spring的XML配置文件
@ComponentScan("annotation")
//自动扫描 annotation 包下使用的注解，并注册为 Bean
/*相当于在 Spring 的 XML 配置文件中使用<context:component-scan base-package="Bean
所在的包路径"/> 语句*/
public class ConfigAnnotation {
}
```

❻ 创建测试类

在模块 ch1_2 的 annotation 包下创建测试类 TestAnnotation，具体代码如下：

```
package annotation;
import org.springframework.context.annotation.AnnotationConfigApplicationContext;
import annotation.controller.TestController;
public class TestAnnotation {
    public static void main(String[] args) {
        //初始化 Spring 容器 ApplicationContext
        AnnotationConfigApplicationContext appCon =
            new AnnotationConfigApplicationContext(ConfigAnnotation.class);
        TestController tc = appCon.getBean(TestController.class);
        tc.save();
        appCon.close();
    }
}
```

❼ 运行结果

运行测试类 TestAnnotation 的 main 方法，运行结果如图 1.11 所示。

图 1.11 ch1_2 模块的运行结果

1.4　Spring AOP

Spring AOP 是 Spring 框架体系结构中非常重要的功能模块之一，该模块提供了面向切面编程的实现。面向切面编程在事务处理、日志记录、安全控制等操作中被广泛使用。

▶ 1.4.1　Spring AOP 的基本概念

❶ AOP 的概念

AOP（Aspect-Oriented Programming）即面向切面编程。它与 OOP（Object-Oriented Programming，面向对象编程）相辅相成，提供了与 OOP 不同的抽象软件结构的视角。在 OOP 中，以类作为程序的基本单元，而 AOP 中的基本单元是 Aspect（切面）。Struts 2 的拦截器的设计基于的就是 AOP 的思想，是一个比较经典的应用。

在业务处理代码中通常都有日志记录、性能统计、安全控制、事务处理、异常处理等操作，尽管使用 OOP 可以通过封装或继承的方式实现代码的重用，但仍然有同样的代码分散到各个方法中，因此采用 OOP 处理日志记录等操作不仅增加了开发者的工作量，而且提高了升级和维护的困难。为了解决此类问题，AOP 思想应运而生。AOP 采取横向抽取机制，即将分散在各个方法中的重复代码提取出来，然后在程序编译或运行阶段再将这些抽取出来的代码应用到需要执行的地方。这种横向抽取机制采用传统的 OOP 是无法实现的，因为 OOP 实现的是父子关系的纵向重用。AOP 不是 OOP 的替代品，而是 OOP 的补充，它们相辅相成。

在 AOP 中，横向抽取机制的类与切面的关系如图 1.12 所示。

图 1.12　AOP 中类与切面的关系

从图 1.12 可以看出，通过切面 Aspect 分别在业务类 1 和业务类 2 中加入了日志记录、性能统计、安全控制、事务处理、异常处理等操作。

❷ AOP 的术语

在 Spring AOP 框架中涉及以下常用术语。

1）切面

切面（Aspect）是指封装横切到系统功能（例如事务处理）的类。

2）连接点

连接点（Joinpoint）是指程序运行中的一些时间点，例如方法的调用或异常的抛出。

3）切入点

切入点（Pointcut）是指需要处理的连接点。在 Spring AOP 中，所有的方法执行都是连接点，而切入点是一个描述信息，它修饰的是连接点，通过切入点确定哪些连接点需要被处理。切面、连接点和切入点的关系如图 1.13 所示。

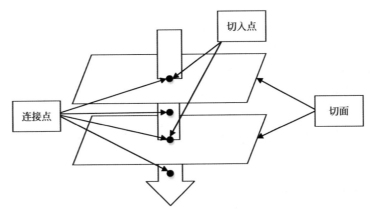

图 1.13　切面、连接点和切入点的关系

4）通知（增强处理）

通知是由切面添加到特定连接点（满足切入点规则）的一段代码，即在定义好的切入点处所要执行的程序代码，可以将其理解为切面开启后切面的方法，因此通知是切面的具体实现。

5）引入

引入（Introduction）允许在现有的实现类中添加自定义的方法和属性。

6）目标对象

目标对象（Target Object）是指所有被通知的对象。如果 AOP 框架使用运行时代理的方式（动态的 AOP）来实现切面，那么通知对象总是一个代理对象。

7）代理

代理（Proxy）是通知应用到目标对象之后被动态创建的对象。

8）织入

织入（Weaving）是将切面代码插入目标对象上，从而生成代理对象的过程。根据不同的实现技术，AOP 织入有 3 种方式：编译器织入，需要有特殊的 Java 编译器；类装载器织入，需要有特殊的类装载器；动态代理织入，在运行期为目标类添加通知生成子类的方式。Spring AOP 框架默认采用动态代理织入，而 AspectJ（基于 Java 语言的 AOP 框架）采用编译器织入和类装载器织入。

1.4.2 基于注解开发 AspectJ

基于注解开发 AspectJ 要比基于 XML 配置开发 AspectJ 便捷许多，所以在实际开发中推荐使用注解方式。在讲解 AspectJ 之前先了解一下 Spring 的通知类型。根据在目标类方法中的连接点位置，Spring 通知可以分为以下 6 种类型。

❶ 环绕通知

环绕通知是在目标方法执行前和执行后实施增强，可以应用于日志记录、事务处理等功能。

❷ 前置通知

前置通知是在目标方法执行前实施增强，可以应用于权限管理等功能。

❸ 后置返回通知

后置返回通知是在目标方法成功执行后实施增强，可以应用于关闭流、删除临时文件等功能。

❹ 后置（最终）通知

后置通知是在目标方法执行后实施增强，与后置返回通知不同的是，该通知不管是否发生异常都要执行，可以应用于释放资源。

❺ 异常通知

异常通知是在方法抛出异常后实施增强，可以应用于异常处理、日志记录等功能。

❻ 引入通知

引入通知是在目标类中添加一些新的方法和属性，可以应用于修改目标类（增强类）。

有关 AspectJ 注解的知识如表 1.1 所示。

表 1.1 AspectJ 注解

注解名称	描 述
@Aspect	用于定义一个切面，注解在切面类上
@Pointcut	用于定义切入点表达式。在使用时需要定义一个切入点方法。该方法是一个返回值为 void，且方法体为空的普通方法
@Before	用于定义前置通知。在使用时通常为其指定 value 属性值，该值可以是已有的切入点，也可以直接定义切入点表达式
@AfterReturning	用于定义后置返回通知。在使用时通常为其指定 value 属性值，该值可以是已有的切入点，也可以直接定义切入点表达式
@Around	用于定义环绕通知。在使用时通常为其指定 value 属性值，该值可以是已有的切入点，也可以直接定义切入点表达式
@AfterThrowing	用于定义异常通知。在使用时通常为其指定 value 属性值，该值可以是已有的切入点，也可以直接定义切入点表达式。另外它还有一个 throwing 属性用于访问目标方法抛出的异常，该属性值与异常通知方法中同名的形参一致
@After	用于定义后置（最终）通知。在使用时通常为其指定 value 属性值，该值可以是已有的切入点，也可以直接定义切入点表达式

下面通过一个实例讲解基于注解开发 AspectJ 的过程。

【例 1-3】基于注解开发 AspectJ 的过程。

其具体实现步骤如下：

❶ 使用 IntelliJ IDEA 创建模块 ch1_3 并导入 JAR 包

在项目 ch1 中创建一个名为 ch1_3 的模块，同时给模块 ch1_3 添加 Web Application，并将 Spring 的 4 个基础包、Spring Commons Logging Bridge 对应的 JAR 包 spring-jcl-6.0.0.jar、AOP 实现 JAR 包 spring-aop-6.0.0.jar、AspectJ 实现 JAR 包 spring-aspects-6.0.0.jar 以及 AspectJ 框架所提供的规范包 aspectjweaver-1.9.9.1.jar（https://mvnrepository.com/artifact/org.aspectj/aspectjweaver/1.9.9.1）复制到 ch1_3 的 WEB-INF/lib 目录中，添加为模块依赖。

❷ 创建接口及实现类

在模块 ch1_3 的 src 目录下创建一个名为 aspectj.dao 的包，并在该包中创建接口 TestDao 和接口实现类 TestDaoImpl。该实现类作为目标类，在切面类中对其方法进行增强处理。使用 @Repository 注解将目标类 aspectj.dao.TestDaoImpl 注解为目标对象。

TestDao 的代码如下：

```java
package aspectj.dao;
public interface TestDao {
    public void save();
    public void modify();
    public void delete();
}
```

TestDaoImpl 的代码如下：

```java
package aspectj.dao;
import org.springframework.stereotype.Repository;
@Repository("testDao")
public class TestDaoImpl implements TestDao{
    @Override
    public void save() {
        System.out.println("保存");
    }
    @Override
    public void modify() {
        System.out.println("修改");
    }
    @Override
    public void delete() {
        System.out.println("删除");
    }
}
```

❸ 创建切面类

在模块 ch1_3 的 src 目录下创建 aspectj.annotation 包，并在该包中创建切面类 MyAspect。在该类中首先使用@Aspect 注解定义一个切面类，由于该类在 Spring 中是作为组件使用的，所以还需要使用@Component 注解。然后使用@Pointcut 注解切入点表达式，并通过定义方法来表示切入点名称。最后在每个通知方法上添加相应的注解，并将切入点名称作为参数传递给需要执行增强的通知方法。

MyAspect 的代码如下：

```java
package aspectj.annotation;
import org.aspectj.lang.JoinPoint;
import org.aspectj.lang.ProceedingJoinPoint;
import org.aspectj.lang.annotation.After;
import org.aspectj.lang.annotation.AfterReturning;
import org.aspectj.lang.annotation.AfterThrowing;
import org.aspectj.lang.annotation.Around;
import org.aspectj.lang.annotation.Aspect;
import org.aspectj.lang.annotation.Before;
import org.aspectj.lang.annotation.Pointcut;
```

```java
import org.springframework.stereotype.Component;
/**
 * 切面类，在此类中编写各种类型的通知
 */
@Aspect           //@Aspect 声明一个切面
@Component    //@Component 让此切面成为 Spring 容器管理的 Bean
public class MyAspect {
    /**
     * 定义切入点，通知增强哪些方法。
"execution(* aspectj.dao.*.*(..))"是定义切入点表达式，
该切入点表达式的意思是匹配 aspectj.dao 包中任意类的任意方法的执行。
其中execution()是表达式的主体，第一个*表示的是返回类型，使用*代表所有类型；
aspectj.dao 表示的是需要匹配的包名，后面第二个*表示的是类名，使用*代表匹配包中所有的类；
第三个*表示的是方法名，使用*表示所有方法；后面(..)表示方法的参数，其中".."表示任意参数。
另外，注意第一个*与包名之间有一个空格
     */
    @Pointcut("execution(* aspectj.dao.*.*(..))")
    private void myPointCut() {
    }
    /**
     * 前置通知，使用 JoinPoint 接口作为参数获得目标对象的信息
     */
    @Before("myPointCut()")           //myPointCut()是切入点的定义方法
    public void before(JoinPoint jp) {
        System.out.print("前置通知：模拟权限控制");
        System.out.println("，目标类对象：" + jp.getTarget()
            + ",被增强处理的方法：" + jp.getSignature().getName());
    }
    /**
     * 后置返回通知
     */
    @AfterReturning("myPointCut()")
    public void afterReturning(JoinPoint jp) {
        System.out.print("后置返回通知：" + "模拟删除临时文件");
        System.out.println(",被增强处理的方法："+jp.getSignature().getName());
    }
    /**
     * 环绕通知
     * ProceedingJoinPoint 是 JoinPoint 的子接口，代表可以执行的目标方法
     * 返回值的类型必须是 Object
     * 必须有一个参数是 ProceedingJoinPoint 类型
     * 必须 throws Throwable
     */
    @Around("myPointCut()")
    public Object around(ProceedingJoinPoint pjp) throws Throwable{
        //开始
        System.out.println("环绕开始：执行目标方法前，模拟开启事务");
        //执行当前目标方法
        Object obj = pjp.proceed();
        //结束
        System.out.println("环绕结束：执行目标方法后，模拟关闭事务");
        return obj;
    }
    /**
     * 异常通知
     */
```

```java
@AfterThrowing(value="myPointCut()",throwing="e")
public void except(Throwable e) {
    System.out.println("异常通知: " + "程序执行异常" + e.getMessage());
}
/**
 * 后置（最终）通知
 */
@After("myPointCut()")
public void after() {
    System.out.println("最终通知：模拟释放资源");
}
}
```

❹ 创建配置类

在模块 ch1_3 的 src 目录下创建 aspectj.config 包，并在该包中创建配置类 AspectjAOPConfig。在该类中使用@Configuration 注解声明此类为配置类；使用@ComponentScan("aspectj")注解自动扫描 aspectj 包下使用的注解；使用@EnableAspectJAutoProxy 注解开启 Spring 对 AspectJ 的支持。

AspectjAOPConfig 的代码如下：

```java
package aspectj.config;
import org.springframework.context.annotation.ComponentScan;
import org.springframework.context.annotation.Configuration;
import org.springframework.context.annotation.EnableAspectJAutoProxy;
@Configuration                    //声明一个配置类
@ComponentScan("aspectj")         //自动扫描 aspectj 包下使用的注解
@EnableAspectJAutoProxy           //开启 Spring 对 AspectJ 的支持
public class AspectjAOPConfig {
}
```

❺ 创建测试类

在模块 ch1_4 的 aspectj.config 包中创建测试类 AOPTest。

AOPTest 的代码如下：

```java
package aspectj.config;
import org.springframework.context.annotation.AnnotationConfigApplicationContext;
import aspectj.dao.TestDao;
public class AOPTest {
    public static void main(String[] args) {
        //初始化 Spring 容器 ApplicationContext
        AnnotationConfigApplicationContext appCon =
            new AnnotationConfigApplicationContext(AspectjAOPConfig.class);
        //从容器中获取增强后的目标对象
        TestDao testDaoAdvice = appCon.getBean(TestDao.class);
        //执行方法
        testDaoAdvice.save();
        System.out.println("=================");
        testDaoAdvice.modify();
        System.out.println("=================");
        testDaoAdvice.delete();
        appCon.close();
    }
}
```

❻ 运行测试类

运行测试类 AOPTest 的 main 方法，运行结果如图 1.14 所示。

[图片：ch1_3 模块的运行结果截图]

图 1.14　ch1_3 模块的运行结果

扫一扫

视频讲解

1.5　Spring Bean

在 Spring 的应用中，Spring IoC 容器可以创建、装配和配置应用组件对象，这里的组件对象称为 Bean。

▶ 1.5.1　Bean 的实例化

在面向对象编程中，如果想使用某个对象，需要事先实例化该对象。同样，在 Spring 框架中，如果想使用 Spring 容器中的 Bean，也需要实例化 Bean。Spring 框架实例化 Bean 有 3 种方式，即构造方法实例化、静态工厂实例化和实例工厂实例化（其中最常用的实例化方法是构造方法实例化），使用实例工厂实例化具有解耦、降低代码复用等优点。

下面通过一个实例来演示 Bean 的实例化过程。

【例 1-4】Bean 的实例化过程。

其具体实现步骤如下：

❶ 使用 IntelliJ IDEA 创建模块 ch1_4 并导入 JAR 包

在项目 ch1 中创建一个名为 ch1_4 的模块，同时给模块 ch1_4 添加 Web Application，并将 Spring 的 4 个基础包、Spring Commons Logging Bridge 对应的 JAR 包 spring-jcl-6.0.0.jar，以及 AOP 实现 JAR 包 spring-aop-6.0.0.jar 复制到 ch1_4 的 WEB-INF/lib 目录中，添加为模块依赖。

❷ 创建实例化 Bean 的类

在模块 ch1_4 的 src 目录下创建一个名为 instance 的包，并在该包中创建 BeanClass、BeanInstanceFactory 以及 BeanStaticFactory 等实例化 Bean 的类。

BeanClass 的代码如下：

```
package instance;
public class BeanClass {
    public String message;
    public BeanClass() {
```

```java
        message = "构造方法实例化Bean";
    }
    publicBeanClass(String s) {
        message = s;
    }
}
```

BeanInstanceFactory 的代码如下：

```java
package instance;
public class BeanInstanceFactory {
    publicBeanClass createBeanClassInstance() {
        return newBeanClass("调用实例工厂方法实例化Bean");
    }
}
```

BeanStaticFactory 的代码如下：

```java
package instance;
public class BeanStaticFactory {
    private staticBeanClass beanInstance = new BeanClass("调用静态工厂方法实例化Bean");
    public staticBeanClass createInstance() {
        return beanInstance;
    }
}
```

❸ 创建配置类

在模块 ch1_4 的 src 目录下创建 config 包，并在该包中创建配置类 JavaConfig，在该配置类中使用@Bean 定义 3 个 Bean，具体代码如下：

```java
package config;
import org.springframework.context.annotation.Bean;
import org.springframework.context.annotation.Configuration;
import instance.BeanClass;
import instance.BeanInstanceFactory;
import instance.BeanStaticFactory;
@Configuration
public class JavaConfig {
    /**
     * 构造方法实例化
     */
    @Bean(value="beanClass")          //value 可以省略
    publicBeanClass getBeanClass() {
        return new BeanClass();
    }
    /**
     * 静态工厂实例化
     */
    @Bean(value="beanStaticFactory")
    publicBeanClass getBeanStaticFactory() {
        return BeanStaticFactory.createInstance();
    }
    /**
     * 实例工厂实例化
     */
    @Bean(value="beanInstanceFactory")
    publicBeanClass getBeanInstanceFactory() {
        BeanInstanceFactory bi = new BeanInstanceFactory();
        return bi.createBeanClassInstance();
    }
}
```

❹ 创建测试类

在模块 ch1_4 的 config 包中创建测试类 TestBean，并在该类中测试配置类定义的 Bean，具体代码如下：

```
package config;
import org.springframework.context.annotation.AnnotationConfigApplicationContext;
import instance.BeanClass;
public class TestBean {
    public static void main(String[] args) {
        //初始化 Spring 容器 ApplicationContext
        AnnotationConfigApplicationContext appCon =
            new AnnotationConfigApplicationContext(JavaConfig.class);
        BeanClass b1 = (BeanClass)appCon.getBean("beanClass");
        System.out.println(b1+ b1.message);
        BeanClass b2 = (BeanClass)appCon.getBean("beanStaticFactory");
        System.out.println(b2+ b2.message);
        BeanClass b3 = (BeanClass)appCon.getBean("beanInstanceFactory");
        System.out.println(b3+ b3.message);
        appCon.close();
    }
}
```

❺ 运行测试类

运行测试类 TestBean 的 main 方法，运行结果如图 1.15 所示。

图 1.15　ch1_4 模块的运行结果

▶ 1.5.2　Bean 的作用域

在 Spring 中不仅可以完成 Bean 的实例化，还可以为 Bean 指定作用域。在 Spring 中为 Bean 的实例定义了如表 1.2 所示的作用域，通过@Scope 注解来实现。

表 1.2　Bean 的作用域

作用域的名称	描述
singleton	默认的作用域，使用 singleton 定义的 Bean 在 Spring 容器中只有一个 Bean 实例
prototype	Spring 容器每次获取 prototype 定义的 Bean，容器都将创建一个新的 Bean 实例
request	在一次 HTTP 请求中容器将返回一个 Bean 实例，不同的 HTTP 请求返回不同的 Bean 实例。该作用域仅在 Web Spring 应用程序上下文中使用
session	在一个 HTTP Session 中容器将返回同一个 Bean 实例。该作用域仅在 Web Spring 应用程序上下文中使用
application	为每个 ServletContext 对象创建一个实例，即同一个应用共享一个 Bean 实例。该作用域仅在 Web Spring 应用程序上下文中使用
websocket	为每个 WebSocket 对象创建一个 Bean 实例。该作用域仅在 Web Spring 应用程序上下文中使用

在表 1.2 所示的 6 种作用域中，singleton 和 prototype 是最常用的两种，后面 4 种作用域仅在 Web Spring 应用程序上下文中使用。下面通过一个实例来演示 Bean 的作用域。

【例 1-5】Bean 的作用域。

其具体实现步骤如下：

❶ 使用 IntelliJ IDEA 创建模块 ch1_5 并导入 JAR 包

在项目 ch1 中创建一个名为 ch1_5 的模块，同时给模块 ch1_5 添加 Web Application，并导入与模块 ch1_4 相同的 JAR 包到 WEB-INF/lib 目录中，添加为模块依赖。

❷ 编写不同作用域的 Bean

在模块 ch1_5 的 src 目录下创建 service 包，并在该包中创建 SingletonService 和 PrototypeService 类。在 SingletonService 类中，Bean 的作用域为默认作用域 singleton；在 PrototypeService 类中，Bean 的作用域为 prototype。

SingletonService 的代码如下：

```
package service;
import org.springframework.stereotype.Service;
@Service              //默认为 singleton, 相当于@Scope("singleton")
public class SingletonService {
}
```

PrototypeService 的代码如下：

```
package service;
import org.springframework.context.annotation.Scope;
import org.springframework.stereotype.Service;
@Service
@Scope("prototype")
public class PrototypeService {
}
```

❸ 创建配置类

在模块 ch1_5 的 src 目录下创建 config 包，并在该包中创建配置类 ScopeConfig，具体代码如下：

```
package config;
import org.springframework.context.annotation.ComponentScan;
import org.springframework.context.annotation.Configuration;
@Configuration
@ComponentScan("service")
public class ScopeConfig {
}
```

❹ 创建测试类

在模块 ch1_5 的 config 包中创建测试类 TestScope，并在该测试类中分别获得 SingletonService 和 PrototypeService 的两个 Bean 实例，具体代码如下：

```
package config;
import org.springframework.context.annotation.AnnotationConfigApplicationContext;
import service.PrototypeService;
import service.SingletonService;
public class TestScope {
    public static void main(String[] args) {
        //初始化 Spring 容器 ApplicationContext
        AnnotationConfigApplicationContext appCon =
            new AnnotationConfigApplicationContext(ScopeConfig.class);
        SingletonService ss1 = appCon.getBean(SingletonService.class);
        SingletonService ss2 = appCon.getBean(SingletonService.class);
```

```
            System.out.println(ss1);
            System.out.println(ss2);
            PrototypeService ps1 = appCon.getBean(PrototypeService.class);
            PrototypeService ps2 = appCon.getBean(PrototypeService.class);
            System.out.println(ps1);
            System.out.println(ps2);
            appCon.close();
        }
    }
```

❺ 运行测试类

运行测试类 TestScope 的 main 方法，运行结果如图 1.16 所示。

图 1.16 ch1_5 模块的运行结果

从图 1.16 所示的运行结果可以得知，两次获取 SingletonService 的 Bean 实例时，IoC 容器返回两个相同的 Bean 实例；而两次获取 PrototypeService 的 Bean 实例时，IoC 容器返回两个不同的 Bean 实例。

▶ 1.5.3 Bean 的初始化和销毁

在实际工程应用中经常需要在 Bean 使用之前或之后做一些必要的操作，Spring 对 Bean 的生命周期的操作提供了支持。用户可以使用@Bean 注解的 initMethod 和 destroyMethod 属性（相当于 XML 配置的 init-method 和 destroy-method）对 Bean 进行初始化和销毁。下面通过一个实例来演示 Bean 的初始化和销毁。

【例 1-6】Bean 的初始化和销毁。

其具体实现步骤如下：

❶ 使用 IntelliJ IDEA 创建模块 ch1_6 并导入 JAR 包

在项目 ch1 中创建一个名为 ch1_6 的模块，同时给模块 ch1_6 添加 Web Application，并导入与模块 ch1_5 相同的 JAR 包到 WEB-INF/lib 目录中，添加为模块依赖。

❷ 创建 Bean 的类

在模块 ch1_6 的 src 目录下创建 service 包，并在该包中创建 MyService 类，具体代码如下：

```
package service;
public class MyService {
    public void initService() {
        System.out.println(this.getClass().getName() + "执行自定义的初始化方法");
    }
    public MyService() {
        System.out.println("执行构造方法，创建对象。");
    }
    public void destroyService() {
        System.out.println(this.getClass().getName() +"执行自定义的销毁方法");
    }
}
```

❸ 创建配置类

在模块 ch1_6 的 src 目录下创建 config 包，并在该包中创建配置类 JavaConfig，具体代码如下：

```
package config;
import org.springframework.context.annotation.Bean;
import org.springframework.context.annotation.Configuration;
import service.MyService;
@Configuration
public class JavaConfig {
    //initMethod 和 destroyMethod 指定 MyService 类的 initService 和 destroyService 方法
    //在构造方法之后、销毁之前执行
    @Bean(initMethod="initService",destroyMethod="destroyService")
    public MyService getMyService() {
        return new MyService();
    }
}
```

❹ 创建测试类

在模块 ch1_6 的 config 包中创建测试类 TestInitAndDestroy，具体代码如下：

```
package config;
import org.springframework.context.annotation.AnnotationConfigApplicationContext;
import service.MyService;
public class TestInitAndDestroy {
    public static void main(String[] args) {
        //初始化 Spring 容器 ApplicationContext
        AnnotationConfigApplicationContext appCon =
            new AnnotationConfigApplicationContext(JavaConfig.class);
        System.out.println("获得对象前");
        MyService ms = appCon.getBean(MyService.class);
        System.out.println("获得对象后" + ms);
        appCon.close();
    }
}
```

❺ 运行测试类

运行测试类 TestInitAndDestroy 的 main 方法，运行结果如图 1.17 所示。

图 1.17　ch1_6 模块的运行结果

本章小结

本章讲解了 Spring IoC、AOP 以及 Bean 等基础知识，目的是让读者在学习 Spring Boot 之前对 Spring 有一个简要的了解。

习题 1

1. Spring 的核心容器由哪些模块组成？
2. 什么是 Spring IoC？什么是依赖注入？
3. 在 Java 配置类中如何开启 Spring 对 AspectJ 的支持？
4. 什么是 Spring AOP？它与 OOP 是什么关系？

第 2 章　Spring MVC 基础

学习目的与要求

本章重点讲解 Spring MVC 的工作原理、控制器以及 JSON 数据交互。通过本章的学习，要求读者了解 Spring MVC 的工作原理，掌握 Spring MVC 应用的开发步骤。

本章主要内容

* Spring MVC 的工作原理
* Spring MVC 的工作环境
* 基于注解的控制器
* JSON 数据交互
* Spring MVC 的基本配置

MVC 思想是将一个应用分成 3 个基本部分，即 Model（模型）、View（视图）和 Controller（控制器），让这 3 个部分以最低的耦合进行协同工作，从而提高应用的可扩展性及可维护性。Spring MVC 是一款优秀的基于 MVC 思想的应用框架，它是 Spring 提供的一个实现了 Web MVC 设计模式的轻量级 Web 框架。

2.1　Spring MVC 的工作原理

Spring MVC 框架是高度可配置的，包含多种视图技术，例如 JSP 技术、Velocity、Tiles、iText 和 POI。Spring MVC 框架并不关心开发者使用的视图技术，也不会强迫开发者只使用 JSP 技术，本章使用的视图是 JSP。

Spring MVC 框架主要由 DispatcherServlet、处理器映射器、控制器、视图解析器、视图组成，其工作原理如图 2.1 所示。

从图 2.1 可总结出 Spring MVC 的工作流程如下：

（1）将客户端请求提交到 DispatcherServlet。

（2）由 DispatcherServlet 控制器寻找一个或多个 HandlerMapping，找到处理请求的 Controller。

（3）DispatcherServlet 将请求提交到 Controller。

（4）Controller 调用业务逻辑处理后返回 ModelAndView。

（5）DispatcherServlet 寻找一个或多个 ViewResolver 视图解析器，找到 ModelAndView 指定的视图。

（6）视图负责将结果显示到客户端。

在图 2.1 中包含 4 个 Spring MVC 接口，即 DispatcherServlet、HandlerMapping、Controller 和 ViewResolver。

图 2.1　Spring MVC 的工作原理图

　　Spring MVC 所有的请求都经过 DispatcherServlet 统一分发。DispatcherServlet 在将请求分发给 Controller 之前，需要借助于 Spring MVC 提供的 HandlerMapping 定位到具体的 Controller。

　　HandlerMapping 接口负责完成客户请求到 Controller 的映射。

　　Controller 接口将处理用户请求，这和 Java Servlet 扮演的角色是一致的。一旦 Controller 处理完用户请求，将返回 ModelAndView 对象给 DispatcherServlet 前端控制器，ModelAndView 中包含了模型（Model）和视图（View）。从宏观角度考虑，DispatcherServlet 是整个 Web 应用的控制器；从微观角度考虑，Controller 是单个 HTTP 请求处理过程中的控制器，而 ModelAndView 是 HTTP 请求处理过程中返回的模型（Model）和视图（View）。

　　ViewResolver 接口（视图解析器）在 Web 应用中负责查找 View 对象，从而将相应结果渲染给客户。

2.2　Spring MVC 的工作环境

▶ 2.2.1　Spring MVC 所需要的 JAR 包

　　在第 1 章 Spring 开发环境的基础上导入 Spring MVC 的相关 JAR 包，即可开发 Spring MVC 应用。

　　对于 Spring MVC 框架的初学者而言，在开发 Spring MVC 应用时，只需要将 Spring 的 4 个基础 JAR 包、Spring Commons Logging Bridge 对应的 JAR 包 spring-jcl-6.0.0.jar、AOP 实现 JAR 包 spring-aop-6.0.0.jar 以及两个与 Web 相关的 JAR 包（spring-web-6.0.0.jar 和 spring-webmvc-6.0.0.jar）复制到 Web 应用的 WEB-INF/lib 目录下即可。

在 Tomcat 10 运行 Spring MVC 6.0 应用时，DispatcherServlet 接口依赖性能监控 micrometer-observation 和 micrometer-commons 两个包进行请求的分发，因此 Spring MVC 应用所需要添加的 JAR 包如图 2.2 所示。

图 2.2　Spring MVC 应用所需要添加的 JAR 包

▶ 2.2.2　使用 IntelliJ IDEA 开发 Spring MVC 应用

本小节通过一个实例来演示 Spring MVC 入门程序的实现过程。

【例 2-1】Spring MVC 入门程序的实现过程。

其具体实现步骤如下：

❶ 使用 IDEA 创建 Web 应用并添加相关依赖

1）向项目或模块的/WEB-INF/lib 目录中添加依赖

首先在 IDEA 中创建一个名为 ch2 的项目，在该项目中创建一个名为 ch2_1 的模块，同时为模块 ch2_1 添加 Web Application，然后将如图 2.2 所示的 JAR 包复制到模块 ch2_1 的 /WEB-INF/lib 目录中，并添加为模块依赖。

2）为项目或模块添加 Tomcat 依赖

选择 File→Project Structure 命令，打开如图 2.3 所示的 Project Structure 界面。

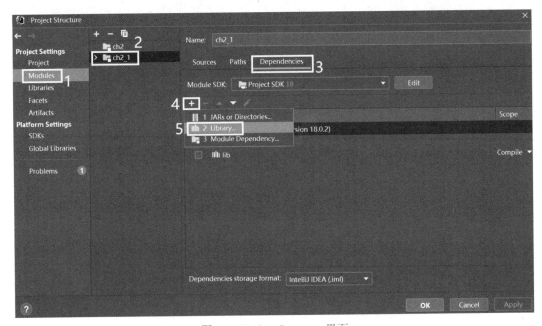

图 2.3　Project Structure 界面

按照图 2.3 中的操作顺序打开如图 2.4 所示的 Choose Libraries 界面。

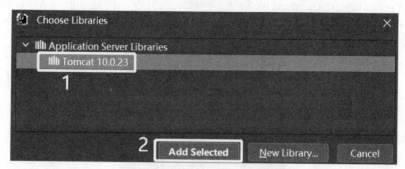

图 2.4 Choose Libraries 界面

按照图 2.4 中的操作顺序返回如图 2.5 所示的 Project Structure 界面。

图 2.5 返回的 Project Structure 界面

按照图 2.5 中的操作顺序即可将 Tomcat 的相关 JAR 包添加到模块 ch2_1 中，为后续 Web 开发奠定基础。

❷ 在 web.xml 文件中部署 DispatcherServlet

在开发 Spring MVC 应用时需要在 web.xml 中部署 DispatcherServlet，代码如下：

```
<?xml version="1.0" encoding="UTF-8"?>
<web-app xmlns:xsi="http://www.w3.org/2001/XMLSchema-instance"
    xmlns="https://jakarta.ee/xml/ns/jakartaee"
    xmlns:web="http://xmlns.jcp.org/xml/ns/javaee"
    xsi:schemaLocation="https://jakarta.ee/xml/ns/jakartaee
```

```xml
        https://jakarta.ee/xml/ns/jakartaee/web-app_5_0.xsd" id="WebApp_ID"
version="5.0">
    <display-name>ch2_1</display-name>
    <welcome-file-list>
        <welcome-file>index.html</welcome-file>
        <welcome-file>index.jsp</welcome-file>
        <welcome-file>index.htm</welcome-file>
        <welcome-file>default.html</welcome-file>
        <welcome-file>default.jsp</welcome-file>
        <welcome-file>default.htm</welcome-file>
    </welcome-file-list>
    <!--部署DispatcherServlet-->
    <servlet>
        <servlet-name>springmvc</servlet-name>
        <servlet-class>org.springframework.web.servlet.DispatcherServlet
            </servlet-class>
        <!--表示容器在启动时立即加载servlet-->
        <load-on-startup>1</load-on-startup>
    </servlet>
    <servlet-mapping>
        <servlet-name>springmvc</servlet-name>
        <!--处理所有URL-->
        <url-pattern>/</url-pattern>
    </servlet-mapping>
</web-app>
```

上述 DispatcherServlet 的 servlet 对象 springmvc 初始化时将在应用程序的 WEB-INF 目录下查找一个配置文件，该配置文件的命名规则是 "servletName-servlet.xml"，例如 springmvc-servlet.xml。

❸ 创建 Web 应用首页

在模块 ch2_1 的 web 目录下有一个应用首页 index.jsp。index.jsp 的代码如下：

```jsp
<%@ page language="java" contentType="text/html; charset=UTF-8" pageEncoding=
            "UTF-8"%>
<!DOCTYPE html>
<html>
<head>
<meta charset="UTF-8">
<title>Insert title here</title>
</head>
<body>
    没注册的用户，请<a href="index/register">注册</a>! <br>
    已注册的用户，去<a href="index/login">登录</a>!
</body>
</html>
```

❹ 创建 Controller 类

在模块 ch2_1 的 src 目录下创建 controller 包，并在该包中创建基于注解的名为 "IndexController" 的控制器类，在该类中有两个处理请求方法，分别处理首页中的 "注册" 和 "登录" 超链接请求。

```java
package controller;
import org.springframework.stereotype.Controller;
import org.springframework.web.bind.annotation.RequestMapping;
/** "@Controller" 表示 IndexController 的实例是一个控制器
 * @Controller 相当于@Controller("indexController")
```

```
 * 或@Controller(value = "indexController")
 */
@Controller
@RequestMapping("/index")
public class IndexController {
    @RequestMapping("/login")
    public String login() {
        /**login 代表逻辑视图名称,需要根据 Spring MVC 配置
         * 文件中 internalResourceViewResolver 的前缀和后缀找到对应的物理视图
         */
        return "login";
    }
    @RequestMapping("/register")
    public String register() {
        return "register";
    }
}
```

❺ 创建 Spring MVC 的配置文件

在 Spring MVC 中使用扫描机制找到应用中所有基于注解的控制器类,因此为了让控制器类能被 Spring MVC 框架扫描到,需要在配置文件中声明 spring-context,并使用 `<context:component-scan/>`元素指定控制器类的基本包(请确保所有控制器类都在基本包及其子包下),另外需要在配置文件中定义 Spring MVC 的视图解析器(ViewResolver),示例代码如下:

```xml
<bean class="org.springframework.web.servlet.view.InternalResourceViewResolver"
      id="internalResourceViewResolver">
    <!--前缀-->
    <property name="prefix" value="/WEB-INF/jsp/" />
    <!--后缀-->
    <property name="suffix" value=".jsp" />
</bean>
```

上述视图解析器配置了前缀和后缀两个属性,因此控制器类中的视图路径仅需提供 register 和 login,视图解析器将会自动添加前缀和后缀。

在模块 ch2_1 的 WEB-INF 目录下创建名为 springmvc-servlet.xml 的配置文件,其代码如下:

```xml
<?xml version="1.0" encoding="UTF-8"?>
<beans xmlns="http://www.springframework.org/schema/beans"
    xmlns:xsi="http://www.w3.org/2001/XMLSchema-instance"
    xmlns:context="http://www.springframework.org/schema/context"
    xmlns:mvc="http://www.springframework.org/schema/mvc"
    xsi:schemaLocation="
        http://www.springframework.org/schema/beans
        http://www.springframework.org/schema/beans/spring-beans.xsd
        http://www.springframework.org/schema/context
        http://www.springframework.org/schema/context/spring-context.xsd
        http://www.springframework.org/schema/mvc
        http://www.springframework.org/schema/mvc/spring-mvc.xsd">
    <!--使用扫描机制扫描控制器类-->
    <context:component-scan base-package="controller"/>
    <mvc:annotation-driven/>
    <!--使用 resources 过滤掉不需要 dispatcher servlet 的资源(即静态资源,例如 CSS、
        JS、HTML、images)。在使用 resources 时必须使用 annotation-driven,否则
        resources 元素会阻止任意控制器被调用
    -->
```

第 2 章　Spring MVC 基础

```
    <!--允许 css 目录下的所有文件可见-->
    <mvc:resources location="/css/" mapping="/css/**"></mvc:resources>
        <!--配置视图解析器-->
        <bean class="org.springframework.web.servlet.view.InternalResourceViewResolver"
              id="internalResourceViewResolver">
            <!--前缀-->
            <property name="prefix" value="/WEB-INF/jsp/" />
            <!--后缀-->
            <property name="suffix" value=".jsp" />
        </bean>
</beans>
```

❻ 完成应用的其他页面

IndexController 控制器的 register 方法处理成功后，跳转到"/WEB-INF/jsp/register.jsp"视图；IndexController 控制器的 login 方法处理成功后，跳转到"/WEB-INF/jsp/login.jsp"视图，因此模块 ch2_1 的"/WEB-INF/jsp"目录下应该有"register.jsp"和"login.jsp"页面，这里省略这两个 JSP 页面的代码。

❼ 发布并运行 Spring MVC 应用

在 IDEA 中第一次运行 Spring MVC 应用时需要将应用发布到 Tomcat，具体步骤如下：

在 IDEA 主界面中单击如图 2.6 所示的向下三角符号，选择 Edit Configurations 命令，打开如图 2.7 所示的服务器选择界面。

图 2.6　选择 Edit Configurations 命令

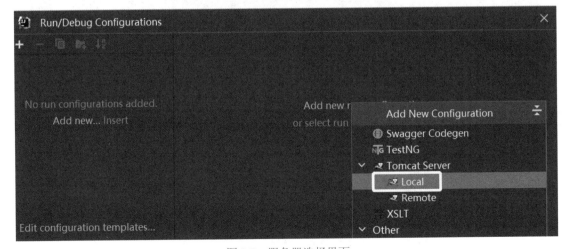

图 2.7　服务器选择界面

在图 2.7 中选择 Tomcat Server 下的 Local 选项打开 Deployment 界面，如图 2.8 所示。

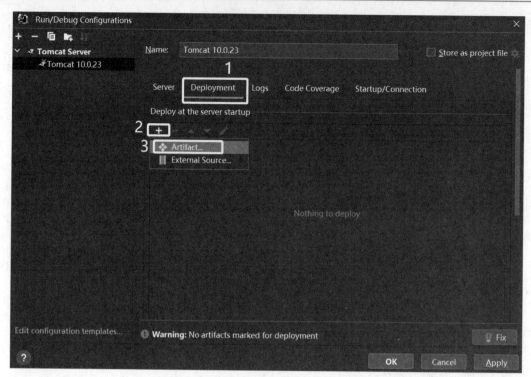

图 2.8 Deployment 界面

按照图 2.8 中的操作顺序打开如图 2.9 所示的界面，并修改应用的上下文路径。

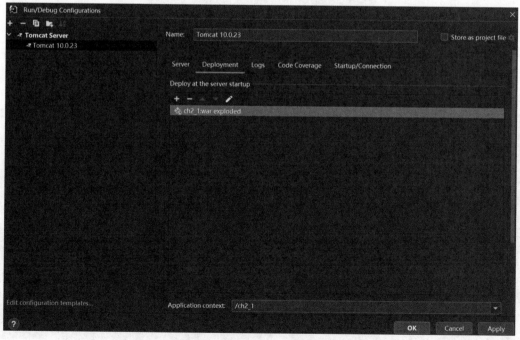

图 2.9 修改应用的上下文路径

单击图 2.9 中的 OK 按钮，即可将模块 ch2_1 发布到 Tomcat。在发布成功后，回到如图 2.10 所示的 IDEA 主界面。

第 2 章　Spring MVC 基础

图 2.10　IDEA 主界面

单击图 2.10 中的三角符号启动 Tomcat，即可运行 Web 应用 ch2_1，如图 2.11 所示。

图 2.11　Web 应用 ch2_1 的 index.jsp 页面

在如图 2.11 所示的页面中，用户单击"注册"超链接时，根据 springmvc-servlet.xml 文件中的映射，将请求转发给 IndexController 控制器处理，处理后跳转到"/WEB-INF/jsp/register.jsp"视图。同理，用户单击"登录"超链接时，控制器处理后跳转到"/WEB-INF/jsp/login.jsp"视图。

▶ 2.2.3　基于 Java 配置的 Spring MVC 应用

在例 2-1 中使用 web.xml 和 springmvc-servlet.xml 配置文件进行 Web 配置和 Spring MVC 配置，但 Spring Boot 推荐使用 Java 配置的方式进行项目配置，因此本小节通过一个实例来演示 Spring MVC 应用的 Java 配置。

【例 2-2】Spring MVC 应用的 Java 配置。

其具体实现步骤如下：

❶ 创建模块 ch2_2 并导入 JAR 包

首先在项目 ch2 中创建一个名为 ch2_2 的模块，同时为模块 ch2_2 添加 Web Application，然后将如图 2.2 所示的 JAR 包复制到模块 ch2_2 的/WEB-INF/lib 目录中，并添加为模块依赖，最后参照 2.2.2 节为模块 ch2_2 添加 Tomcat 依赖。

❷ 复制 JSP 和 Java 文件

将模块 ch2_1 中的 JSP 和 Java 文件按照相同目录复制到模块 ch2_2 中。

❸ 创建 Spring MVC 的 Java 配置（相当于 springmvc-servlet.xml 文件）

在模块 ch2_2 的 src 目录中创建名为 config 的包，并在该包中创建 Spring MVC 的 Java 配置类 SpringMVCConfig。在该配置类中使用@Configuration 注解声明该类为 Java 配置类；使用@EnableWebMvc 注解开启默认配置，例如 ViewResolver；使用@ComponentScan 注解扫描注解的类；使用@Bean 注解配置视图解析器。该类需要实现 WebMvcConfigurer 接口来配置 Spring MVC。其具体代码如下：

```
package config;
import org.springframework.context.annotation.Bean;
import org.springframework.context.annotation.ComponentScan;
import org.springframework.context.annotation.Configuration;
import org.springframework.web.servlet.config.annotation.EnableWebMvc;
import org.springframework.web.servlet.config.annotation.ResourceHandlerRegistry;
import org.springframework.web.servlet.config.annotation.WebMvcConfigurer;
import org.springframework.web.servlet.view.InternalResourceViewResolver;
@Configuration
@EnableWebMvc
```

```java
@ComponentScan("controller")
public class SpringMVCConfig implements WebMvcConfigurer {
    /**
     * 配置视图解析器
     */
    @Bean
    public InternalResourceViewResolver getViewResolver() {
        InternalResourceViewResolver viewResolver =
            new InternalResourceViewResolver();
        viewResolver.setPrefix("/WEB-INF/jsp/");
        viewResolver.setSuffix(".jsp");
        return viewResolver;
    }
    /**
     * 配置静态资源
     */
    @Override
    public void addResourceHandlers(ResourceHandlerRegistry registry) {
        registry.addResourceHandler("/html/**").addResourceLocations("/html/");
    }
}
```

❹ 创建 Web 的 Java 配置（相当于 web.xml 文件）

在模块 ch2_2 的 config 包中创建 Web 的 Java 类 WebConfig。该类需要实现 WebApplicationInitializer 接口代替 web.xml 文件的配置。实现该接口将会自动启动 Servlet 容器。在 WebConfig 类中需要使用 AnnotationConfigWebApplicationContext 注册 Spring MVC 的 Java 配置类 SpringMVCConfig，并和当前 ServletContext 关联。最后，在该类中需要注册 Spring MVC 的 DispatcherServlet。其具体代码如下：

```java
package config;
import jakarta.servlet.ServletContext;
import jakarta.servlet.ServletException;
import jakarta.servlet.ServletRegistration.Dynamic;
import org.springframework.web.WebApplicationInitializer;
import org.springframework.web.context.support.
        AnnotationConfigWebApplicationContext;
import org.springframework.web.servlet.DispatcherServlet;
public class WebConfig implements WebApplicationInitializer{
    @Override
    public void onStartup(ServletContext arg0) throws ServletException {
        AnnotationConfigWebApplicationContext ctx
        = new AnnotationConfigWebApplicationContext();
        ctx.register(SpringMVCConfig.class);
        //注册 Spring MVC 的 Java 配置类 SpringMVCConfig
        ctx.setServletContext(arg0);      //和当前 ServletContext 关联
        /**
         * 注册 Spring MVC 的 DispatcherServlet
         */
        Dynamic servlet=arg0.addServlet("dispatcher",new DispatcherServlet(ctx));
        servlet.addMapping("/");
        servlet.setLoadOnStartup(1);
    }
}
```

❺ 发布并运行 Spring MVC 应用

参照 2.2.2 节，按照图 2.12 中的操作顺序打开 Edit Configuration Settings 界面发布并运行 ch2_2 应用。

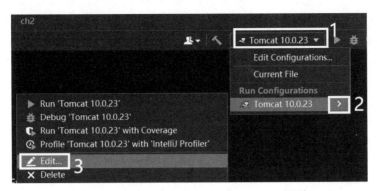

图 2.12　打开 Edit Configuration Settings 界面

扫一扫

视频讲解

2.3　基于注解的控制器

在使用 Spring MVC 进行 Web 应用开发时，Controller 是 Web 应用的核心。Controller 实现类包含了对用户请求的处理逻辑，是用户请求和业务逻辑之间的"桥梁"，是 Spring MVC 框架的核心部分。

▶ 2.3.1　Controller 注解类型

在 Spring MVC 中使用 org.springframework.stereotype.Controller 注解类型声明某类的实例是一个控制器，例如 2.2.2 节中的 IndexController 控制器类。注意，不要忘了在 Spring MVC 的配置文件中使用<context:component-scan/>元素（见例 2-1）或在 Spring MVC 配置类中使用@ComponentScan（见例 2-2）指定控制器类的基本包，进而扫描所有注解的控制器类。

▶ 2.3.2　RequestMapping 注解类型

在基于注解的控制器类中可以为每个请求编写对应的处理方法。那么如何将请求与处理方法一一对应呢？需要使用 org.springframework.web.bind.annotation.RequestMapping 注解类型将请求与处理方法一一对应。

❶ 方法级别注解

方法级别注解的示例代码如下：

```
package controller;
import org.springframework.stereotype.Controller;
import org.springframework.web.bind.annotation.RequestMapping;
@Controller
public class IndexController {
    @RequestMapping(value = "/index/login")
    public String login() {
        /**login 代表逻辑视图名称，需要根据 Spring MVC 配置中
        *internalResourceViewResolver 的前缀和后缀找到对应的物理视图
        */
        return "login";
    }
    @RequestMapping(value = "/index/register")
    public String register() {
        return "register";
    }
}
```

在上述示例中有两个 RequestMapping 注解语句，它们都作用在处理方法上。注解的 value 属性将请求 URI 映射到方法，value 属性是 RequestMapping 注解的默认属性，如果只有一个 value 属性，则可以省略该属性。用户可以使用以下 URL 访问 login 方法（请求处理方法）。

```
http://localhost:xxx/yyyy/index/login
```

❷ 类级别注解

类级别注解的示例代码如下：

```java
package controller;
import org.springframework.stereotype.Controller;
import org.springframework.web.bind.annotation.RequestMapping;
@Controller
@RequestMapping("/index")
public class IndexController {
    @RequestMapping("/login")
    public String login() {
        return "login";
    }
    @RequestMapping("/register")
    public String register() {
        return "register";
    }
}
```

在类级别注解的情况下，控制器类中的所有方法都将映射为类级别的请求。用户可以使用以下 URL 访问 login 方法。

```
http://localhost:xxx/yyy/index/login
```

为了方便维护程序，建议开发者采用类级别注解，将相关处理放在同一个控制器类中，例如对商品的增、删、改、查处理方法都可以放在一个名为 GoodsOperate 的控制器类中。

@RequestMapping 注解的 value 属性表示请求路径；method 属性表示请求方式。如果方法上的@RequestMapping 注解没有设置 method 属性，则 get 和 post 请求都可以访问；如果方法上的@RequestMapping 注解设置了 method 属性，则只能是相应的请求方式可以访问。

@RequestMapping 还有特定于 HTTP 请求方式的组合注解，具体如下。

1）@GetMapping

@GetMapping 相当于@RequestMapping(method=RequestMethod.GET)，用于处理 get 请求。使用@RequestMapping 编写是@RequestMapping(value = "requestpath", method = RequestMethod.GET)；使用@GetMapping 可简写为@GetMapping("requestpath")。该注解通常在查询数据时使用。

2）@PostMapping

@PostMapping 相当于@RequestMapping(method=HttpMethod.POST)，用于处理 post 请求。使用@RequestMapping 编写是@RequestMapping(value = "requestpath", method = RequestMethod.POST)；使用@PostMapping 可简写为@PostMapping("requestpath")。该注解通常在新增数据时使用。

3）@PutMapping、@PatchMapping

@PutMapping、@PatchMapping 相当于@RequestMapping(method = RequestMethod.PUT/PATCH)，用于处理 put 和 patch 请求。使用@RequestMapping 编写是@RequestMapping(value = "requestpath", method = RequestMethod.PUT/PATCH)；使用 @PutMapping 可简写为@PutMapping("requestpath")。两者都是更新，@PutMapping 为全局更新，@PatchMapping 是

对 put 方式的一种补充，put 是对整体的更新，patch 是对局部的更新。该注解通常在更新数据时使用。

4）@DeleteMapping

@DeleteMapping 相当于@RequestMapping(method = RequestMethod.DELETE)，用于处理 delete 请求。使用@RequestMapping 编写是@RequestMapping(value = "requestpath", method = RequestMethod.DELETE)；使用@DeleteMapping 可简写为@DeleteMapping("requestpath")。该注解通常在删除数据时使用。

▶ 2.3.3 编写请求处理方法

在控制类中每个请求处理方法可以有多个不同类型的参数，以及一个多种类型的返回结果。

❶ 请求处理方法中常出现的参数类型

如果需要在请求处理方法中使用 Servlet API 类型，那么可以将这些类型作为请求处理方法的参数类型。Servlet API 参数类型的示例代码如下：

```
package controller;
import jakarta.servlet.http.HttpServletRequest;
import jakarta.servlet.http.HttpSession;
import org.springframework.stereotype.Controller;
import org.springframework.web.bind.annotation.RequestMapping;
@Controller
@RequestMapping("/index")
public class IndexController {
    @RequestMapping("/login")
    public String login(HttpSession session, HttpServletRequest request) {
        session.setAttribute("skey", "session 范围的值");
        request.setAttribute("rkey", "request 范围的值");
        return "login";
    }
}
```

除了 Servlet API 参数类型以外，还有输入/输出流、表单实体类、注解类型、与 Spring 框架相关的类型等，这些类型将在后续章节中使用时再详细介绍。org.springframework.ui.Model 类型是一个包含 Map 的 Spring 框架类型，特别重要。每次调用请求处理方法时，Spring MVC 都将创建 org.springframework.ui.Model 对象。Model 参数类型的示例代码如下：

```
package controller;
import org.springframework.stereotype.Controller;
import org.springframework.ui.Model;
import org.springframework.web.bind.annotation.RequestMapping;
@Controller
@RequestMapping("/index")
public class IndexController {
    @RequestMapping("/register")
    public String register(Model model) {
        /*在视图中可以使用 EL 表达式${success}取出 model 中的值,有关 EL 的知识请参考相关内容*/
        model.addAttribute("success", "注册成功");
        return "register";
    }
}
```

❷ 请求处理方法常见的返回类型

最常见的返回类型就是代表逻辑视图名称的 String 类型，例如前面章节中的请求处理方法。除了 String 类型以外，还有 Model、View 以及其他任意的 Java 类型。

2.3.4 Controller 接收请求参数的常见方式

Controller 接收请求参数的方式有很多种,有的适合 get 请求方式,有的适合 post 请求方式,有的两者都适合。下面介绍几种常见的方式,读者可以根据实际情况选择合适的接收方式。

❶ 通过实体 Bean 接收请求参数

通过一个实体 Bean 接收请求参数,适用于 get 和 post 提交请求方式。需要注意的是,Bean 的属性名称必须与请求参数名称相同。下面通过一个实例讲解"通过实体 Bean 接收请求参数"。

【例 2-3】通过实体 Bean 接收请求参数。

其具体实现步骤如下:

1) 创建模块 ch2_3 并导入 JAR 包

首先在项目 ch2 中创建一个名为 ch2_3 的模块,同时为模块 ch2_3 添加 Web Application,然后除如图 2.2 所示的 JAR 包以外,还需要将 Java 增强库 lombok-1.18.24.jar 复制到模块 ch2_3 的/WEB-INF/lib 目录中,并添加为模块依赖,最后参照 2.2.2 节为模块 ch2_3 添加 Tomcat 依赖。

2) 创建视图文件

在模块 ch2_3 的/WEB-INF/jsp/目录下有 register.jsp、login.jsp 和 main.jsp 文件,这里省略 main.jsp 的代码。

register.jsp 的核心代码如下:

```html
<body>
<form action="user/register" method="post" name="registForm">
    <table border=1>
        <tr>
            <td>姓名:</td>
            <td>
                <input type="text" name="uname" value="${user.uname}"/>
            </td>
        </tr>
        <tr>
            <td>密码:</td>
            <td><input type="password" name="upass"/></td>
        </tr>
        <tr>
            <td>确认密码:</td>
            <td><input type="password" name="reupass"/></td>
        </tr>
        <tr>
            <td colspan="2" align="center">
                <input type="submit" value="注册"/>
            </td>
        </tr>
    </table>
</form>
</body>
```

login.jsp 的核心代码如下:

```html
<body>
    <form action="user/login" method="post">
    <table>
        <tr>
            <td align="center" colspan="2">登录</td>
```

```
            </tr>
            <tr>
                <td>姓名: </td>
                <td><input type="text" name="uname"></td>
            </tr>
            <tr>
                <td>密码: </td>
                <td><input type="password" name="upass"></td>
            </tr>
            <tr>
                <td colspan="2">
                    <input type="submit" value="提交" >
                    <input type="reset" value="重置" >
                </td>
            </tr>
        </table>
        ${messageError}
    </form>
</body>
```

在模块 ch2_3 的 web 目录下有 index.jsp 文件，核心代码如下：

```
<body>
    没注册的用户，请<a href="index/register">注册</a>! <br>
    已注册的用户，去<a href="index/login">登录</a>!
</body>
```

3) 创建 POJO 实体类

在模块 ch2_3 的 src 目录下创建 pojo 包，并在该包中创建实体类 UserForm，代码如下：

```
package pojo;
import lombok.Data;
@Data
public class UserForm {
    private String uname;         //与请求参数名称相同
    private String upass;
    private String reupass;
}
```

4) 创建控制器类

在模块 ch2_3 的 src 目录下创建 controller 包，并在该包中创建控制器类 UserController 和 IndexController。

IndexController 的代码如下：

```
package controller;
import org.springframework.stereotype.Controller;
import org.springframework.web.bind.annotation.GetMapping;
import org.springframework.web.bind.annotation.RequestMapping;
@Controller
@RequestMapping("/index")
public class IndexController {
    @GetMapping("/login")
    public String login() {
        return "login";
    }
    @GetMapping("/register")
    public String register() {
        return "register";
    }
}
```

UserController 的代码如下：

```java
package controller;
import jakarta.servlet.http.HttpSession;
import org.apache.commons.logging.Log;
import org.apache.commons.logging.LogFactory;
import org.springframework.stereotype.Controller;
import org.springframework.ui.Model;
import org.springframework.web.bind.annotation.PostMapping;
import org.springframework.web.bind.annotation.RequestMapping;
import pojo.UserForm;
@Controller
@RequestMapping("/user")
public class UserController {
    //得到一个用来记录日志的对象，这样在打印信息的时候能够标记打印的是哪个类的信息
    private static final Log logger = LogFactory.getLog(UserController.class);
    /**
     * 处理登录
     * 使用 UserForm 对象（实体 Bean）user 接收登录页面提交的请求参数
     */
    @PostMapping("/login")
    public String login(UserForm user, HttpSession session, Model model) {
        if("zhangsan".equals(user.getUname())
                && "123456".equals(user.getUpass())) {
            session.setAttribute("u", user);
            logger.info("成功");
            return "main";        //登录成功，跳转到 main.jsp
        }else{
            logger.info("失败");
            model.addAttribute("messageError", "用户名或密码错误");
            return "login";
        }
    }

    /**
     *处理注册
     *使用 UserForm 对象（实体 Bean）user 接收注册页面提交的请求参数
     */
    @PostMapping("/register")
    public String register(UserForm user, Model model) {
        if("zhangsan".equals(user.getUname())
                && "123456".equals(user.getUpass())) {
            logger.info("成功");
            return "login";         //注册成功，跳转到 login.jsp
        }else{
            logger.info("失败");
            //在 register.jsp 页面上可以使用 EL 表达式取出 model 的 uname 值
            model.addAttribute("uname", user.getUname());
            return "register";        //返回 register.jsp
        }
    }
}
```

5）创建 Web 与 Spring MVC 的配置类

将例 2-2 中模块 ch2_2 的 Web 配置类 WebConfig 和 Spring MVC 配置类 SpringMVCConfig 复制到模块 ch2_3 的相同位置。

6）发布并运行应用

参照例 2-2 发布并运行 ch2_3 应用。

❷ 通过处理方法的形参接收请求参数

通过处理方法的形参接收请求参数也就是直接把表单参数写在控制器类相应方法的形参中，即形参名称与请求参数名称完全相同。该接收参数方式适用于 get 和 post 提交请求方式。这里可以将例 2-3 的控制器类 UserController 中 register 方法的代码修改如下：

```java
@PostMapping("/register")
/**
 * 通过形参接收请求参数，形参名称与请求参数名称完全相同
 */
public String register(String uname, String upass, Model model) {
    if("zhangsan".equals(uname)
        && "123456".equals(upass)) {
        logger.info("成功");
        return "login";        //注册成功，跳转到 login.jsp
    }else{
        logger.info("失败");
        //在 register.jsp 页面上可以使用 EL 表达式取出 model 的 uname 值
        model.addAttribute("uname", uname);
        return "register";        //返回 register.jsp
    }
}
```

❸ 通过@RequestParam 接收请求参数

通过@RequestParam 接收请求参数适用于 get 和 post 提交请求方式。这里可以将例 2-3 的控制器类 UserController 中 register 方法的代码修改如下：

```java
@PostMapping("/register")
/**
 * 通过@RequestParam 接收请求参数
 */
public String register(@RequestParam String uname, @RequestParam String
                    upass, Model model) {
    if("zhangsan".equals(uname)
        && "123456".equals(upass)) {
        logger.info("成功");
        return "login";        //注册成功，跳转到 login.jsp
    }else{
        logger.info("失败");
        //在 register.jsp 页面上可以使用 EL 表达式取出 model 的 uname 值
        model.addAttribute("uname", uname);
        return "register";        //返回 register.jsp
    }
}
```

通过@RequestParam 接收请求参数与通过处理方法的形参接收请求参数的区别是，当请求参数与接收参数名称不一致时，通过处理方法的形参接收请求参数不会报 400 错误，而通过@RequestParam 接收请求参数会报 400 错误。

❹ 通过@ModelAttribute 接收请求参数

当将@ModelAttribute 注解放在处理方法的形参上时，用于将多个请求参数封装到一个实体对象，从而简化数据绑定流程，而且自动暴露为模型数据，在展示视图页面时使用。通过实体 Bean 接收请求参数只是将多个请求参数封装到一个实体对象，并不能暴露为模型数据，需要使用 model.addAttribute 语句才能暴露为模型数据。

通过@ModelAttribute 注解接收请求参数适用于 get 和 post 提交请求方式。这里可以将

例 2-3 的控制器类 UserController 中 register 方法的代码修改如下：

```
@PostMapping("/register")
public String register(@ModelAttribute("user") UserForm user) {
    if("zhangsan".equals(user.getUname())
            && "123456".equals(user.getUpass())){
        logger.info("成功");
        return "login";        //注册成功，跳转到login.jsp
    }else{
        logger.info("失败");
        //使用@ModelAttribute("user")与model.addAttribute("user", user)的功能相同
    //在register.jsp页面上可以使用EL表达式${user.uname}取出ModelAttribute的
      uname值
        return "register";     //返回register.jsp
    }
}
```

▶ 2.3.5 重定向与转发

　　重定向是将用户对当前处理的请求定向到另一个视图（例如 JSP）或处理请求，以前的请求（request）中存放的信息全部失效，并进入一个新的 request 作用域；转发是将用户对当前处理的请求转发给另一个视图或处理请求，以前的 request 中存放的信息不会失效。

　　转发是服务器行为，重定向是客户端行为，具体工作流程如下。

　　转发过程：客户浏览器发送 HTTP 请求，Web 服务器接受此请求，调用内部的一个方法在容器内部完成请求处理和转发动作，将目标资源发送给客户。在这里转发的路径必须是同一个 Web 容器下的 URL，其不能转向到其他的 Web 路径上去，中间传递的是自己容器内的 request。在客户浏览器的地址栏中显示的仍然是其第一次访问的路径，也就是说客户是感觉不到服务器做了转发的。转发行为是浏览器只做了一次访问请求。

　　重定向过程：客户浏览器发送 HTTP 请求，Web 服务器接受后发送 302 状态码响应及对应新的 location 给客户浏览器，客户浏览器发现是 302 响应，则自动再发送一个新的 HTTP 请求，请求 URL 是新的 location 地址，服务器根据此请求寻找资源并发送给客户。在这里 location 可以重定向到任意 URL，既然是浏览器重新发出了请求，则就没有什么 request 传递的概念了。在客户浏览器的地址栏中显示的是其重定向的路径，客户可以观察到地址的变化。重定向行为是浏览器做了至少两次的访问请求。

　　在 Spring MVC 框架中，控制器类中处理方法的 return 语句默认就是转发实现，只不过实现的是转发到视图，示例代码如下：

```
@RequestMapping("/register")
public String register() {
    return "register";         //转发到register.jsp
}
```

　　在 Spring MVC 框架中，重定向与转发的示例代码如下：

```
package controller;
import org.springframework.stereotype.Controller;
import org.springframework.web.bind.annotation.RequestMapping;
@Controller
@RequestMapping("/index")
public class IndexController {
    @RequestMapping("/login")
    public String login() {
```

```
            //转发到一个请求方法（同一个控制器类中，可省略/index/）
            return "forward:/index/isLogin";
    }
    @RequestMapping("/isLogin")
    public String isLogin() {
            //重定向到一个请求方法
            return "redirect:/index/isRegister";
    }
    @RequestMapping("/isRegister")
    public String isRegister() {
            //转发到一个视图
            return "register";
    }
}
```

在 Spring MVC 框架中，不管是重定向还是转发，都需要符合视图解析器的配置，如果直接重定向到一个不需要 DispatcherServlet 的资源，例如：

```
return "redirect:/html/my.html";
```

在 Spring MVC 配置文件中，需要使用 mvc:resources 配置：

```
<mvc:resources location="/html/" mapping="/html/**"></mvc:resources>
```

在 Spring MVC 配置类中，需要实现 WebMvcConfigurer 的接口方法 public void addResourceHandlers(ResourceHandlerRegistry registry)，示例代码如下：

```
@Override
public void addResourceHandlers(ResourceHandlerRegistry registry) {
    registry.addResourceHandler("/html/**").addResourceLocations("/html/");
}
```

2.3.6 应用@Autowired 进行依赖注入

在前面学习的控制器中并没有体现 MVC 的 M 层，这是因为控制器既充当 C 层又充当 M 层。这样设计程序的系统结构很不合理，应该将 M 层从控制器中分离出来。Spring MVC 框架本身就是一个非常优秀的 MVC 框架，它具有依赖注入的优点。这里可以通过 org.springframework.beans.factory.annotation.Autowired 注解类型将依赖注入一个属性（成员变量）或方法，例如：

```
@Autowired
public UserService userService;
```

在 Spring MVC 中，为了能被作为依赖注入，服务层的类必须使用 org.springframework.stereotype.Service 注解类型注明为@Service（一个服务），另外还需要在配置文件中使用<context:component-scan base-package="基本包"/>元素或者在配置类中使用@ComponentScan("基本包")来扫描依赖基本包。下面将例 2-3 的 ch2_3 应用的"登录"和"注册"的业务逻辑处理分离出来，使用 Service 层实现。

首先创建 service 包，在该包中创建 UserService 接口和 UserServiceImpl 实现类。

UserService 接口的具体代码如下：

```
package service;
import pojo.UserForm;
public interface UserService {
    boolean login(UserForm user);
    boolean register(UserForm user);
}
```

UserServiceImpl 实现类的具体代码如下：

```java
package service;
import org.springframework.stereotype.Service;
import pojo.UserForm;
//注解为一个服务
@Service
public class UserServiceImpl implements UserService{
    @Override
    public boolean login(UserForm user) {
        if("zhangsan".equals(user.getUname())
                && "123456".equals(user.getUpass()))
            return true;
        return false;
    }
    @Override
    public boolean register(UserForm user) {
        if("zhangsan".equals(user.getUname())
                && "123456".equals(user.getUpass()))
            return true;
        return false;
    }
}
```

然后将配置类中的@ComponentScan("controller")修改如下：

```java
@ComponentScan(basePackages = {"controller","service"})        //扫描基本包
```

最后修改控制器类 UserController，核心代码如下：

```java
@Controller
@RequestMapping("/user")
public class UserController {
    private static final Log logger = LogFactory.getLog(UserController.class);
    //将服务层依赖注入属性 userService
    @Autowired
    public UserService userService;
    /**
     * 处理登录
     */
    @PostMapping("/login")
    public String login(UserForm user, HttpSession session, Model model) {
        if(userService.login(user)){
            session.setAttribute("u", user);
            logger.info("成功");
            return "main";
        }else{
            logger.info("失败");
            model.addAttribute("messageError", "用户名或密码错误");
            return "login";
        }
    }
    /**
     *处理注册
     */
    @PostMapping("/register")
    public String register(@ModelAttribute("user") UserForm user) {
        if(userService.register(user)){
            logger.info("成功");
            return "login";
        }else{
            logger.info("失败");
```

```
            return "register";
        }
    }
}
```

▶ 2.3.7 @ModelAttribute

通过 org.springframework.web.bind.annotation.ModelAttribute 注解类型可以实现以下两个功能。

❶ 绑定请求参数到实体对象（表单的命令对象）

该用法和 2.3.4 节中的"通过@ModelAttribute 接收请求参数"一样：

```
@PostMapping("/register")
public String register(@ModelAttribute("user") UserForm user) {
    if("zhangsan".equals(user.getUname())
            && "123456".equals(user.getUpass())){
        return "login";
    }else{
        return "register";
    }
}
```

在上述代码中"@ModelAttribute("user") UserForm user"语句的功能有两个，一个是将请求参数的输入封装到 user 对象中；另一个是创建 UserForm 实例，以"user"为键值存储在 Model 对象中，与"model.addAttribute("user", user)"语句的功能一样。如果没有指定键值，即"@ModelAttribute UserForm user"，那么在创建 UserForm 实例时，以"userForm"为键值存储在 Model 对象中，与"model.addAttribute("userForm", user)"语句的功能一样。

❷ 注解一个非请求处理方法

在控制器类中，被@ModelAttribute 注解的一个非请求处理方法，将在每次调用该控制器类的请求处理方法前被调用。这种特性可以用来控制登录权限，当然控制登录权限的方法有很多，例如拦截器、过滤器等。

使用该特性控制登录权限的示例代码如下：

```
package controller;
import jakarta.servlet.http.HttpSession;
import org.springframework.web.bind.annotation.ModelAttribute;
public class BaseController {
    @ModelAttribute
    public void isLogin(HttpSession session) throws Exception {
        if(session.getAttribute("user") == null){
            throw new Exception("没有权限");
        }
    }
}
package controller;
import org.springframework.stereotype.Controller;
import org.springframework.web.bind.annotation.RequestMapping;
@Controller
@RequestMapping("/admin")
public class ModelAttributeController extends BaseController{
    @RequestMapping("/add")
    public String add(){
        return "addSuccess";
    }
    @RequestMapping("/update")
```

```
    public String update(){
        return "updateSuccess";
    }
    @RequestMapping("/delete")
    public String delete(){
        return "deleteSuccess";
    }
}
```

上述 ModelAttributeController 类中的 add、update、delete 请求处理方法在执行时首先执行父类 BaseController 中的 isLogin 方法判断登录权限。

扫一扫

视频讲解

2.4 JSON 数据交互

Spring MVC 在数据绑定的过程中需要对传递数据的格式和类型进行转换，它既可以转换 String 等类型的数据，也可以转换 JSON 等其他类型的数据。本节将针对 Spring MVC 中 JSON 类型的数据交互进行讲解。

2.4.1 JSON 数据结构

JSON（JavaScript Object Notation，JS 对象标记）是一种轻量级的数据交换格式。和 XML 一样，JSON 也是基于纯文本的数据格式。它有对象和数组两种数据结构。

❶ 对象结构

对象结构以"{"开始，以"}"结束，中间部分由 0 个或多个以英文","分隔的 key-value 对构成，key 和 value 之间以英文":"分隔。对象结构的语法结构如下：

```
{
    key1:value1,
    key2:value2,
    …
}
```

其中，key 必须为 String 类型，value 可以是 String、Number、Object、Array 等数据类型。例如，一个 person 对象包含姓名、密码、年龄等信息，其 JSON 的表示形式如下：

```
{
    "pname":"陈恒",
    "password":"123456",
    "page":40
}
```

❷ 数组结构

数组结构以"["开始，以"]"结束，中间部分由 0 个或多个以英文","分隔的值的列表组成。数组结构的语法结构如下：

```
[
    value1,
    value2,
    …
]
```

上述两种（对象、数组）数据结构也可以分别组合构成更加复杂的数据结构。例如，一个 student 对象包含 sno、sname、hobby 和 college 对象，其 JSON 的表示形式如下：

```
{
    "sno":"201802228888",
```

```
    "sname":"陈恒",
    "hobby":["篮球","足球"],
    "college":{
        "cname":"清华大学",
        "city":"北京"
    }
}
```

▶ 2.4.2　JSON 数据转换

为了实现浏览器与控制器类之间的 JSON 数据交互，Spring MVC 提供了 MappingJackson2HttpMessageConverter 实现类默认处理 JSON 格式请求响应。该实现类利用 Jackson 开源包读/写 JSON 数据，将 Java 对象转换为 JSON 对象和 XML 文档，同时也可以将 JSON 对象和 XML 文档转换为 Java 对象。

Jackson 开源包及其描述如下。

（1）jackson-annotations.jar：JSON 转换注解包。

（2）jackson-core.jar：JSON 转换核心包。

（3）jackson-databind.jar：JSON 转换的数据绑定包。

以上 3 个 Jackson 的开源包在编写本书时最新版本是 2.14.1，读者可以通过地址"http://mvnrepository.com/artifact/com.fasterxml.jackson.core"下载得到。

在使用注解开发时需要用到两个重要的 JSON 格式转换注解，分别是@RequestBody 和@ResponseBody。

（1）@RequestBody：用于将请求体中的数据绑定到方法的形参中，该注解应用在方法的形参上。

（2）@ResponseBody：用于直接返回 JSON 对象，该注解应用在方法上。

下面通过一个实例来演示 JSON 数据交互过程，在该实例中针对返回实体对象、ArrayList 集合、Map<String, Object>集合以及 List<Map<String, Object>>集合分别处理。

【例 2-4】JSON 数据交互过程。

其具体实现步骤如下：

❶ 创建模块 ch2_4 并导入相关的 JAR 包

首先在项目 ch2 中创建一个名为 ch2_4 的模块，同时为模块 ch2_4 添加 Web Application，然后除导入模块 ch2_3 的 JAR 包以外，还需要将与 JSON 相关的 3 个 JAR 包（jackson-annotations-2.14.1.jar、jackson-databind-2.14.1.jar 和 jackson-core-2.14.1.jar）复制到模块 ch2_4 的 WEB-INF/lib 目录中，并添加为模块依赖，最后参照 2.2.2 节为模块 ch2_4 添加 Tomcat 依赖。

❷ 创建 Web 和 Spring MVC 配置类

在模块 ch2_4 的 src 目录下创建名为 config 的包，并在该包中创建 Web 配置类 WebConfig 和 Spring MVC 配置类 SpringMVCConfig。

由于 WebConfig 的代码和 ch2_3 应用的相同，这里不再赘述。

在 SpringMVCConfig 配置类中需要配置静态资源，核心代码如下：

```
@Configuration
@EnableWebMvc
@ComponentScan("controller")
public class SpringMVCConfig implements WebMvcConfigurer {
    /**
```

```java
     * 配置视图解析器
     */
    @Bean
    public InternalResourceViewResolver getViewResolver() {
        InternalResourceViewResolver viewResolver =
            new InternalResourceViewResolver();
        viewResolver.setPrefix("/WEB-INF/jsp/");
        viewResolver.setSuffix(".jsp");
        return viewResolver;
    }
    /**
     * 配置静态资源
     */
    @Override
    public void addResourceHandlers(ResourceHandlerRegistry registry) {
        registry.addResourceHandler("/js/**").addResourceLocations("/js/");
    }
}
```

❸ 创建 JSP 页面并引入 jQuery

首先从 jQuery 官方网站 "http://jquery.com/download/" 下载 jQuery 插件 jquery-3.6.0.min.js，将其复制到模块 ch2_4 的 web/js 目录下，然后在 JSP 页面中通过 <script type="text/javascript" src ="js/jquery-3.6.0.min.js"></script> 代码将 jQuery 引入当前页面。

在模块 ch2_4 的 web 目录下创建 JSP 文件 index.jsp，在该页面中使用 Ajax 向控制器异步提交数据，代码如下：

```jsp
<%@ page language="java" contentType="text/html; charset=UTF-8" pageEncoding=
            "UTF-8"%>
<%
String path = request.getContextPath();
String basePath = request.getScheme()+"://"+request.getServerName()+":"+
            request.getServerPort()+path+"/";
%>
<!DOCTYPE html>
<html>
<head>
<base href="<%=basePath%>">
<meta charset="UTF-8">
<title>Insert title here</title>
<script type="text/javascript" src ="js/jquery-3.6.0.min.js"></script>
<script type="text/javascript">
    function testJson() {
        //获取输入的值 pname 为 id
        var pname = $("#pname").val();
        var password = $("#password").val();
        var page = $("#page").val();
        $.ajax({
            //请求路径
            url: "testJson",
            //请求类型
            type: "post",
            //data 表示发送的数据
            data: JSON.stringify({pname:pname,password:password,page:page}),
            //定义发送请求的数据格式为 JSON 字符串
            contentType: "application/json;charset=UTF-8",
            //定义回调响应的数据格式为 JSON 字符串，该属性可以省略
            dataType: "json",
```

```javascript
                    //成功响应的结果
                    success : function(data){
                        if(data != null){
                            //返回一个 Person 对象
                    //alert("输入的用户名:" + data.pname + ",密码: " + data.password + ",
                    //年龄: " + data.page);
                            //ArrayList<Person>对象
                            /**for(var i = 0; i < data.length; i++){
                                alert(data[i].pname);
                            }**/
                            //返回一个 Map<String, Object>对象
                            //alert(data.pname);              //pname 为 key
                            //返回一个 List<Map<String, Object>>对象
                            for(var i = 0; i < data.length; i++){
                                alert(data[i].pname);
                            }
                        }
                    }
                });
            }
</script>
</head>
<body>
    <form action="">
        用户名:<input type="text" name="pname" id="pname"/><br>
        密码:<input type="password" name="password" id="password"/><br>
        年龄:<input type="text" name="page" id="page"/><br>
        <input type="button" value="测试" onclick="testJson()"/>
    </form>
</body>
</html>
```

❹ 创建实体类

在模块 ch2_4 的 src 目录下创建名为 pojo 的包,并在该包中创建 Person 实体类,代码如下:

```java
package pojo;
import lombok.Data;
@Data
public class Person {
    private String pname;
    private String password;
    private Integer page;
}
```

❺ 创建控制器类

在模块 ch2_4 的 src 目录下创建名为 controller 的包,并在该包中创建 TestController 控制器类,在处理方法中使用@ResponseBody 和@RequestBody 注解进行 JSON 数据交互,核心代码如下:

```java
@Controller
public class TestController {
    /**
     * 接收页面请求的 JSON 数据,并返回 JSON 格式结果
     */
    @RequestMapping("/testJson")
    @ResponseBody
    public List<Map<String, Object>> testJson(@RequestBody Person user) {
        //打印接收的 JSON 格式数据
```

```
            System.out.println("pname=" + user.getPname() +
                    ", password=" + user.getPassword() + ",page=" + user.getPage());
            //返回 Person 对象
            //return user;
            /**ArrayList<Person> allp = new ArrayList<Person>();
            Person p1 = new Person();
            p1.setPname("陈恒1");
            p1.setPassword("123456");
            p1.setPage(80);
            allp.add(p1);
            Person p2 = new Person();
            p2.setPname("陈恒2");
            p2.setPassword("78910");
            p2.setPage(90);
            allp.add(p2);
            //返回 ArrayList<Person>对象
            return allp;
            **/
            Map<String, Object> map = new HashMap<String, Object>();
            map.put("pname", "陈恒2");
            map.put("password", "123456");
            map.put("page", 25);
            //返回一个 Map<String, Object>对象
            //return map;
            //返回一个 List<Map<String, Object>>对象
            List<Map<String,Object>> allp = new ArrayList<Map<String,Object>>();
            allp.add(map);
            Map<String, Object> map1 = new HashMap<String, Object>();
            map1.put("pname", "陈恒3");
            map1.put("password", "54321");
            map1.put("page", 55);
            allp.add(map1);
            return allp;
        }
    }
```

❻ 测试应用

参照例 2-2 发布并运行 ch2_4 应用。

扫一扫

视频讲解

2.5 Spring MVC 的基本配置

Spring MVC 的定制配置需要配置类实现 WebMVCConfigurer 接口，并在配置类中使用 @EnableWebMvc 注解来开启对 Spring MVC 的配置支持，这样开发者就可以重写接口方法完成常用的配置。

▶ 2.5.1 静态资源配置

应用的静态资源（CSS、JS、图片）等需要直接访问，这时需要开发者在配置类中重写 public void addResourceHandlers(ResourceHandlerRegistry registry)接口方法来实现。配置的示例代码如下：

```
@Configuration
@EnableWebMvc
@ComponentScan(basePackages = {"controller","service"})
public class SpringMVCConfig implements WebMvcConfigurer {
```

```
/**
 * 配置静态资源
 */
@Override
public void addResourceHandlers(ResourceHandlerRegistry registry) {
    registry.addResourceHandler("/html/**").addResourceLocations("/html/");
    //addResourceHandler 指的是对外暴露的访问路径
    //addResourceLocations 指的是静态资源存放的位置
}
```

根据上述配置,可以直接访问 Web 应用(例如 ch2_3 应用)的/html/目录下的静态资源。假设有如图 2.13 所示的静态资源,可以通过"http://localhost:8080/ch2_3/html/NewFile.html"直接访问静态资源文件 NewFile.html。

图 2.13　静态资源

▶ 2.5.2　拦截器配置

Spring 的拦截器(Interceptor)实现对每一个请求处理的前后进行相关的业务处理,类似于 Servlet 的过滤器(Filter)。开发者如果需要自定义 Spring 的拦截器,可以通过以下两个步骤完成。

❶ 创建自定义拦截器类

自定义拦截器类需要实现 HandlerInterceptor 接口或继承 HandlerInterceptorAdapter 类,示例代码如下:

```
package interceptor;
import jakarta.servlet.http.HttpServletRequest;
import jakarta.servlet.http.HttpServletResponse;
import org.springframework.web.servlet.HandlerInterceptor;
import org.springframework.web.servlet.ModelAndView;
public class MyInterceptor implements HandlerInterceptor{
    /**
     * 重写 preHandle 方法在请求发生前执行
     */
    @Override
    public boolean preHandle(HttpServletRequest request, HttpServletResponse
                        response, Object handler)
            throws Exception {
        System.out.println("preHandle 方法在请求发生前执行");
```

```
        return true;
    }
    /**
     * 重写postHandle方法在请求完成后执行
     */
    @Override
    public void postHandle(HttpServletRequest request, HttpServletResponse
                       response, Object handler,
        ModelAndView modelAndView) throws Exception {
        System.out.println("postHandle方法在请求完成后执行");
    }
}
```

❷ 配置拦截器

在配置类中需要首先配置拦截器Bean，然后重写addInterceptors方法注册拦截器，示例代码如下：

```
@Configuration
@EnableWebMvc
@ComponentScan(basePackages = {"controller","service"})
public class SpringMVCConfig implements WebMvcConfigurer {
    /**
     * 配置拦截器Bean
     */
    @Bean
    public MyInterceptor myInterceptor() {
        return new MyInterceptor();
    }
    /**
     * 重写addInterceptors方法注册拦截器
     */
    @Override
    public void addInterceptors(InterceptorRegistry registry) {
        registry.addInterceptor(myInterceptor()).addPathPatterns("/api/**");
    }
}
```

▶ 2.5.3 文件上传配置

org.springframework.web.multipart.MultipartResolver 是一个解析包括文件上传在内的 multipart 请求接口。从 Spring 6.0 和 Servlet 5.0+开始，基于 Apache Commons FileUpload 组件的 MultipartResolver 接口实现类 CommonsMultipartResolver 被弃用，改用基于 Servlet 容器的 MultipartResolver 接口实现类 StandardServletMultipartResolver，进行 multipart 请求解析。

在 Spring MVC 框架中上传文件时，将文件的相关信息及操作封装到 MultipartFile 接口对象中，因此开发者只需要使用 MultipartFile 类型声明模型类的一个属性，即可对被上传文件进行操作。该接口具有如下方法。

（1）byte[] getBytes()：以字节数组的形式返回文件的内容。

（2）String getContentType()：返回文件内容的类型。

（3）InputStream getInputStream()：返回一个 InputStream，从中读取文件的内容。

（4）String getName()：返回请求参数的名称。

（5）String getOriginalFilename()：返回客户端提交的原始文件名称。

（6）long getSize()：返回文件的大小，单位为字节。

（7）boolean isEmpty()：判断被上传文件是否为空。

（8）void transferTo(File destination)：将上传文件保存到目标目录下。

在 Spring MVC 应用中上传文件时需要完成以下配置或设置：

（1）在 Spring MVC 配置文件中，使用 Spring 的 org.springframework.web.multipart.support.StandardServletMultipartResolver 类配置一个名为 multipartResolver（MultipartResolver 接口）的 Bean，用于文件的上传。

（2）必须将文件上传表单的 method 设置为 post，并将 enctype 设置为 multipart/form-data。只有这样设置，浏览器才能将所选文件的二进制数据发送给服务器。

（3）需要通过 Servlet 容器配置启用 Servlet multipart 请求解析。可以在 Web 配置类中的 Servlet 声明部分进行配置启用 Servlet multipart 请求解析，Web 配置类的示例代码如下：

```java
public class WebConfig implements WebApplicationInitializer{
    @Override
    public void onStartup(ServletContext arg0) throws ServletException {
        AnnotationConfigWebApplicationContext ctx
        = new AnnotationConfigWebApplicationContext();
        ctx.register(SpringMVCConfig.class);
        ctx.setServletContext(arg0);
        Dynamic servlet = arg0.addServlet("dispatcher",
            new DispatcherServlet(ctx));
        servlet.setMultipartConfig(getMultiPartConfig());
        servlet.addMapping("/");
        servlet.setLoadOnStartup(1);
    }
    private MultipartConfigElement getMultiPartConfig() {
        String location = "";
        long maxFileSize = 20848820;    //允许上传文件大小的最大值，默认为-1（不限制）
        long maxRequestSize = 418018841;
        //multipart/form-data 请求允许的最大值，默认为-1（不限制）
        int fileSizeThreshold = 1048576;        //文件将写入磁盘的阈值大小，默认为 0
        return new MultipartConfigElement(
            location,
            maxFileSize,
            maxRequestSize,
            fileSizeThreshold
        );
    }
}
```

下面通过一个实例讲解如何上传多个文件。

【例 2-5】上传多个文件。

其具体实现步骤如下：

❶ 创建模块 ch2_5 并导入相关的 JAR 包

首先在项目 ch2 中创建一个名为 ch2_5 的模块，同时为模块 ch2_5 添加 Web Application，然后导入与 Spring MVC 相关的 JAR 包，包括 Spring 的 4 个基础 JAR 包、Spring Commons Logging Bridge 对应的 JAR 包 spring-jcl-6.0.0.jar、AOP 实现 JAR 包 spring-aop-6.0.0.jar、DispatcherServlet 接口所依赖的性能监控包（micrometer-observation.jar 和 micrometer-commons.jar）、Java 增强库（lombok-1.18.24.jar）以及两个与 Web 相关的 JAR 包（spring-web-6.0.0.jar 和 spring-webmvc-6.0.0.jar）。另外，本实例还使用 JSTL 标签展示页面，因此还需要从 Tomcat 的 webapps\examples\WEB-INF\lib 目录中将 JSTL 的相关 JAR 包复制到应用的 WEB-INF/lib 目录下，并添加为模块依赖。最后参照 2.2.2 节为模块 ch2_5 添加 Tomcat 依赖。

❷ 创建多文件选择页面

在模块 ch2_5 的 web 目录下创建 JSP 页面 multiFiles.jsp。在该页面中使用表单（不要忘记 enctype 属性值为 multipart/form-data）上传多个文件，具体代码如下：

```jsp
<%@ page language="java" contentType="text/html; charset=UTF-8" pageEncoding=
            "UTF-8"%>
<%
String path = request.getContextPath();
String basePath = request.getScheme()+"://"+request.getServerName()+":"+
            request.getServerPort()+path+"/";
%>
<!DOCTYPE html>
<html>
<head>
<base href="<%=basePath%>">
<meta charset="UTF-8">
<title>Insert title here</title>
</head>
<body>
<form action="multifile" method="post" enctype="multipart/form-data">
    选择文件1:<input type="file" name="myfile"> <br>
    文件描述1:<input type="text" name="description"> <br>
    选择文件2:<input type="file" name="myfile"> <br>
    文件描述2:<input type="text" name="description"> <br>
    选择文件3:<input type="file" name="myfile"> <br>
    文件描述3:<input type="text" name="description"> <br>
<input type="submit" value="提交">
</form>
</body>
</html>
```

❸ 创建 POJO 类

在模块 ch2_5 的 src 目录下创建 pojo 包，并在 pojo 包中创建实体类 MultiFileDomain。在上传多个文件时，需要用 POJO 类 MultiFileDomain 封装文件信息，MultiFileDomain 类的具体代码如下：

```java
package pojo;
import java.util.List;
import org.springframework.web.multipart.MultipartFile;
import lombok.Data;
@Data
public class MultiFileDomain {
    private List<String> description;
    private List<MultipartFile> myfile;
}
```

❹ 创建控制器类

在模块 ch2_5 的 src 目录下创建 controller 包，并在 controller 包中创建控制器类 MultiFilesController，具体代码如下：

```java
package controller;
import java.io.File;
import java.util.List;
import jakarta.servlet.http.HttpServletRequest;
import org.apache.commons.logging.Log;
import org.apache.commons.logging.LogFactory;
import org.springframework.stereotype.Controller;
import org.springframework.web.bind.annotation.ModelAttribute;
import org.springframework.web.bind.annotation.RequestMapping;
```

```java
import org.springframework.web.multipart.MultipartFile;
import pojo.MultiFileDomain;
@Controller
public class MultiFilesController {
    //得到一个用来记录日志的对象,这样在打印信息时能够标记打印的是哪个类的信息
    private static final Log logger = LogFactory.getLog(MultiFilesController.class);
    /**
     * 多文件上传
     */
    @RequestMapping("/multifile")
    public String multiFileUpload(@ModelAttribute MultiFileDomain
        multiFileDomain, HttpServletRequest request){
        String realpath = request.getServletContext().getRealPath("uploadfiles");
        File targetDir = new File(realpath);
        if(!targetDir.exists()){
            targetDir.mkdirs();
        }
        List<MultipartFile> files = multiFileDomain.getMyfile();
        for (int i = 0; i < files.size(); i++) {
            MultipartFile file = files.get(i);
            String fileName = file.getOriginalFilename();
            File targetFile = new File(realpath,fileName);
            //上传
            try {
                file.transferTo(targetFile);
            } catch (Exception e) {
                e.printStackTrace();
            }
        }
        logger.info("成功");
        return "showMulti";
    }
}
```

❺ 创建 Web 与 Spring MVC 配置类

在模块 ch2_5 的 src 目录下创建名为 config 的包,并在该包中创建 Web 配置类 WebConfig 和 Spring MVC 配置类 SpringMVCConfig。

在 Web 配置类 WebConfig 中通过 Servlet 容器配置启用 Servlet multipart 请求解析,核心代码见上述配置示例。

在 SpringMVCConfig 配置类中需要配置 MultipartResolver 进行文件的上传,核心代码如下:

```java
@Configuration
@EnableWebMvc
@ComponentScan(basePackages = {"controller"})
public class SpringMVCConfig implements WebMvcConfigurer {
    /**
     * 配置视图解析器
     */
    @Bean
    public InternalResourceViewResolver getViewResolver() {
        InternalResourceViewResolver viewResolver =
            new InternalResourceViewResolver();
        viewResolver.setPrefix("/WEB-INF/jsp/");
        viewResolver.setSuffix(".jsp");
        return viewResolver;
    }
    /**
     * MultipartResolver 配置
     */
```

```
        @Bean(name = "multipartResolver")
        public MultipartResolver multipartResolver() {
            StandardServletMultipartResolver multipartResolver =
                new StandardServletMultipartResolver();
            return multipartResolver;
        }
}
```

❻ 创建成功显示页面

在模块 ch2_5 的 WEB-INF/jsp 目录下创建多文件上传成功显示页面 showMulti.jsp，具体代码如下：

```
<%@ page language="java" contentType="text/html; charset=UTF-8" pageEncoding=
        "UTF-8"%>
<%@ taglib uri="http://java.sun.com/jsp/jstl/core" prefix="c" %>
<%
String path = request.getContextPath();
String basePath = request.getScheme()+"://"+request.getServerName()+":"+
    request.getServerPort()+path+"/";
%>
<!DOCTYPE html>
<html>
<head>
<base href="<%=basePath%>">
<meta charset="UTF-8">
<title>Insert title here</title>
</head>
<body>
    <table>
        <tr>
            <td>详情</td><td>文件名</td>
        </tr>
        <!--同时取两个数组的元素-->
        <c:forEach items="${multiFileDomain.description}" var="description"
                    varStatus="loop">
        <tr>
            <td>${description}</td>
            <td>${multiFileDomain.myfile[loop.count-1].originalFilename}</td>
        </tr>
        </c:forEach>
        <!--fileDomain.getMyfile().getOriginalFilename()-->
    </table>
</body>
</html>
```

❼ 发布并运行应用

发布 ch2_5 应用到 Tomcat 服务器，并启动 Tomcat 服务器。然后通过地址"http://localhost:8080/ch2_5/multiFiles.jsp"运行多文件选择页面，运行结果如图 2.14 所示。

图 2.14　多文件选择页面

第 2 章 Spring MVC 基础

在图 2.14 中选择文件,并输入文件描述,然后单击"提交"按钮上传多个文件,如果成功则显示如图 2.15 所示的结果。

图 2.15 多文件成功上传结果

本章小结

本章简单地介绍了 Spring MVC 框架基础,包括 Spring MVC 的工作流程、控制器、JSON 数据交互以及 Spring MVC 的基本配置等内容。第 1 章和本章只是简单地介绍了 Spring 和 Spring MVC 的基础知识,对于它们的更多知识,请参考编者的另一本教程《Java EE 框架整合开发入门到实战——Spring+Spring MVC+MyBatis(微课版)》。

扫一扫

自测题

习 题 2

1. 在开发 Spring MVC 应用时如何配置 DispatcherServlet?又如何配置 Spring MVC?
2. 简述 Spring MVC 的工作流程。
3. @ModelAttribute 可实现哪些功能?
4. 在 Spring MVC 中,JSON 类型的数据如何交互?请按照返回实体对象、ArrayList 集合、Map<String, Object>集合以及 List<Map<String, Object>>集合举例说明。
5. 在 Spring MVC 中,需要使用(   )注解将请求与处理方法一一对应。
   A.@RequestMapping　　B.@Controller　　C.@RequestBody　　D.@ResponseBody

第 3 章 Spring Boot 入门

学习目的与要求

本章首先介绍什么是 Spring Boot，然后介绍 Spring Boot 应用的开发环境，最后介绍如何快速构建一个 Spring Boot 应用。通过本章的学习，要求读者掌握如何构建 Spring Boot 应用的开发环境以及 Spring Boot 应用。

本章主要内容

- ❖ Spring Boot 概述
- ❖ 快速构建 Spring Boot 应用

从前两章的学习可知 Spring 框架非常优秀，但"配置过多"，造成开发效率低、部署流程复杂以及集成难度大等问题。为了解决上述问题，Spring Boot 应运而生。编者在编写本书时 Spring Boot 的最新正式版是 3.0.2，建议读者在测试本书示例代码时使用 3.0.2 或更高版本。

3.1 Spring Boot 概述

▶ 3.1.1 什么是 Spring Boot

Spring Boot 是由 Pivotal 团队提供的全新框架，其设计目的是用来简化新 Spring 应用的初始搭建以及开发过程。使用 Spring Boot 框架可以做到专注于 Spring 应用的开发，无须过多关注样板化的配置。

在 Spring Boot 框架中使用"约定优于配置"（Convention Over Configuration，COC）的理念。针对企业应用开发提供了符合各种场景的 spring-boot-starter 自动配置依赖模块，这些模块都是基于"开箱即用"的原则，进而使企业应用开发更加快捷和高效。可以说 Spring Boot 是开发者和 Spring 框架的中间层，目的是帮助开发者管理应用的配置，提供应用开发中常见配置的默认处理（即约定优于配置），简化 Spring 应用的开发和运维，降低开发人员对框架的关注度，使开发人员把更多精力放在业务逻辑代码上。通过"约定优于配置"的原则，Spring Boot 致力于在蓬勃发展的快速应用开发领域成为领导者。

▶ 3.1.2 Spring Boot 的优点

Spring Boot 之所以能够应运而生，是因为它具有以下优点。

（1）使编码变得简单：推荐使用注解。

（2）使配置变得快捷：自动配置、快速构建项目、快速集成第三方技术的能力。

（3）使部署变得简便：内嵌 Tomcat、Jetty 等 Web 容器。

（4）使监控变得容易：自带项目监控。

3.1.3 Spring Boot 的主要特性

Spring Boot 的主要特性如下。

（1）约定优于配置：Spring Boot 遵循"约定优于配置"的原则，只需很少的配置，大多数情况直接使用默认配置即可。

（2）独立运行的 Spring 应用：Spring Boot 可以以 JAR 包的形式独立运行。使用 java -jar 命令或者在项目的主程序中执行 main() 方法运行 Spring Boot 应用（项目）。

（3）内嵌 Web 容器：内嵌 Servlet 容器，Spring Boot 可以选择内嵌 Tomcat、Jetty 等 Web 容器，无须以 WAR 包形式部署应用。

（4）提供 starter 简化 Maven 配置：Spring Boot 提供了一系列的 starter pom 简化 Maven 的依赖加载，基本上可以做到自动化配置，高度封装，开箱即用。

（5）自动配置 Spring：Spring Boot 根据项目依赖（在类路径中的 JAR 包、类）自动配置 Spring 框架，极大地减少了项目的配置。

（6）提供准生产的应用监控：Spring Boot 提供基于 HTTP、SSH、Telnet 对运行的项目进行跟踪监控。

（7）无代码生成和 XML 配置：Spring Boot 不是借助于代码生成来实现的，而是通过条件注解来实现的，提倡使用 Java 配置和注解配置相结合的配置方式，方便、快捷。

3.2 第一个 Spring Boot 应用

3.2.1 Maven 简介

Apache Maven 是一个软件项目管理工具，基于项目对象模型（Project Object Model, POM）的理念，通过一段核心描述信息来管理项目构建、报告和文档信息。在 Java 项目中，Maven 主要完成两件工作：①统一开发规范与工具；②统一管理 JAR 包。

Maven 统一管理项目开发所需要的 JAR 包，但这些 JAR 包将不再包含在项目内（即不在 lib 目录下），而是存放于仓库当中。仓库主要包括中央仓库和本地仓库。

❶ 中央仓库

中央仓库存放开发过程中的所有 JAR 包，例如 JUnit，它们都可以通过互联网从中央仓库中下载，中央仓库的地址为 http://mvnrepository.com。

❷ 本地仓库

本地仓库指本地计算机中的仓库。官方下载 Maven 的本地仓库配置在"%MAVEN_HOME%\conf\settings.xml"文件中，找到"localRepository"即可。

Maven 项目首先从本地仓库中获取所需要的 JAR 包，当无法获取指定的 JAR 包时，本地仓库将从远程仓库（中央仓库）中下载 JAR 包，并存入本地仓库以备将来使用。

3.2.2 Maven 的 pom.xml

Maven 是基于项目对象模型的理念管理项目的，所以 Maven 的项目都有一个 pom.xml 配置文件来管理项目的依赖以及项目的编译等功能。

在 Maven Web 项目中重点关注以下元素：

❶ properties 元素

在<properties></properties>之间可以定义变量，以便在<dependency></dependency>中引用，示例代码如下：

```xml
<properties>
    <!--Spring 的版本号-->
    <spring.version>6.0.0</spring.version>
</properties>
<dependencies>
    <dependency>
        <groupId>org.springframework</groupId>
        <artifactId>spring-core</artifactId>
        <version>${spring.version}</version>
    </dependency>
</dependencies>
```

❷ dependencies 元素

<dependencies></dependencies>元素包含多个项目依赖需要使用的<dependency></dependency>元素。

❸ dependency 元素

<dependency></dependency>元素内部通过<groupId></groupId>、<artifactId></artifactId>、<version></version>3 个子元素确定唯一的依赖，也可以称为 3 个坐标，示例代码如下：

```xml
<dependency>
    <!--groupId 组织的唯一标识-->
    <groupId>org.springframework</groupId>
    <!--artifactId 项目的唯一标识-->
    <artifactId>spring-core</artifactId>
    <!--version 项目的版本号-->
    <version>${spring.version}</version>
</dependency>
```

❹ scope 子元素

在<dependency></dependency>元素中有时使用<scope></scope>子元素管理依赖的部署。<scope></scope>子元素可以使用以下 5 个值。

（1）compile（编译范围）：compile 是默认值，即默认范围。如果依赖没有提供范围，那么该依赖的范围就是编译范围。编译范围的依赖在所有的 classpath 中可用，同时也会被打包发布。

（2）provided（已提供范围）：provided 表示已提供范围，只有当 JDK 或者容器已提供该依赖才可以使用。已提供范围的依赖不是传递性的，也不会被打包发布。

（3）runtime（运行时范围）：runtime 范围依赖在运行和测试系统时需要，在编译时不需要。

（4）test（测试范围）：test 范围依赖在一般的编译和运行时都不需要，它们只有在测试编译和测试运行阶段可用，不会随项目发布。

（5）system（系统范围）：system 范围与 provided 范围类似，但需要显式提供包含依赖的 JAR 包，Maven 不会在 Repository 中查找它。

▶ 3.2.3 使用 IntelliJ IDEA 快速构建 Spring Boot 应用

IntelliJ IDEA 简称 IDEA，是 Java 编程语言的集成开发环境。IDEA 在业界被公认为较好

第 3 章 Spring Boot 入门

的 Java 开发工具，尤其是在智能代码助手、代码自动提示、重构、Java EE 支持、各类版本工具（Git、SVN 等）、JUnit、CVS 整合、代码分析、创新的 GUI 设计等方面的功能，可以说是超常的。读者可以登录官网"https://www.jetbrains.com/idea/"，根据操作系统下载相应的 IntelliJ IDEA。本书采用的是 Windows Ultimate 版 ideaIU-2022.2.1（内置 Java 运行环境 Open JDK）。

下面详细讲解如何使用 IDEA 集成开发工具快速构建一个 Spring Boot Web 应用，具体实现步骤如下：

❶ 新建 Spring Project

打开 IDEA，通过选择 File→New→Project 命令打开 New project 对话框，在该对话框的左侧选中 Spring Initializr 选项，在右侧输入项目信息，如图 3.1 所示。

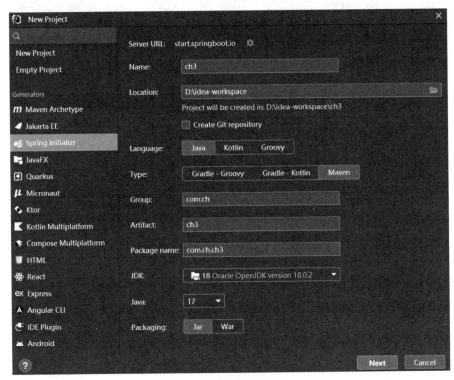

图 3.1　输入项目信息

❷ 选择项目依赖

在图 3.1 中输入项目信息后单击 Next 按钮，打开如图 3.2 所示的选择项目依赖界面。在图 3.2 中选择项目依赖（例如 Spring Web）后单击 Create 按钮，即可完成 Spring Boot Web 应用的创建。

❸ 编写测试代码

在应用 ch3 的 src/main/java 目录下创建 com.ch.ch3.test 包，并在该包中创建 TestController 类，代码如下：

```
package com.ch.ch3.test;
import org.springframework.web.bind.annotation.GetMapping;
import org.springframework.web.bind.annotation.RestController;
@RestController
public class TestController {
    @GetMapping("/hello")
```

```
    public String hello() {
        return "您好，Spring Boot!";
    }
}
```

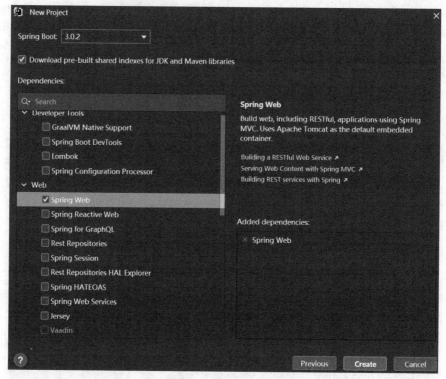

图 3.2　选择项目依赖

在上述代码中使用的@RestController 注解是一个组合注解，相当于 Spring MVC 中的@Controller 和@ResponseBody 注解的组合，具体应用如下：

（1）如果只是使用@RestController 注解 Controller，则 Controller 中的方法无法返回 JSP、HTML 等视图，返回的内容就是 return 的内容。

（2）如果需要返回到指定页面，则需要用@Controller 注解。如果需要返回 JSON、XML 或自定义 mediaType 内容到页面，则需要在对应的方法上添加@ResponseBody 注解。

❹ 应用程序的 App 类

在应用 ch3 的 com.ch.ch3 包中自动生成了应用程序的 App 类 Ch3Application。这里省略 Ch3Application 的代码。

❺ 运行 main 方法启动 Spring Boot 应用

运行 Ch3Application 类的 main 方法，控制台信息如图 3.3 所示。

图 3.3　启动 Spring Boot 应用后的控制台信息

从控制台信息可以看到 Tomcat 的启动过程、Spring MVC 的加载过程。注意 Spring Boot 3.0 内嵌了 Tomcat 10，因此 Spring Boot Web 应用不需要开发者配置与启动 Tomcat。

❻ 测试 Spring Boot 应用

在启动 Spring Boot 应用后，默认访问地址为"http://localhost:8080/"，将项目路径直接设为根路径，这是 Spring Boot 的默认设置，因此用户可以通过"http://localhost:8080/hello"测试应用（hello 与测试类 TestController 中的@GetMapping("/hello")对应），测试效果如图 3.4 所示。

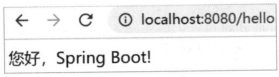

图 3.4 访问 Spring Boot 应用

本章小结

本章首先简单介绍了 Spring Boot 应运而生的缘由，然后演示了如何使用 IDEA 快速构建 Spring Boot 应用。IDEA 在业界被公认为较好的 Java 开发工具，本书后续章节都采用 IDEA 编写示例代码。开发者使用何种集成开发工具（IDE）构建 Spring Boot 应用，可根据实际工程需要选择合适的 IDE。

习题 3

1. Spring、Spring MVC、Spring Boot 三者有什么联系？为什么还要学习 Spring Boot？
2. 在 IDEA 中如何快速构建 Spring Boot 的 Web 应用？

第 4 章　Spring Boot 核心

学习目的与要求

本章将详细介绍 Spring Boot 的核心注解、基本配置、自动配置原理以及条件注解。通过本章的学习，要求读者掌握 Spring Boot 的核心注解与基本配置，理解 Spring Boot 的自动配置原理与条件注解。

本章主要内容

- Spring Boot 的基本配置
- 读取应用配置
- Spring Boot 的自动配置原理
- Spring Boot 的条件注解

在 Spring Boot 产生之前，Spring 项目会存在多个配置文件，比如 web.xml、application.xml，应用程序自身也需要多个配置文件，同时需要编写程序读取这些配置文件。现在 Spring Boot 简化了 Spring 项目配置的管理和读取，仅需要一个 application.properties 文件，并提供了多种读取配置文件的方式。本章将学习 Spring Boot 的基本配置与运行原理。

4.1　Spring Boot 的基本配置

▶ 4.1.1　启动类和核心注解@SpringBootApplication

Spring Boot 应用通常都有一个名为*Application 的程序入口类，该入口类需要使用 Spring Boot 的核心注解@SpringBootApplication 标注为应用的启动类。另外，该入口类有一个标准的 Java 应用程序的 main 方法，在 main 方法中通过 "SpringApplication.run(*Application.class, args);" 启动 Spring Boot 应用。

Spring Boot 的核心注解@SpringBootApplication 是一个组合注解，主要组合了@SpringBootConfiguration、@EnableAutoConfiguration 和@ComponentScan 注解，源代码可以从 spring-boot-autoconfigure-x.y.z.jar 依赖包中查看。

❶ @SpringBootConfiguration 注解

@SpringBootConfiguration 是 Spring Boot 应用的配置注解，该注解也是一个组合注解，源代码可以从 spring-boot-x.y.z.jar 依赖包中查看。在 Spring Boot 应用中推荐使用@SpringBootConfiguration 注解代替@Configuration 注解。

❷ @EnableAutoConfiguration 注解

@EnableAutoConfiguration 注解可以让 Spring Boot 根据当前应用项目所依赖的 JAR 包自动配置项目的相关配置。例如，在 Spring Boot 项目的 pom.xml 文件中添加了 spring-boot-starter-web 依赖，Spring Boot 项目会自动添加 Tomcat 和 Spring MVC 的依赖，同时对 Tomcat

和 Spring MVC 进行自动配置。打开 pom.xml 文件，单击窗口右侧的 Maven 即可查看 spring-boot-starter-web 的相关依赖，如图 4.1 所示。

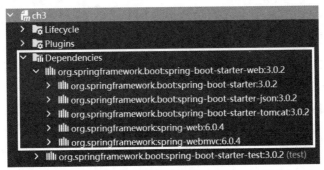

图 4.1　spring-boot-starter-web 的相关依赖

❸ @ComponentScan 注解

该注解的功能是让 Spring Boot 自动扫描@SpringBootApplication 所在类的同级包以及它的子包中的配置，所以建议将@SpringBootApplication 注解的入口类放置在项目包中（Group Id+Artifact Id 组合的包名），并将用户自定义的程序放置在项目包及其子包中，这样可以保证 Spring Boot 自动扫描项目所有包中的配置。

▶ 4.1.2　Spring Boot 的全局配置文件

Spring Boot 的全局配置文件（application.properties 或 application.yml）位于 Spring Boot 应用的 src/main/resources 目录下。

❶ 设置端口号

全局配置文件主要用于修改项目的默认配置。例如，在 Spring Boot 应用 ch13_1 的 src/main/resources 目录下找到名为 application.properties 的全局配置文件，添加如下配置内容：

```
server.port=8888
```

可以将内嵌的 Tomcat 的默认端口改为 8888。

❷ 设置 Web 应用的上下文路径

如果开发者想设置一个 Web 应用程序的上下文路径，可以在 application.properties 文件中配置如下内容：

```
server.servlet.context-path=/XXX
```

这时应该通过"http://localhost:8080/XXX/testStarters"访问如下控制器类中的请求处理方法：

```
@GetMapping("/testStarters")
public String index() {
}
```

❸ 配置文档

在 Spring Boot 的全局配置文件中可以配置与修改多个参数，如果读者想了解参数的详细说明和描述，可以查看官方文档说明（https://docs.spring.io/spring-boot/docs/current/reference/htmlsingle/#appendix.application-properties）。

▶ 4.1.3　Spring Boot 的 Starters

Spring Boot 提供了许多简化企业级开发的"开箱即用"的 Starters。Spring Boot 项目只要使用所需要的 Starters，Spring Boot 即可自动关联项目开发所需要的相关依赖。例如在应用的 pom.xml 文件中添加如下依赖配置：

```
<dependency>
    <groupId>org.springframework.boot</groupId>
    <artifactId>spring-boot-starter-web</artifactId>
</dependency>
```

Spring Boot 将自动关联 Web 开发的相关依赖，例如 Tomcat、spring-webmvc 等，进而对 Web 开发予以支持，并实现相关技术的自动配置。

通过访问"https://docs.spring.io/spring-boot/docs/current/reference/htmlsingle/#using.build-systems.starters"官网，可以查看 Spring Boot 官方提供的 Starters。

除了 Spring Boot 官方提供的 Starters 以外，用户还可以通过访问"https://github.com/spring-projects/spring-boot/blob/master/spring-boot-project/spring-boot-starters/README.adoc"网站查看第三方为 Spring Boot 贡献的 Starters。

4.2 读取应用配置

Spring Boot 提供了 3 种方式读取项目的 application.properties 配置文件的内容，这 3 种方式分别为使用 Environment 类、使用@Value 注解以及使用@ConfigurationProperties 注解。

▶ 4.2.1 使用 Environment 类

Environment 是一个通用的读取应用程序运行时的环境变量的类，可以通过 key-value 方式读取 application.properties、命令行输入参数、系统属性、操作系统环境变量等。下面通过一个实例来演示如何使用 Environment 类读取 application.properties 配置文件的内容。

【例 4-1】使用 Environment 类读取 application.properties 配置文件的内容。

其具体实现步骤如下：

❶ 创建 Spring Boot Web 应用 ch4

参考 3.2.3 节，使用 IDEA 快速创建 Spring Boot Web 应用 ch4，同时给应用 ch4 添加如图 4.2 所示的依赖。在 4.2.3 节使用@ConfigurationProperties 注解读取配置时需要 Spring Configuration Processor 依赖。注意，在使用 Lombok 时需要事先给 IDEA 安装 Lombok 插件。

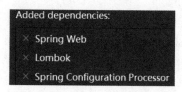

图 4.2　应用 ch4 的依赖

❷ 添加配置文件内容

在应用 ch4 的 src/main/resources 目录下找到全局配置文件 application.properties，并添加如下内容：

```
test.msg=read config
```

❸ 创建控制器类 EnvReaderConfigController

在应用 ch4 的 src/main/java 目录下创建名为 com.ch.ch4.controller 的包（com.ch.ch4 包的

子包，保障注解全部被扫描），并在该包下创建控制器类 EnvReaderConfigController。在控制器类 EnvReaderConfigController 中使用@Autowired 注解依赖注入 Environment 类的对象，具体代码如下：

```
package com.ch.ch4.controller;
import org.springframework.beans.factory.annotation.Autowired;
import org.springframework.core.env.Environment;
import org.springframework.web.bind.annotation.GetMapping;
import org.springframework.web.bind.annotation.RestController;
@RestController
public class EnvReaderConfigController{
    @Autowired
    private Environment env;
    @GetMapping("/testEnv")
    public String testEnv() {
        return "方法一：" + env.getProperty("test.msg") ;
        //test.msg 为配置文件 application.properties 中的 key
    }
}
```

❹ 启动 Spring Boot 应用

运行 Ch4Application 类的 main 方法，启动 Spring Boot 应用。

❺ 测试应用

在启动 Spring Boot 应用后，默认访问地址为"http://localhost:8080/"，将项目路径直接设为根路径，这是 Spring Boot 的默认设置，因此可以通过"http://localhost:8080/testEnv"测试应用（testEnv 与控制器类 ReaderConfigController 中的@GetMapping("/testEnv")对应），测试效果如图 4.3 所示。

图 4.3　使用 Environment 类读取 application.properties 文件的内容

4.2.2　使用@Value 注解

使用@Value 注解读取配置文件内容的示例代码如下：

```
@Value("${test.msg}")    //test.msg 为配置文件 application.properties 中的 key
    private String msg; //通过@Value 注解将配置文件中 key 对应的 value 赋值给变量 msg
```

下面通过实例讲解如何使用@Value 注解读取配置文件内容。

【例 4-2】使用@Value 注解读取配置文件内容。

其具体实现步骤如下：

❶ 创建控制器类 ValueReaderConfigController

在 ch4 应用的 com.ch.ch4.controller 包中创建名为 ValueReaderConfigController 的控制器类，在该控制器类中使用@Value 注解读取配置文件内容，具体代码如下：

```
package com.ch.ch4.controller;
import org.springframework.beans.factory.annotation.Value;
import org.springframework.web.bind.annotation.GetMapping;
import org.springframework.web.bind.annotation.RestController;
```

```java
@RestController
public class ValueReaderConfigController {
    @Value("${test.msg}")
    private String msg;
    @GetMapping("/testValue")
    public String testValue() {
        return "方法二: " + msg;
    }
}
```

❷ 启动并测试应用

首先运行 Ch4Application 类的 main 方法，启动 Spring Boot 应用，然后通过 "http://localhost:8080/testValue" 测试应用。

4.2.3 使用@ConfigurationProperties 注解

使用@ConfigurationProperties 首先建立配置文件与对象的映射关系，然后在控制器方法中使用@Autowired 注解将对象注入。

下面通过实例讲解如何使用@ConfigurationProperties 读取配置文件内容。

【例 4-3】使用@ConfigurationProperties 读取配置文件内容。

其具体实现步骤如下：

❶ 添加配置文件内容

在 ch4 项目的 src/main/resources 目录下找到全局配置文件 application.properties，并添加如下内容：

```
# nest Simple properties
obj.sname=chenheng
obj.sage=88
#List properties
obj.hobby[0]=running
obj.hobby[1]=basketball
#Map Properties
obj.city.cid=dl
obj.city.cname=dalian
```

❷ 建立配置文件与对象的映射关系

在 ch4 项目的 src/main/java 目录下创建名为 com.ch.ch4.model 的包，并在包中创建实体类 StudentProperties，在该类中使用@ConfigurationProperties 注解建立配置文件与对象的映射关系，具体代码如下：

```java
package com.ch.ch4.model;
import java.util.List;
import java.util.Map;
import org.springframework.boot.context.properties.ConfigurationProperties;
import org.springframework.stereotype.Component;
import lombok.Data;
@Component          //使用 Component 注解声明一个组件，被控制器依赖注入
@ConfigurationProperties(prefix = "obj")    //obj 为配置文件中 key 的前缀
@Data
public class StudentProperties {
    private String sname;
    private int sage;
    private List<String> hobby;
    private Map<String, String> city;
    @Override
```

```java
    public String toString() {
        return "StudentProperties [sname=" + sname
            + ", sage=" + sage
            + ", hobby0=" + hobby.get(0)
            + ", hobby1=" + hobby.get(1)
            + ", city=" + city + "]";
    }
}
```

❸ 创建控制器类 ConfigurationPropertiesController

在 ch4 项目的 com.ch.ch4.controller 包中创建名为 ConfigurationPropertiesController 的控制器类，在该控制器类中使用@Autowired 注解依赖注入 StudentProperties 对象，具体代码如下：

```java
package com.ch.ch4.controller;
import org.springframework.beans.factory.annotation.Autowired;
import org.springframework.web.bind.annotation.GetMapping;
import org.springframework.web.bind.annotation.RestController;
import com.ch.ch4.model.StudentProperties;
@RestController
public class ConfigurationPropertiesController {
    @Autowired
    StudentProperties studentProperties;
    @GetMapping("/testConfigurationProperties")
    public String testConfigurationProperties() {
        return studentProperties.toString();
    }
}
```

❹ 启动并测试应用

首先运行 Ch4Application 类的 main 方法，启动 Spring Boot 应用，然后通过 "http://localhost:8080/testConfigurationProperties" 测试应用。

▶ 4.2.4 使用@PropertySource 注解

如果开发者希望读取项目的其他配置文件，而不是全局配置文件 application.properties，应该如何实现呢？可以使用@PropertySource 注解找到项目的其他配置文件，然后结合 4.2.1～4.2.3 小节中的任意一种方式读取。

下面通过实例讲解如何使用@PropertySource + @Value 读取其他配置文件的内容。

【例 4-4】使用@PropertySource + @Value 读取其他配置文件的内容。

其具体实现步骤如下：

❶ 创建配置文件

在 ch4 的 src/main/resources 目录下创建配置文件 ok.properties 和 test.properties，并在 ok.properties 文件中添加如下内容：

```
your.msg=hello.
```

在 test.properties 文件中添加如下内容：

```
my.msg=test PropertySource
```

❷ 创建控制器类 PropertySourceValueReaderOtherController

在 ch4 项目的 com.ch.ch4.controller 包中创建名为 PropertySourceValueReaderOtherController 的控制器类，在该控制器类中首先使用@PropertySource 注解找到其他配置文件，然后使用 @Value 注解读取配置文件的内容，具体代码如下：

```java
package com.ch.ch4.controller;
import org.springframework.beans.factory.annotation.Value;
```

```
import org.springframework.context.annotation.PropertySource;
import org.springframework.web.bind.annotation.GetMapping;
import org.springframework.web.bind.annotation.RestController;
@RestController
@PropertySource({"test.properties","ok.properties"})
public class PropertySourceValueReaderOtherController {
    @Value("${my.msg}")
    private String mymsg;
    @Value("${your.msg}")
    private String yourmsg;
    @GetMapping("/testProperty")
    public String testProperty() {
        return "其他配置文件test.properties: " + mymsg + "<br>"
                + "其他配置文件ok.properties: " + yourmsg;
    }
}
```

❸ 启动并测试应用

首先运行 Ch4Application 类的 main 方法，启动 Spring Boot 应用，然后通过 "http://localhost:8080/testProperty" 测试应用，测试结果如图 4.4 所示。

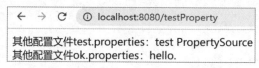

图 4.4 读取其他配置文件的内容

4.3 日志配置

在默认情况下，Spring Boot 项目使用 LogBack 实现日志，使用 apache Commons Logging 作为日志接口，因此代码中通常如下使用日志：

```
package com.ch.ch4.controller;
import org.apache.commons.logging.Log;
import org.apache.commons.logging.LogFactory;
import org.springframework.web.bind.annotation.GetMapping;
import org.springframework.web.bind.annotation.RestController;
@RestController
public class LogTestController {
    private Log log = LogFactory.getLog(LogTestController.class);
    @GetMapping("/testLog")
    public String testLog() {
        log.info("测试日志");
        return "测试日志";
    }
}
```

通过地址 "http://localhost:8080/testLog" 运行上述控制器类代码，可以在控制台输出如图 4.5 所示的日志。

图 4.5 默认日志

日志级别有 ERROR、WARN、INFO、DEBUG 和 TRACE。Spring Boot 默认的日志级别为 INFO，日志信息可以输出到控制台。开发者可以自己设定 Spring Boot 项目的日志输出级别，例如在 application.properties 配置文件中加入以下配置：

```
#设定日志的默认级别为 info
logging.level.root=info
#设定 org 包下的日志级别为 warn
logging.level.org=warn
#设定 com.ch.ch4 包下的日志级别为 debug
logging.level.com.ch.ch4=debug
```

Spring Boot 项目默认并没有输出日志到文件，但开发者可以在 application.properties 配置文件中指定输出日志到文件，配置示例如下：

```
logging.file=my.log
```

日志输出到 my.log 文件，该日志文件位于 Spring Boot 项目运行的当前目录（项目工程目录下）。开发者也可以指定日志文件目录，配置示例如下：

```
logging.file=C:/log/my.log
```

这样将在 C:/log 目录下生成一个名为 my.log 的日志文件。不管日志文件位于何处，当日志文件的大小达到 10MB 时将自动生成一个新日志文件。

Spring Boot 使用内置的 LogBack 支持对控制台日志输出和文件输出进行格式控制，比如开发者可以在 application.properties 配置文件中添加如下配置：

```
logging.pattern.console=%level %date{yyyy-MM-dd HH:mm:ss:SSS} %logger{50}.%M %L:%m%n
logging.pattern.file=%level %date{ISO8601} %logger{50}.%M %L:%m%n
```

logging.pattern.console：指定控制台日志格式。

logging.pattern.file：指定日志文件格式。

%level：指定输出日志级别。

%date：指定日志发生的时间。ISO8601 表示标准日期，相当于 yyyy-MM-dd HH:mm:ss:SSS。

%logger：指定输出 Logger 的名字，格式为包名+类名，{n}限定了输出长度。

%M：指定日志发生时的方法名。

%L：指定日志调用时所在的代码行，适用于开发调试，在线上运行时不建议使用此参数，因为获取代码行对性能有消耗。

%m：表示日志消息。

%n：表示日志换行。

4.4 Spring Boot 的自动配置原理

从 4.1.1 小节可知 Spring Boot 使用核心注解@SpringBootApplication 将一个带有 main 方法的类标注为应用的启动类。@SpringBootApplication 注解最主要的功能之一是为 Spring Boot 开启了一个@EnableAutoConfiguration 注解的自动配置功能。

@EnableAutoConfiguration 注解主要利用了一个类名为 AutoConfigurationImportSelector 的选择器向 Spring 容器自动配置一些组件。@EnableAutoConfiguration 注解的源代码可以从 spring-boot-autoconfigure-x.y.z.jar（org.springframework.boot.autoconfigure）依赖包中查看，核心代码如下：

扫一扫

视频讲解

```java
@Import(AutoConfigurationImportSelector.class)
public @interface EnableAutoConfiguration {
String ENABLED_OVERRIDE_PROPERTY = "spring.boot.enableautoconfiguration";
    Class<?>[] exclude() default {};
    String[] excludeName() default {};
}
```

AutoConfigurationImportSelector（源代码位于 org.springframework.boot.autoconfigure 包）类中有一个名为 selectImports 的方法，该方法规定了向 Spring 容器自动配置的组件。

selectImports 方法的代码如下：

```java
@Override
    public String[] selectImports(AnnotationMetadata annotationMetadata) {
        //判断@EnableAutoConfiguration注解有没有开启，默认开启
        if (!isEnabled(annotationMetadata)) {
            return NO_IMPORTS;
        }
        //获得自动配置
        AutoConfigurationEntry autoConfigurationEntry =
        getAutoConfigurationEntry (annotationMetadata);
        return StringUtils.toStringArray(autoConfigurationEntry.getConfigurations());
}
```

在 selectImports 方法中调用 getAutoConfigurationEntry 方法获得自动配置，进入该方法，查看到的源代码如下：

```java
    protected AutoConfigurationEntry getAutoConfigurationEntry(AnnotationMetadata
        annotationMetadata) {
        if (!isEnabled(annotationMetadata)) {
            return EMPTY_ENTRY;
        }
        AnnotationAttributes attributes = getAttributes(annotationMetadata);
        //获取自动配置数据
        List<String> configurations =
            getCandidateConfigurations(annotationMetadata,attributes);
        //去重
        configurations = removeDuplicates(configurations);
        //去除一些多余的类
        Set<String> exclusions = getExclusions(annotationMetadata, attributes);
        checkExcludedClasses(configurations, exclusions);
        configurations.removeAll(exclusions);
        //过滤掉一些条件没有满足的配置
        configurations = getConfigurationClassFilter().filter(configurations);
        fireAutoConfigurationImportEvents(configurations, exclusions);
        return new AutoConfigurationEntry(configurations, exclusions);
}
```

在 getAutoConfigurationEntry 方法中调用 getCandidateConfigurations 方法获取自动配置数据，进入该方法，查看到的源代码如下：

```java
    protected List<String> getCandidateConfigurations(AnnotationMetadata metadata,
        AnnotationAttributes attributes) {
        List<String> configurations = ImportCandidates.load(AutoConfiguration.class,
            getBeanClassLoader()).getCandidates();
        Assert.notEmpty(configurations, "No auto configuration classes found
            in "+" META-INF/spring/org.springframework.boot.autoconfigure.
            AutoConfiguration.imports. If you are using a custom packaging,
            make sure that file is correct.");
            return configurations;
        }
```

在 getCandidateConfigurations 方法中调用了 ImportCandidates 类的静态方法 load,进入该方法,查看到的源代码如下:

```java
public static ImportCandidates load(Class<?> annotation, ClassLoader classLoader) {
    Assert.notNull(annotation, "'annotation' must not be null");
    ClassLoader classLoaderToUse = decideClassloader(classLoader);
    String location = String.format(LOCATION, annotation.getName());
    Enumeration<URL> urls = findUrlsInClasspath(classLoaderToUse, location);
    List<String> importCandidates = new ArrayList<>();
    while (urls.hasMoreElements()) {
        URL url = urls.nextElement();
        importCandidates.addAll(readCandidateConfigurations(url));
    }
    return new ImportCandidates(importCandidates);
}
```

在 load 方法中可以看到加载了一个字符串常量"LOCATION",该常量的源代码如下:

```java
private static final String LOCATION = "META-INF/spring/%s.imports";
```

从上述源代码中可以看出,最终 Spring Boot 是通过加载所有(in multiple JAR files) META-INF/spring/XXX.imports 配置文件进行自动配置的,所以@SpringBootApplication 注解通过使用@EnableAutoConfiguration 注解自动配置的原理是从 classpath 中搜索所有 META-INF/spring/XXX.imports 配置文件,并将其中 org.springframework.boot.autoconfigure 对应的配置项通过 Java 反射机制进行实例化,然后汇总并加载到 Spring 的 IoC 容器。

在 Spring Boot 项目的 Maven Dependencies 的 spring-boot-autoconfigure-x.y.z.jar 目录下可以找到 META-INF/spring/org.springframework.boot.autoconfigure.AutoConfiguration.imports 配置文件,该文件定义了许多自动配置。

4.5 Spring Boot 的条件注解

打开 META-INF/spring/org.springframework.boot.autoconfigure.AutoConfiguration.imports 配置文件中的任意一个 AutoConfiguration,一般都可以找到条件注解。例如,打开 org.springframework.boot.autoconfigure.aop.AopAutoConfiguration 的源代码,可以看到@ConditionalOnClass 和@ConditionalOnProperty 等条件注解。

通过 org.springframework.boot.autoconfigure.aop.AopAutoConfiguration 的源代码可以看出 Spring Boot 的自动配置是使用 Spring 的@Conditional 注解实现的,因此本节将介绍相关的条件注解,并讲述如何自定义 Spring 的条件注解。

▶ 4.5.1 条件注解

所谓 Spring 的条件注解,就是应用程序的配置类在满足某些特定条件时才会被自动启用此配置类的配置项。Spring Boot 的条件注解位于 spring-boot-autoconfigure-x.y.z.jar 的 org.springframework.boot.autoconfigure.condition 包下,具体如表 4.1 所示。

表 4.1 Spring Boot 的条件注解

注 解 名	条件实现类	条 件
@ConditionalOnBean	OnBeanCondition	Spring 容器中存在指定的实例 Bean
@ConditionalOnClass	OnClassCondition	类加载器(类路径)中存在对应的类

续表

注 解 名	条件实现类	条 件
@ConditionalOnCloudPlatform	OnCloudPlatformCondition	是否在云平台
@ConditionalOnExpression	OnExpressionCondition	判断 SpEL 表达式是否成立
@ConditionalOnJava	OnJavaCondition	指定 Java 版本是否符合要求
@ConditionalOnJndi	OnJndiCondition	在 JNDI（Java 命名和目录接口）存在的条件下查找指定的位置
@ConditionalOnMissingBean	OnBeanCondition	Spring 容器中不存在指定的实例 Bean
@ConditionalOnMissingClass	OnClassCondition	类加载器（类路径）中不存在对应的类
@ConditionalOnNotWebApplication	OnWebApplicationCondition	当前应用程序不是 Web 程序
@ConditionalOnProperty	OnPropertyCondition	应用环境中属性是否存在指定的值
@ConditionalOnResource	OnResourceCondition	是否存在指定的资源文件
@ConditionalOnSingleCandidate	OnBeanCondition	Spring 容器中是否存在且只存在一个对应的实例 Bean
@ConditionalOnWarDeployment	OnWarDeploymentCondition	当前应用程序是传统 WAR 部署
@ConditionalOnWebApplication	OnWebApplicationCondition	当前应用程序是 Web 程序

表 4.1 中的条件注解都组合了@Conditional 元注解，只是针对不同的条件去实现。

▶ 4.5.2 自定义条件

Spring 的@Conditional 注解根据满足某特定条件创建一个特定的 Bean。例如，当某 JAR 包在类路径下时自动配置多个 Bean。这就是根据特定条件控制 Bean 的创建行为，这样开发者就可以利用这个特性进行一些自动配置。那么开发者如何自己构造条件呢？在 Spring 框架中可以通过实现 Condition 接口并重写 matches 方法来构造条件。下面通过实例讲解条件的构造过程。

【例 4-5】自定义条件。要求：如果类路径 classpath（src/main/resources）下存在 test.properties 文件，输出"test.properties 文件存在。"；否则输出"test.properties 文件不存在！"。

其具体实现步骤如下：

❶ 构造条件

在 Spring Boot 应用 ch4 的 src/main/java 目录下创建 com.ch.ch4.conditional 包，并在该包中分别创建条件实现类 MyCondition（存在文件 test.properties）和 YourCondition（不存在文件 test.properties）。

MyCondition 的代码如下：

```java
package com.ch.ch4.conditional;
import org.springframework.context.annotation.Condition;
import org.springframework.context.annotation.ConditionContext;
import org.springframework.core.type.AnnotatedTypeMetadata;
public class MyCondition implements Condition{
    @Override
    public boolean matches(ConditionContext context, AnnotatedTypeMetadata
        metadata) {
        return context.getResourceLoader().getResource("classpath:
            test.properties").exists();
    }
}
```

YourCondition 的代码如下：

```java
package com.ch.ch4.conditional;
import org.springframework.context.annotation.Condition;
import org.springframework.context.annotation.ConditionContext;
import org.springframework.core.type.AnnotatedTypeMetadata;
public class YourCondition implements Condition{
    @Override
    public boolean matches(ConditionContext context, AnnotatedTypeMetadata
        metadata) {
        return !context.getResourceLoader().getResource("classpath:
            test.properties").exists();
    }
}
```

❷ 创建不同条件下 Bean 的类

在 com.ch.ch4.conditional 包中创建接口 MessagePrint，并分别创建该接口的实现类 MyMessagePrint 和 YourMessagePrint。

MessagePrint 的代码如下：

```java
package com.ch.ch4.conditional;
public interface MessagePrint {
    public String showMessage();
}
```

MyMessagePrint 的代码如下：

```java
package com.ch.ch4.conditional;
public class MyMessagePrint implements MessagePrint{
    @Override
    public String showMessage() {
        return "test.properties 文件存在。";
    }
}
```

YourMessagePrint 的代码如下：

```java
package com.ch.ch4.conditional;
public class YourMessagePrint implements MessagePrint{
    @Override
    public String showMessage() {
        return "test.properties 文件不存在！";
    }
}
```

❸ 创建配置类

在 com.ch.ch4.conditional 包中创建配置类 ConditionConfig，并在该配置类中使用@Bean 和@Conditional 实例化符合条件的 Bean。

ConditionConfig 的代码如下：

```java
package com.ch.ch4.conditional;
import org.springframework.context.annotation.Bean;
import org.springframework.context.annotation.Conditional;
import org.springframework.context.annotation.Configuration;
@Configuration
public class ConditionConfig {
    @Bean
    @Conditional(MyCondition.class)
    public MessagePrint myMessage() {
```

```
        return new MyMessagePrint();
    }
    @Bean
    @Conditional(YourCondition.class)
    public MessagePrint yourMessage() {
        return new YourMessagePrint();
    }
}
```

❹ 创建测试类

在 com.ch.ch4.conditional 包中创建测试类 TestMain，具体代码如下：

```
package com.ch.ch4.conditional;
import org.springframework.context.annotation.AnnotationConfigApplicationContext;
public class TestMain {
    private static AnnotationConfigApplicationContext context;
    public static void main(String[] args) {
        context = new AnnotationConfigApplicationContext(ConditionConfig.class);
        MessagePrint mp = context.getBean(MessagePrint.class);
        System.out.println(mp.showMessage());
    }
}
```

❺ 运行

当 Spring Boot 应用 ch4 的 src/main/resources 目录下存在 test.properties 文件时，运行测试类，结果如图 4.6 所示；当 Spring Boot 应用 ch4 的 src/main/resources 目录下不存在 test.properties 文件时，运行测试类，结果如图 4.7 所示。

图 4.6　存在 test.properties 文件　　　　图 4.7　不存在 test.properties 文件

扫一扫

视频讲解

▶ 4.5.3　自定义 Starters

从 4.1.3 小节可知，第三方为 Spring Boot 贡献了许多 Starters。那么作为开发者是否也可以贡献自己的 Starters？在学习 Spring Boot 的自动配置机制以后，答案是肯定的。下面通过实例讲解如何自定义 Starters。

【例 4-6】自定义一个 Starter（spring_boot_mystarters）。要求：当类路径中存在 MyService 类时自动配置该类的 Bean，并可以将相应 Bean 的属性在 application.properties 中配置。

其具体实现步骤如下：

❶ 新建 Spring Boot 项目 spring_boot_mystarters

首先在 IntelliJ IDEA 中通过选择 File→New→Project 命令打开如图 4.8 所示的 New Project 对话框，然后在该对话框中输入项目名称 spring_boot_mystarters，最后单击 Next 和 Create 按钮。

第 4 章 Spring Boot 核心

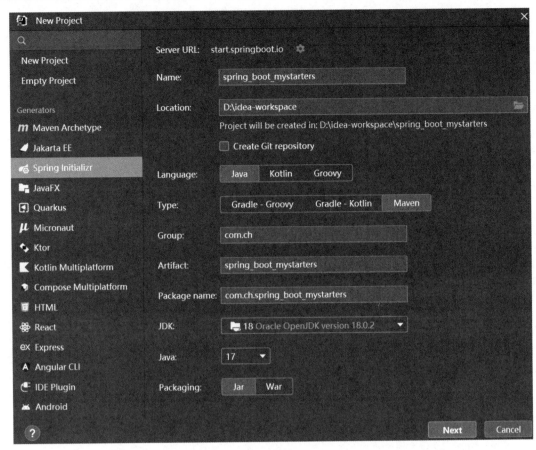

图 4.8 新建 Spring Boot 项目 spring_boot_mystarters

❷ 修改 POM 文件

修改 Spring Boot 项目 spring_boot_mystarters 的 POM 文件，增加 Spring Boot 自身的自动配置作为依赖。另外，在使用@ConfigurationProperties 注解读取配置时需要 Spring Configuration Processor 依赖。POM 文件的代码具体如下：

```xml
<?xml version="1.0" encoding="UTF-8"?>
<project xmlns="http://maven.apache.org/POM/4.0.0"
    xmlns:xsi="http://www.w3.org/2001/XMLSchema-instance"
        xsi:schemaLocation="http://maven.apache.org/POM/4.0.0
    https://maven.apache.org/xsd/maven-4.0.0.xsd">
        <modelVersion>4.0.0</modelVersion>
        <parent>
            <groupId>org.springframework.boot</groupId>
            <artifactId>spring-boot-starter-parent</artifactId>
            <version>3.0.2</version>
            <relativePath/> <!--从存储库中查找父级项目-->
        </parent>
        <groupId>com.ch</groupId>
        <artifactId>spring_boot_mystarters</artifactId>
        <version>0.0.1-SNAPSHOT</version>
        <name>spring_boot_mystarters</name>
        <description>spring_boot_mystarters</description>
        <properties>
            <java.version>17</java.version>
```

```xml
        </properties>
        <dependencies>
            <dependency>
                <groupId>org.springframework.boot</groupId>
                <artifactId>spring-boot-starter</artifactId>
            </dependency>
            <dependency>
                <groupId>org.springframework.boot</groupId>
                <artifactId>spring-boot-autoconfigure</artifactId>
            </dependency>
            <dependency>
                <groupId>org.springframework.boot</groupId>
                <artifactId>spring-boot-configuration-processor</artifactId>
                <optional>true</optional>
            </dependency>
            <dependency>
                <groupId>org.springframework.boot</groupId>
                <artifactId>spring-boot-starter-test</artifactId>
                <scope>test</scope>
            </dependency>
        </dependencies>
        <build>
            <plugins>
                <plugin>
                    <groupId>org.springframework.boot</groupId>
                    <artifactId>spring-boot-maven-plugin</artifactId>
                    <configuration>
                        <skip>true</skip>
                    </configuration>
                </plugin>
            </plugins>
        </build>
</project>
```

❸ 创建属性配置类 MyProperties

在项目 spring_boot_mystarters 的 com.ch.spring_boot_mystarters 包中创建属性配置类 MyProperties。在使用 spring_boot_mystarters 的 Spring Boot 项目的配置文件 application.properties 中可以使用 my.msg=设置属性，若不设置，默认为 my.msg=默认值。属性配置类 MyProperties 的代码如下：

```java
package com.ch.spring_boot_mystarters;
import org.springframework.boot.context.properties.ConfigurationProperties;
//在application.properties中使用my.msg=设置属性
@ConfigurationProperties(prefix="my")
public class MyProperties {
    private String msg = "默认值";
    public String getMsg() {
        return msg;
    }
    public void setMsg(String msg) {
        this.msg = msg;
    }
}
```

❹ 创建判断依据类 MyService

在项目 spring_boot_mystarters 的 com.ch.spring_boot_mystarters 包中创建判断依据类

MyService。本例自定义的 Starters 将根据该类的存在与否来创建该类的 Bean，该类可以是第三方类库的类。判断依据类 MyService 的代码如下：

```java
package com.ch.spring_boot_mystarters;
public class MyService {
    private String msg;
    public String sayMsg() {
        return "my " + msg;
    }
    public String getMsg() {
        return msg;
    }
    public void setMsg(String msg) {
        this.msg = msg;
    }
}
```

❺ 创建自动配置类 MyAutoConfiguration

在项目 spring_boot_mystarters 的 com.ch.spring_boot_mystarters 包中创建自动配置类 MyAutoConfiguration。在该类中使用@EnableConfigurationProperties 注解开启属性配置类 MyProperties 提供参数；使用@ConditionalOnClass 注解判断类加载器（类路径）中是否存在 MyService 类；使用@ConditionalOnMissingBean 注解判断当容器中不存在 MyService 的 Bean 时自动配置这个 Bean。自动配置类 MyAutoConfiguration 的代码如下：

```java
package com.ch.spring_boot_mystarters;
import org.springframework.beans.factory.annotation.Autowired;
import org.springframework.boot.autoconfigure.condition.ConditionalOnClass;
import org.springframework.boot.autoconfigure.condition.ConditionalOnMissingBean;
import org.springframework.boot.autoconfigure.condition.ConditionalOnProperty;
import org.springframework.boot.context.properties.EnableConfigurationProperties;
import org.springframework.context.annotation.Bean;
import org.springframework.context.annotation.Configuration;
@Configuration
//开启属性配置类 MyProperties 提供参数
@EnableConfigurationProperties(MyProperties.class)
//类加载器（类路径）中是否存在对应的类
@ConditionalOnClass(MyService.class)
//应用环境中属性是否存在指定的值
@ConditionalOnProperty(prefix = "my", value = "enabled", matchIfMissing = true)
public class MyAutoConfiguration {
    @Autowired
    private MyProperties myProperties;
    @Bean
    //当容器中不存在 MyService 的 Bean 时自动配置这个 Bean
    @ConditionalOnMissingBean(MyService.class)
    public MyService myService() {
        MyService myService = new MyService();
        myService.setMsg(myProperties.getMsg());
        return myService;
    }
}
```

❻ 注册配置

在项目 spring_boot_mystarters 的 src/main/resources 目录下新建文件夹 META-INF/spring，并在该文件夹下创建名为 com.ch.spring_boot_mystarters.MyAutoConfiguration.imports 的文件。

在 com.ch.spring_boot_mystarters.MyAutoConfiguration.imports 文件中添加如下内容注册自动配置类 MyAutoConfiguration：

```
com.ch.spring_boot_mystarters.MyAutoConfiguration
```

在上述文件内容中，若有多个自动配置类，换行即可注册另一个自动配置类。

至此，经过上述 6 个步骤后，自定义 Starters（spring_boot_mystarters）已经完成。开发者可以将 spring_boot_mystarters 安装到 Maven 的本地库，或者将 JAR 包发布到 Maven 的私服上。

下面讲解如何将 spring_boot_mystarters 安装到 Maven 的本地库，具体做法是使用 IntelliJ IDEA 打开 spring_boot_mystarters 项目，然后进一步打开右侧的 Maven，双击 Lifecycle 中的 install，即可将 spring_boot_mystarters 安装到 Maven 的本地库，如图 4.9 所示。

图 4.9　将 spring_boot_mystarters 安装到 Maven 的本地库

在成功地将 spring_boot_mystarters 安装到 Maven 的本地库之后，在项目 spring_boot_mystarters 的 target 目录下可以看到 spring_boot_mystarters-0.0.1-SNAPSHOT.jar 文件。

下面讲解如何在另一个 Spring Boot 应用中使用 spring_boot_mystarters-0.0.1-SNAPSHOT.jar 作为项目依赖。

【例 4-7】创建 Spring Boot 的 Web 应用 ch4_1，并在 ch4_1 中使用 spring_boot_mystarters 作为项目依赖。

其具体实现步骤如下：

❶ 创建 Spring Boot 的 Web 应用 ch4_1

使用 IntelliJ IDEA 快速创建 Spring Boot 的 Web 应用 ch4_1。

❷ 添加 spring_boot_mystarters 的依赖

在 Web 应用 ch4_1 的 pom.xml 文件中添加 spring_boot_mystarters 的依赖，代码如下：

```xml
<dependency>
    <groupId>com.ch</groupId>
    <artifactId>spring_boot_mystarters</artifactId>
    <version>0.0.1-SNAPSHOT</version>
</dependency>
```

在添加依赖以后，可以在 Maven 的依赖中查看 spring_boot_mystarters 依赖，如图 4.10 所示。

第 4 章 Spring Boot 核心

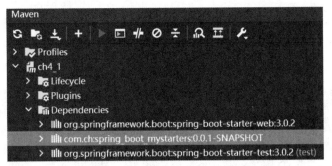

图 4.10 查看 spring_boot_mystarters 依赖

❸ 修改程序入口类 Ch41Application，测试 spring_boot_mystarters

Ch41Application 类修改后的代码如下：

```
package com.ch.ch4_1;
import com.ch.spring_boot_mystarters.MyService;
import org.springframework.beans.factory.annotation.Autowired;
import org.springframework.boot.SpringApplication;
import org.springframework.boot.autoconfigure.SpringBootApplication;
import org.springframework.web.bind.annotation.GetMapping;
import org.springframework.web.bind.annotation.RestController;
@RestController
//扫描 com.ch.spring_boot_mystarters 包
@SpringBootApplication(scanBasePackages = {"com.ch.spring_boot_mystarters"})
public class Ch41Application {
    @Autowired
    private MyService myService;
    public static void main(String[] args) {
        SpringApplication.run(Ch41Application.class, args);
    }
    @GetMapping("/testStarters")
    public String index() {
        return myService.sayMsg();
    }
}
```

运行 Ch41Application 应用程序，启动 Web 应用，通过访问"http://localhost:8080/testStarters"测试 spring_boot_mystarters，运行效果如图 4.11 所示。

图 4.11 访问"http://localhost:8080/testStarters"

这时在 Web 应用 ch4_1 的 application.properties 文件中配置 msg 的内容：

```
my.msg=starter pom
```

然后运行 Ch41Application 应用程序，重新启动 Web 应用，再次访问"http://localhost:8080/testStarters"，运行效果如图 4.12 所示。

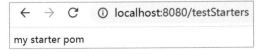

图 4.12 配置 msg 后的效果

另外，可以在 Web 应用 ch4_1 的 application.properties 文件中配置 debug 属性（debug=true），查看自动配置报告。重新启动 Web 应用，可以在控制台中查看到如图 4.13 所示的自定义的自动配置。

图 4.13　查看自动配置报告

本章小结

本章重点讲解了 Spring Boot 的基本配置、读取应用配置、日志配置、自动配置原理以及条件注解。通过本章的学习，读者应掌握 Spring Boot 的核心注解@SpringBootApplication 的基本用法，理解 Spring Boot 的自动配置原理，了解 Spring Boot 的条件注解。重要的是，开发者可以利用 Spring Boot 的自动配置与条件注解贡献自己的 Starters。

扫一扫

自测题

习题 4

1. 如何读取 Spring Boot 项目的应用配置？请举例说明。

2. 参考例 4-5，编写 Spring Boot 应用程序 practice4_2。要求：以不同的操作系统作为条件，若在 Windows 系统下运行程序，则输出列表命令为 dir；若在 Linux 操作系统下运行程序，则输出列表命令为 ls。

3. 参考例 4-6 和例 4-7，自定义一个 Starter（spring_boot_addstarters）和 Spring Boot 的 Web 应用 practice4_3。在 practice4_3 中，使用 spring_boot_addstarters 计算两个整数的和，通过访问"http://localhost:8080/testAddStarters"返回两个整数的和。在 spring_boot_addstarters 中，首先创建属性配置类 AddProperties（有 Integer 类型的 number1 和 number2 两个属性），在该属性配置类中使用@ConfigurationProperties(prefix="add")注解设置属性前缀为 add；然后创建判断依据类 AddService（有 Integer 类型的 number1 和 number2 两个属性），在 AddService 类中提供 add 方法（计算 number1 与 number2 的和）；接着创建自动配置类 AddAutoConfiguration，当类路径中存在 AddService 类时自动配置该类的 Bean，并可以将相应 Bean 的属性在 application.properties 中配置；最后注册自动配置类 AddAutoConfiguration。

第 5 章　Spring Boot 的 Web 开发

学习目的与要求

本章首先介绍 Spring Boot 的 Web 开发支持，然后介绍 Thymeleaf 模板引擎技术，最后介绍 Spring Boot 的 Web 开发技术（JSON 数据交互、文件的上传与下载、异常统一处理以及对 JSP 的支持）。通过本章的学习，要求读者掌握 Spring Boot 的 Web 开发技术。

本章主要内容

- Thymeleaf 模板引擎
- 使用 Spring Boot 处理 JSON 数据
- Spring Boot 的文件上传与下载
- Spring Boot 的异常处理
- Spring Boot 对 JSP 的支持

Web 开发是一种基于 B/S 架构（即浏览器/服务器）的应用软件开发技术，分为前端（用户接口）和后端（业务逻辑和数据），前端的可视化及用户交互由浏览器实现，即通过浏览器作为客户端，实现客户与服务器远程的数据交互。Spring Boot 的 Web 开发内容主要包括内嵌 Servlet 容器和 Spring MVC。

5.1　Spring Boot 的 Web 开发支持

Spring Boot 提供了 spring-boot-starter-web 依赖模块，该依赖模块包含了 Spring Boot 预定义的 Web 开发常用的依赖包，为 Web 开发者提供了内嵌的 Servlet 容器（Tomcat）以及 Spring MVC 的依赖。如果开发者希望开发 Spring Boot 的 Web 应用程序，可以在 Spring Boot 项目的 pom.xml 文件中添加如下依赖配置：

```xml
<dependency>
    <groupId>org.springframework.boot</groupId>
    <artifactId>spring-boot-starter-web</artifactId>
</dependency>
```

Spring Boot 将自动关联 Web 开发的相关依赖，例如 Tomcat、spring-webmvc 等，进而对 Web 开发予以支持，并实现相关技术的自动配置。

另外，开发者也可以使用 IDEA 集成开发工具快速创建 Spring Initializr，在 New Project 对话框中添加 Spring Boot 的 Web 依赖。

5.2　Thymeleaf 模板引擎

在 Spring Boot 的 Web 应用中建议开发者使用 HTML 完成动态页面。Spring Boot 提供了许多模板引擎，主要包括 FreeMarker、Groovy、Thymeleaf、Velocity 和 Mustache。因为 Thymeleaf

提供了完美的 Spring MVC 支持，所以在 Spring Boot 的 Web 应用中推荐使用 Thymeleaf 作为模板引擎。

Thymeleaf 是一个 Java 类库，是一个 XML/XHTML/HTML 5 的模板引擎，能够处理 HTML、XML、JavaScript 以及 CSS，可以作为 MVC Web 应用的 View 层显示数据。

▶ 5.2.1 Spring Boot 的 Thymeleaf 支持

在 Spring Boot 1.X 版本中，spring-boot-starter-thymeleaf 依赖包含了 spring-boot-starter-web 模块，但是在 Spring 5 中 WebFlux 的出现使 Web 应用的解决方案不再唯一，所以 spring-boot-starter-thymeleaf 依赖不再包含 spring-boot-starter-web 模块，需要开发人员自己选择 spring-boot-starter-web 模块依赖。下面通过一个实例讲解如何创建基于 Thymeleaf 模板引擎的 Spring Boot Web 应用 ch5_1。

【例 5-1】创建基于 Thymeleaf 模板引擎的 Spring Boot Web 应用 ch5_1。

其具体实现步骤如下：

❶ 创建基于 Thymeleaf 模板引擎的 Spring Boot Web 应用 ch5_1

在 IDEA 中选择 File→New→Project 命令，打开 New Project 对话框；在 New Project 对话框中选择和输入相关信息后单击 Next 按钮，进入新的界面；在新的界面中选择 Thymeleaf、Lombok 和 Spring Web 依赖，单击 Create 按钮即可创建应用 ch5_1。

❷ 打开项目目录

创建如图 5.1 所示的基于 Thymeleaf 模板引擎的 Spring Boot Web 应用 ch5_1。

图 5.1 基于 Thymeleaf 模板引擎的 Spring Boot Web 应用 ch5_1

Thymeleaf 模板默认将 JS 脚本、CSS 样式、图片等静态文件放置在 src/main/resources/static 目录下；将视图页面放置在 src/main/resources/templates 目录下。

❸ 设置 Web 应用 ch5_1 的上下文路径

在 ch5_1 的 application.properties 文件中配置如下内容：

```
server.servlet.context-path=/ch5_1
```

第 5 章　Spring Boot 的 Web 开发

❹ 创建控制器类

创建一个名为 com.ch.ch5_1.controller 的包，并在该包中创建控制器类 TestThymeleafController，具体代码如下：

```
package com.ch.ch5_1.controller;
import org.springframework.stereotype.Controller;
import org.springframework.web.bind.annotation.GetMapping;
@Controller
public class TestThymeleafController {
    @GetMapping("/")
    public String test(){
        //根据 Thymeleaf 模板，默认将返回 src/main/resources/templates/index.html
        return "index";
    }
}
```

❺ 新建 index.html 页面

在 src/main/resources/templates 目录下新建 index.html 页面，代码如下：

```
<!DOCTYPE html>
<html>
<head>
<meta charset="UTF-8">
<title>Insert title here</title>
</head>
<body>
测试 Spring Boot 的 Thymeleaf 支持
</body>
</html>
```

❻ 运行测试

首先运行 Ch51Application 主类，然后访问"http://localhost:8080/ch5_1/"，运行结果如图 5.2 所示。

图 5.2　例 5-1 的运行结果

▶ 5.2.2　Thymeleaf 的基础语法

❶ 引入 Thymeleaf

首先将 View 层页面文件的 HTML 标签修改如下：

```
<html xmlns:th="http://www.thymeleaf.org">
```

然后在 View 层页面文件的其他标签中使用 th:*动态处理页面，示例代码如下：

```
<img th:src="'images/' + ${aBook.picture}"/>
```

其中，${aBook.picture}获得数据对象 aBook 的 picture 属性。

❷ 输出内容

使用 th:text 和 th:utext（不对文本转义，正常输出）将文本内容输出到所在标签的 body 中。假如在国际化资源文件 messages_en_US.properties 中有消息文本"test.myText=<strong>Test International Message</strong>"，那么在页面中可以使用以下两种方式获得消息文本：

87

```html
<p th:text="#{test.myText}"></p>
<!--对文本转义，即输出<strong>Test International Message</strong> -->
<p th:utext="#{test.myText}"></p>
<!--对文本不转义，即输出加粗的"Test International Message"-->
```

❸ 基本表达式

1）变量表达式：${...}

变量表达式用于访问容器上下文环境中的变量，示例代码如下：

```html
<span th:text="${information}">
```

2）选择变量表达式：*{...}

选择变量表达式计算的是选定的对象（th:object 属性绑定的对象），示例代码如下：

```html
<div th:object="${session.user}">
    name: <span th: text="*{firstName}"></span><br>
    <!-- firstName 为 user 对象的属性-->
    surname: <span th: text="*{lastName}"></span><br>
    nationality: <span th: text="*{nationality}"></span><br>
</div>
```

3）信息表达式：#{...}

信息表达式一般用于显示页面中的静态文本，将可能需要根据需求整体变动的静态文本放在 properties 文件中以便维护（例如国际化），通常与 th:text 属性一起使用，示例代码如下：

```html
<p th:text="#{test.myText}"></p>
```

❹ 引入 URL

Thymeleaf 模板通过@{...}表达式引入 URL，示例代码如下：

```html
<!--默认访问 src/main/resources/static 下的 css 文件夹-->
<link rel="stylesheet" th:href="@{css/bootstrap.min.css}" />
<!--访问相对路径-->
<a th:href="@{/}">去看看</a>
<!--访问绝对路径-->
<a th:href="@{http://www.tup.tsinghua.edu.cn/index.html(param1='传参')}">去清华大学出版社</a>
<!--默认访问 src/main/resources/static 下的 images 文件夹-->
<img th:src="'images/' + ${aBook.picture}"/>
```

❺ 访问 WebContext 对象中的属性

Thymeleaf 模板通过一些专门的表达式从模板的 WebContext 获取请求参数、请求、会话和应用程序中的属性，具体如下：

${xxx}将返回存储在 Thymeleaf 模板上下文中的变量 xxx 或请求 request 作用域中的属性 xxx。

${param.xxx}将返回一个名为 xxx 的请求参数（可能是多个值）。

${session.xxx}将返回一个名为 xxx 的 HttpSession 作用域中的属性。

${application.xxx}将返回一个名为 xxx 的全局 ServletContext 上下文作用域中的属性。

与 EL 表达式一样，使用${xxx}获得变量值，使用${对象变量名.属性名}获取 JavaBean 属性值。需要注意的是，${}表达式只能在 th 标签内部有效。

❻ 运算符

在 Thymeleaf 模板的表达式中可以使用+、-、*、/、%等各种算术运算符，也可以使用>、<、<=、>=、==、!=等各种逻辑运算符。示例代码如下：

```html
<tr th:class="(${row}== 'even')? 'even' : 'odd'">...</tr>
```

❼ 条件判断

1）if 和 unless

标签只有在 th:if 条件成立时才显示，th:unless 与 th:if 相反，只有在条件不成立时才显示标签内容。示例代码如下：

```html
<a href="success.html" th:if="${user != null}">成功</a>
<a href="success.html" th:unless="${user = null}">成功</a>
```

2）switch 语句

Thymeleaf 模板也支持多路选择 switch 语句结构，默认属性 default 可用"*"表示。示例代码如下：

```html
<div th:switch="${user.role}">
<p th:case="'admin'">User is an administrator</p>
<p th:case="'teacher'">User is a teacher</p>
<p th:case="*">User is a student </p>
</div>
```

❽ 循环

1）基本循环

Thymeleaf 模板使用 th:each="obj,iterStat:${objList}"标签进行迭代循环，迭代对象可以是 java.util.List、java.util.Map 或数组等。示例代码如下：

```html
<!--循环取出集合数据-->
<div class="col-md-4 col-sm-6" th:each="book:${books}">
    <a href="">
        <img th:src="'images/' + ${book.picture}" alt="图书封面" style=
                    "height: 180px; width: 40%;"/>
    </a>
    <div class="caption">
        <h4 th:text="${book.bname}"></h4>
        <p th:text="${book.author}"></p>
        <p th:text="${book.isbn}"></p>
        <p th:text="${book.price}"></p>
        <p th:text="${book.publishing}"></p>
    </div>
</div>
```

2）循环状态的使用

在 th:each 标签中可以使用循环状态变量，该变量有以下属性。

index：当前迭代对象的 index（从 0 开始计数）。

count：当前迭代对象的 index（从 1 开始计数）。

size：迭代对象的大小。

current：当前迭代变量。

even/odd：布尔值，当前循环是否为偶数/奇数（从 0 开始计数）。

first：布尔值，当前循环是否为第一个。

last：布尔值，当前循环是否为最后一个。

使用循环状态变量的示例代码如下：

```html
<!--循环取出集合数据-->
<div class="col-md-4 col-sm-6" th:each="book,bookStat:${books}">
    <a href="">
        <img th:src="'images/' + ${book.picture}" alt="图书封面" style=
                    "height: 180px; width: 40%;"/>
```

```
        </a>
        <div class="caption">
            <!--循环状态 bookStat-->
            <h3 th:text="${bookStat.count}"></h3>
            <h4 th:text="${book.bname}"></h4>
            <p th:text="${book.author}"></p>
            <p th:text="${book.isbn}"></p>
            <p th:text="${book.price}"></p>
            <p th:text="${book.publishing}"></p>
        </div>
</div>
```

5.2.3 Thymeleaf 的常用属性

通过 5.2.2 小节的学习，读者可以发现 Thymeleaf 是通过在 HTML 页面的标签中添加 th:xxx 关键字来实现模板套用，且其属性与 HTML 页面的标签基本类似。下面介绍 Thymeleaf 的几个常用属性。

❶ th:action

th:action 用于定义后台控制器路径，类似<form>标签的 action 属性，示例代码如下：

```
<form th:action="@{/login}">...</form>
```

❷ th:each

th:each 用于集合对象的遍历，功能类似 JSTL 标签<c:forEach>，示例代码如下：

```
<div class="col-md-4 col-sm-6" th:each="gtype:${gtypes}">
    <div class="caption">
        <p th:text="${gtype.id}"></p>
        <p th:text="${gtype.typename}"></p>
    </div>
</div>
```

❸ th:field

th:field 常用于表单参数的绑定，通常与 th:object 一起使用，示例代码如下：

```
<form th:action="@{/login}" th:object="${user}">
    <input type="text" value="" th:field="*{username}"></input>
    <input type="text" value="" th:field="*{role}"></input>
</form>
```

❹ th:href

th:href 用于定义超链接，类似<a>标签的 href 属性。其 value 形式为@{/logout}，示例代码如下：

```
<a th:href="@{/gogo}"></a>
```

❺ th:id

th:id 用于 div 标签中的 id 声明，类似 HTML 标签中的 id 属性，示例代码如下：

```
<div th:id ="stu+(${rowStat.index}+1)"></div>
```

❻ th:if

th:if 用于条件判断，如果为否则标签不显示，示例代码如下：

```
<div th:if="${rowStat.index} == 0">... do something ...</div>
```

❼ th:fragment

th:fragment 声明定义该属性的 div 为模板片段，用于头文件、页尾文件的引入，常与 th:include、th:replace 一起使用。

假如在 ch5_1 的 src/main/resources/templates 目录下声明模板片段文件 footer.html，代码如下：

```html
<!DOCTYPE html>
<html xmlns:th="http://www.thymeleaf.org">
<head>
<meta charset="UTF-8">
<title>Insert title here</title>
</head>
<body>
    <!--声明片段 content-->
    <div th:fragment="content" >
        主体内容
    </div>
    <!--声明片段 copy-->
    <div th:fragment="copy" >
        ©清华大学出版社
    </div>
</body>
</html>
```

那么可以在 ch5_1 的 src/main/resources/templates/index.html 文件中引入模板片段，代码如下：

```html
<!DOCTYPE html>
<html xmlns:th="http://www.thymeleaf.org">
<head>
<meta charset="UTF-8">
<title>Insert title here</title>
</head>
<body>
    测试 Spring Boot 的 Thymeleaf 支持<br>
    引入主体内容模板片段：
    <div th:include="footer::content"></div>
    引入版权所有模板片段：
    <div th:replace="footer::copy" ></div>
</body>
</html>
```

❽ th:object

th:object 用于表单数据对象的绑定，将表单绑定到后台 controller 的一个 JavaBean 参数，常与 th:field 一起使用。下面通过实例讲解表单提交及数据绑定的实现过程。

【例 5-2】表单提交及数据绑定的实现过程。

其具体实现步骤如下：

1）创建实体类

在 Web 应用 ch5_1 的 src/main/java 目录下创建 com.ch.ch5_1.model 包，并在该包中创建实体类 LoginBean，代码如下：

```java
package com.ch.ch5_1.model;
import lombok.Data;
@Data
public class LoginBean {
    String uname;
    String urole;
}
```

2）创建控制器类

在 Web 应用 ch5_1 的 com.ch.ch5_1.controller 包中创建控制器类 LoginController，代码如下：

```java
package com.ch.ch5_1.controller;
import org.springframework.stereotype.Controller;
import org.springframework.ui.Model;
import org.springframework.web.bind.annotation.GetMapping;
import org.springframework.web.bind.annotation.ModelAttribute;
import org.springframework.web.bind.annotation.PostMapping;
import com.ch.ch5_1.model.LoginBean;
@Controller
public class LoginController {
    @GetMapping("/toLogin")
    public String toLogin(Model model) {
        /*loginBean与login.html页面中的th:object="${loginBean}"相同*/
        model.addAttribute("loginBean", new LoginBean());
        return "login";
    }
    @PostMapping("/login")
    public String greetingSubmit(@ModelAttribute LoginBean loginBean) {
        /*@ModelAttribute LoginBean loginBean 接收login.html页面中的表单数据，
并将loginBean对象保存到model中返回给result.html页面显示*/
        System.out.println("测试提交的数据: " + loginBean.getUname());
        return "result";
    }
}
```

3）创建页面表示层

在Web应用ch5_1的src/main/resources/templates目录下创建页面login.html和result.html。

页面login.html的代码如下：

```html
<!DOCTYPE html>
<html xmlns:th="http://www.thymeleaf.org">
<head>
<meta charset="UTF-8">
<title>Insert title here</title>
</head>
<body>
    <h1>Form</h1>
    <form action="#" th:action="@{/login}" th:object="${loginBean}" method="post">
    <!--th:field="*{uname}"中的uname与实体类的属性相同，即绑定loginBean对象-->
        <p>Uname: <input type="text" th:field="*{uname}" th:placeholder=
                        "请输入用户名" /></p>
        <p>Urole: <input type="text" th:field="*{urole}" th:placeholder=
                        "请输入角色" /></p>
        <p><input type="submit" value="Submit" /> <input type="reset" value=
                        "Reset" /></p>
    </form>
</body>
</html>
```

页面result.html的代码如下：

```html
<!DOCTYPE html>
<html xmlns:th="http://www.thymeleaf.org">
<head>
<meta charset="UTF-8">
<title>Insert title here</title>
</head>
<body>
    <h1>Result</h1>
    <p th:text="'Uname: ' + ${loginBean.uname}" />
    <p th:text="'Urole: ' + ${loginBean.urole}" />
```

```
        <a href="toLogin">继续提交</a>
    </body>
</html>
```

4）运行

首先运行 Ch51Application 主类，然后访问 "http://localhost:8080/ch5_1/toLogin"，运行结果如图 5.3 所示。

在图 5.3 的文本框中输入信息后单击 Submit 按钮，打开如图 5.4 所示的页面。

图 5.3　页面 login.html 的运行结果

图 5.4　页面 result.html 的运行结果

❾ th:src

th:src 用于外部资源的引入，类似 \<script\>标签的 src 属性，示例代码如下：

```
<img th:src="'images/' + ${aBook.picture}" />
```

❿ th:text

th:text 用于文本的显示，将文本内容显示到所在标签的 body 中，示例代码如下：

```
<td th:text="${username}"></td>
```

⓫ th:value

th:value 用于标签的赋值，类似标签的 value 属性，示例代码如下：

```
<option th:value="Adult">Adult</option>
<input type="hidden" th:value="${msg}" />
```

⓬ th:style

th:style 用于修改标签 style，示例代码如下：

```
<span th:style="'display:' + @{(${myVar} ? 'none' : 'inline-block')}"> myVar
是一个变量</span>
```

⓭ th:onclick

th:onclick 用于修改单击事件，示例代码如下：

```
<button th:onclick="'getCollect()'"></button>
```

5.2.4　通过 Spring Boot 与 Thymeleaf 实现页面信息的国际化

在 Spring Boot 的 Web 应用中实现页面信息的国际化非常简单，下面通过实例讲解页面信息的国际化的实现过程。

【例 5-3】页面信息的国际化的实现过程。

其具体实现步骤如下：

❶ 编写国际化资源属性文件

1）编写管理员模块的国际化信息

在 ch5_1 的 src/main/resources 目录下创建 i18n/admin 文件夹，并在该文件夹下创建

adminMessages.properties、adminMessages_en_US.properties 和 adminMessages_zh_CN.properties 资源属性文件。adminMessages.properties 表示默认加载的信息；adminMessages_en_US.properties 表示英文信息（en 代表语言代码，US 代表国家地区）；adminMessages_zh_CN.properties 表示中文信息。

adminMessages.properties 的内容如下：

```
test.admin=\u6D4B\u8BD5\u540E\u53F0
admin=\u540E\u53F0\u9875\u9762
```

adminMessages_en_US.properties 的内容如下：

```
test.admin=test admin
admin=admin
```

adminMessages_zh_CN.properties 的内容如下：

```
test.admin=\u6D4B\u8BD5\u540E\u53F0
admin=\u540E\u53F0\u9875\u9762
```

2）编写用户模块的国际化信息

在 ch5_1 的 src/main/resources 目录下创建 i18n/before 文件夹，并在该文件夹下创建 beforeMessages.properties、beforeMessages_en_US.properties 和 beforeMessages_zh_CN.properties 资源属性文件。

beforeMessages.properties 的内容如下：

```
test.before=\u6D4B\u8BD5\u524D\u53F0
before=\u524D\u53F0\u9875\u9762
```

beforeMessages_en_US.properties 的内容如下：

```
test.before=test before
before=before
```

beforeMessages_zh_CN.properties 的内容如下：

```
test.before=\u6D4B\u8BD5\u524D\u53F0
before=\u524D\u53F0\u9875\u9762
```

3）编写公共模块的国际化信息

在 ch5_1 的 src/main/resources 目录下创建 i18n/common 文件夹，并在该文件夹下创建 commonMessages.properties、commonMessages_en_US.properties 和 commonMessages_zh_CN.properties 资源属性文件。

commonMessages.properties 的内容如下：

```
chinese.key=\u4E2D\u6587\u7248
english.key=\u82F1\u6587\u7248
return=\u8FD4\u56DE\u9996\u9875
```

commonMessages_en_US.properties 的内容如下：

```
chinese.key=chinese
english.key=english
return=return
```

commonMessages_zh_CN.properties 的内容如下：

```
chinese.key=\u4E2D\u6587\u7248
english.key=\u82F1\u6587\u7248
return=\u8FD4\u56DE\u9996\u9875
```

❷ 添加配置文件内容，引入资源属性文件

在 ch5_1 应用的配置文件中添加如下内容，引入资源属性文件。

```
spring.messages.basename=i18n/admin/adminMessages,i18n/before/beforeMessages,
i18n/common/commonMessages
```

❸ 重写 localeResolver 方法设定语言区域

在 ch5_1 应用的 com.ch.ch5_1 包中创建配置类 LocaleConfig，该配置类实现 WebMvcConfigurer 接口，并设定语言区域。LocaleConfig 的代码如下：

```
package com.ch.ch5_1;
import java.util.Locale;
import org.springframework.boot.autoconfigure.EnableAutoConfiguration;
import org.springframework.context.annotation.Bean;
import org.springframework.context.annotation.Configuration;
import org.springframework.web.servlet.LocaleResolver;
import org.springframework.web.servlet.config.annotation.InterceptorRegistry;
import org.springframework.web.servlet.config.annotation.WebMvcConfigurer;
import org.springframework.web.servlet.i18n.LocaleChangeInterceptor;
import org.springframework.web.servlet.i18n.SessionLocaleResolver;
@Configuration
@EnableAutoConfiguration
public class LocaleConfig implements WebMvcConfigurer {
    /**
     *根据用户本次会话过程中的语义设定语言区域（例如用户进入首页时选择的语言种类）
     */
    @Bean
    public LocaleResolver localeResolver() {
        SessionLocaleResolver slr = new SessionLocaleResolver();
        //默认语言
        slr.setDefaultLocale(Locale.CHINA);
        return slr;
    }
    /**
     *在使用 SessionLocaleResolver 存储语言区域时必须配置 localeChangeInterceptor 拦截器
     */
    @Bean
    public LocaleChangeInterceptor localeChangeInterceptor() {
        LocaleChangeInterceptor lci = new LocaleChangeInterceptor();
        //选择语言的参数名
        lci.setParamName("locale");
        return lci;
    }
    /**
     * 注册拦截器
     */
    @Override
    public void addInterceptors(InterceptorRegistry registry) {
        registry.addInterceptor(localeChangeInterceptor());
    }
}
```

❹ 创建控制器类 I18nTestController

在 ch5_1 应用的 com.ch.ch5_1.controller 包中创建控制器类 I18nTestController，具体代码如下：

```
package com.ch.ch5_1.controller;
import org.springframework.stereotype.Controller;
import org.springframework.web.bind.annotation.GetMapping;
import org.springframework.web.bind.annotation.RequestMapping;
@Controller
@RequestMapping("/i18n")
```

```java
public class I18nTestController {
    @GetMapping("/first")
    public String testI18n(){
        return "/i18n/first";
    }
    @GetMapping("/admin")
    public String admin(){
        return "/i18n/admin";
    }
    @GetMapping("/before")
    public String before(){
        return "/i18n/before";
    }
}
```

❺ 创建视图页面，并获得国际化信息

在 ch5_1 应用的 src/main/resources/templates 目录下创建文件夹 i18n，并在该文件夹中创建 admin.html、before.html 和 first.html 视图页面，在这些视图页面中使用 th:text="#{xxx}" 获得国际化信息。

admin.html 的代码如下：

```html
<!DOCTYPE html>
<html xmlns:th="http://www.thymeleaf.org">
<head>
<meta charset="UTF-8">
<title>Insert title here</title>
</head>
<body>
    <span th:text="#{admin}"></span><br>
    <a th:href="@{/i18n/first}" th:text="#{return}"></a>
</body>
</html>
```

before.html 的代码如下：

```html
<!DOCTYPE html>
<html xmlns:th="http://www.thymeleaf.org">
<head>
<meta charset="UTF-8">
<title>Insert title here</title>
</head>
<body>
    <span th:text="#{before}"></span><br>
    <a th:href="@{/i18n/first}" th:text="#{return}"></a>
</body>
</html>
```

first.html 的代码如下：

```html
<!DOCTYPE html>
<html xmlns:th="http://www.thymeleaf.org">
<head>
<meta charset="UTF-8">
<title>Insert title here</title>
</head>
<body>
    <a th:href="@{/i18n/first(locale='zh_CN')}" th:text="#{chinese.key}"></a>
    <a th:href="@{/i18n/first(locale='en_US')}" th:text="#{english.key}"></a>
    <br>
    <a th:href="@{/i18n/admin}" th:text="#{test.admin}"></a><br>
    <a th:href="@{/i18n/before}" th:text="#{test.before}"></a><br>
```

```
</body>
</html>
```

❻ 运行

首先运行 Ch51Application 主类，然后访问"http://localhost:8080/ch5_1/i18n/first"，运行效果如图 5.5 所示。

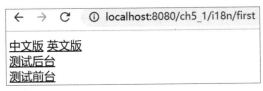

图 5.5　程序入口页面

单击图 5.5 中的"英文版"，打开如图 5.6 所示的页面效果。

图 5.6　英文版效果

扫一扫

视频讲解

5.2.5　Spring Boot 与 Thymeleaf 的表单验证

JSR 是 Java Specification Requests 的缩写，意思是 Java 规范提案。关于数据校验，最新的是 JSR 380，也就是 Bean Validation 2.0。

Bean Validation 是一个通过配置注解来验证数据的框架，它包含 Bean Validation API（规范）和 Hibernate Validator（实现）两部分。

Bean Validation 是 Java 定义的一套基于注解/XML 的数据校验规范，目前已经从 JSR 303 的 Bean Validation 1.0 版本升级到 JSR 349 的 Bean Validation 1.1 版本，再到 JSR 380 的 Bean Validation 2.0 版本。

2018 年，Oracle（甲骨文）公司决定把 Java EE 移交给开源组织 Eclipse 基金会，正式改名为 Jakarta EE，Bean Validation 也就自然命名为 Jakarta Bean Validation。在编写本书时，其最新版本是 Jakarta Bean Validation 3.0。

对于 Jakarta Bean Validation 验证，可以使用它的实现 Hibernate Validator，注意 Hibernate Validator 和 Hibernate 无关，只是使用 Hibernate Validator 进行数据验证。

本小节使用 Hibernate Validator 对表单进行验证，因为 spring-boot-starter-web 不再依赖 hibernate-validator 的 JAR 包，所以在 Spring Boot 的 Web 应用中使用 Hibernate Validator 对表单进行验证时需要加载 Hibernate Validator 所依赖的 JAR 包，示例代码如下：

```
<dependency>
    <groupId>org.hibernate.validator</groupId>
    <artifactId>hibernate-validator</artifactId>
</dependency>
```

在使用 Hibernate Validator 验证表单时，需要利用它的标注类型在实体模型的属性上嵌入约束，标注类型具体如下。

❶ 空检查

@Null：验证对象是否为 null。

@NotNull：验证对象是否不为 null，无法查验长度为 0 的字符串。

@NotBlank：检查约束字符串是否为 null，以及被 trim 后的长度是否大于 0，只针对字符串，且会去掉前后空格。

@NotEmpty：检查约束元素是否为 null 或者是 empty。

示例如下：

```
@NotBlank(message="{goods.gname.required}")
    //goods.gname.required 为属性文件的错误代码
private String gname;
```

❷ boolean 检查

@AssertTrue：验证 boolean 属性是否为 true。

@AssertFalse：验证 boolean 属性是否为 false。

示例如下：

```
@AssertTrue
private boolean isLogin;
```

❸ 长度检查

@Size(min=, max=)：验证对象（Array、Collection、Map、String）的长度是否在给定的范围之内。

@Length(min=, max=)：验证字符串的长度是否在给定的范围之内。

示例如下：

```
@Length(min=1,max=100)
private String gdescription;
```

❹ 日期检查

@Past：验证 Date 和 Calendar 对象是否在当前时间之前。

@Future：验证 Date 和 Calendar 对象是否在当前时间之后。

@Pattern：验证 String 对象是否符合正则表达式的规则。

示例如下：

```
@Past(message="{gdate.invalid}")
private Date gdate;
```

❺ 数值检查

@Min：验证 Number 和 String 对象是否小于或等于指定的值。

@Max：验证 Number 和 String 对象是否大于或等于指定的值。

@DecimalMax：被标注的值必须不大于约束中指定的最大值，这个约束的参数是一个通过 BigDecimal 定义的最大值的字符串表示，小数存在精度。

@DecimalMin：被标注的值必须不小于约束中指定的最小值，这个约束的参数是一个通过 BigDecimal 定义的最小值的字符串表示，小数存在精度。

@Digits：验证 Number 和 String 的构成是否合法。

@Digits(integer=,fraction=)：验证字符串是否符合指定格式的数字，integer 指定整数精度，fraction 指定小数精度。

@Range(min=, max=)：检查数字是否介于 min 和 max 之间。

@Valid：对关联对象进行校验，如果关联对象是一个集合或者数组，那么对其中的元素进行校验，如果是一个 map，则对其中的值部分进行校验。

@CreditCardNumber：信用卡验证。

@Email：验证是否为邮件地址，如果为 null，不进行验证，通过验证。

示例如下：

```
@Range(min=0,max=100,message="{gprice.invalid}")
private double gprice;
```

下面通过实例讲解使用 Hibernate Validator 验证表单的过程。

【例 5-4】使用 Hibernate Validator 验证表单的过程。

其具体实现步骤如下：

❶ 创建表单实体模型

在 ch5_1 应用的 com.ch.ch5_1.model 包中创建表单实体模型类 Goods，在该类中利用 Jakarta Bean Validation 的标注类型对属性进行分组验证，具体代码如下：

```
package com.ch.ch5_1.model;
import jakarta.validation.constraints.NotBlank;
import lombok.Data;
import org.hibernate.validator.constraints.Length;
import org.hibernate.validator.constraints.Range;
@Data
public class Goods {
    //add 组
    public interface Add{}
    //update 组
    public interface Update{}
    @NotBlank(groups = {Add.class, Update.class}, message="商品名必须输入")
    @Length(min=1, max=5, message="商品名长度在 1 到 5 之间")
    private String gname;
    @Range(groups = {Add.class}, min=0, max=100, message="商品价格在 0 到 100 之间")
    private double gprice;
}
```

❷ 创建控制器

在 ch5_1 应用的 com.ch.ch5_1.controller 包中创建控制器类 TestValidatorController，在该类中有两个处理方法，一个是界面初始化处理方法 testValidator，另一个是添加请求处理方法 add，在 add 方法中使用@Validated 注解使验证生效，具体代码如下：

```
package com.ch.ch5_1.controller;
import org.springframework.stereotype.Controller;
import org.springframework.validation.BindingResult;
import org.springframework.validation.annotation.Validated;
import org.springframework.web.bind.annotation.GetMapping;
import org.springframework.web.bind.annotation.ModelAttribute;
import org.springframework.web.bind.annotation.PostMapping;
import com.ch.ch5_1.model.Goods;
@Controller
public class TestValidatorController {
    @GetMapping("/testValidator")
    public String testValidator(@ModelAttribute("goodsInfo") Goods goods){
        goods.setGname("商品名初始化");
        goods.setGprice(0.0);
        return "testValidator";
    }
    @PostMapping(value="/add")
```

```
           //@Validated({Goods.Add.class})验证add组,可以同时验证多组
@Validated({Goods.Add.class, Goods.Update.class})
        public String add(@Validated({Goods.Add.class}) @ModelAttribute
("goodsInfo") Goods goods,BindingResult rs){
            //@ModelAttribute("goodsInfo")与th:object="${goodsInfo}"相对应
            if(rs.hasErrors()){        //验证失败
                return "testValidator";
            }
            //验证成功,可以到任意地方,在这里直接到testValidator界面
            return "testValidator";
        }
    }
```

❸ 创建视图页面

在 ch5_1 应用的 src/main/resources/templates 目录下创建视图页面 testValidator.html,在该视图页面中直接读取到 ModelAttribute 里面注入的数据,然后通过 th:errors="*{xxx}"获得验证错误信息,具体代码如下:

```
<!DOCTYPE html>
<html xmlns:th="http://www.thymeleaf.org">
<head>
<meta charset="UTF-8">
<title>Insert title here</title>
</head>
<body>
    <h2>通过th:object访问对象的方式</h2>
    <div th:object="${goodsInfo}">
        <p th:text="*{gname}"></p>
        <p th:text="*{gprice}"></p>
    </div>
    <h1>表单提交</h1>
    <!--表单提交用户信息,注意表单参数的设置,直接是*{}-->
    <form  th:action="@{/add}" th:object="${goodsInfo}" method="post">
    <div><span>商品名</span><input type="text" th:field=
        "*{gname}"/><span th:errors="*{gname}"></span></div>
    <div><span>商品价格</span><input type="text" th:field=
        "*{gprice}"/><span th:errors="*{gprice}"></span></div>
        <input type="submit"/>
    </form>
</body>
</html>
```

❹ 运行

首先运行 Ch51Application 主类,然后访问 "http://localhost:8080/ch5_1/testValidator",测试效果如图 5.7 所示。

图 5.7　表单验证

5.2.6 基于 Thymeleaf 和 BootStrap 的 Web 开发实例

扫一扫

视频讲解

在本书的后续 Web 应用开发中尽量使用前端开发工具包 BootStrap、JavaScript 框架 jQuery 和 Spring MVC 框架。对于 BootStrap 和 jQuery 的相关知识，请读者自行学习。下面通过一个实例讲解如何创建基于 Thymeleaf 模板引擎的 Spring Boot Web 应用 ch5_2。

【例 5-5】创建基于 Thymeleaf 模板引擎的 Spring Boot Web 应用 ch5_2。

其具体实现步骤如下：

❶ 创建基于 Thymeleaf 模板引擎的 Spring Boot Web 应用 ch5_2

在 IDEA 中选择 File→New→Project 命令，打开 New Project 对话框；在 New Project 对话框中选择和输入相关信息后单击 Next 按钮，进入新的界面；在新的界面中选择 Thymeleaf、Lombok 和 Spring Web 依赖，单击 Create 按钮即可创建应用 ch5_2。

❷ 设置 Web 应用 ch5_2 的上下文路径

在 ch5_2 的 application.properties 文件中配置如下内容：

```
server.servlet.context-path=/ch5_2
```

❸ 创建实体类 Book

在应用 ch5_2 的 src/main/java 目录下创建名为 com.ch.ch5_2.model 的包，并在该包中创建名为 Book 的实体类，此实体类用在模板页面展示数据，具体代码如下：

```java
package com.ch.ch5_2.model;
import lombok.Data;
@Data
public class Book {
    String isbn;
    Double price;
    String bname;
    String publishing;
    String author;
    String picture;
    public Book(String isbn, Double price, String bname, String publishing, String author, String picture) {
        super();
        this.isbn = isbn;
        this.price = price;
        this.bname = bname;
        this.publishing = publishing;
        this.author = author;
        this.picture = picture;
    }
}
```

❹ 创建控制器类 ThymeleafController

在应用 ch5_2 的 src/main/java 目录下创建名为 com.ch.ch5_2.controller 的包，并在该包中创建名为 ThymeleafController 的控制器类，在该控制器类中实例化 Book 类的多个对象，并保存到集合 ArrayList<Book>中，具体代码如下：

```java
package com.ch.ch5_2.controller;
import java.util.ArrayList;
import java.util.List;
import org.springframework.stereotype.Controller;
import org.springframework.ui.Model;
import com.ch.ch5_2.model.Book;
import org.springframework.web.bind.annotation.GetMapping;
```

```java
@Controller
public class ThymeleafController {
    @GetMapping("/")
    public String index(Model model) {
        Book teacherChen = new Book(
            "9787302598503",99.8,
            "SSM + Spring Boot + Vue.js 3 全栈开发从入门到实战",
            "清华大学出版社","陈恒","091883-01.jpg"
        );
        List<Book> chenHeng = new ArrayList<Book>();
        Book b1 = new Book(
            "9787302529118", 69.8,
            "Java Web 开发从入门到实战（微课版）",
            "清华大学出版社", "陈恒","082526-01.jpg"
        );
        chenHeng.add(b1);
        Book b2 = new Book(
            "9787302502968", 69.8,
            "Java EE 框架整合开发入门到实战——Spring+Spring MVC+MyBatis（微课版）",
            "清华大学出版社", "陈恒","079720-01.jpg");
        chenHeng.add(b2);
        model.addAttribute("aBook", teacherChen);
        model.addAttribute("books", chenHeng);
        //根据Thymeleaf模板，默认将返回src/main/resources/templates/index.html
        return "index";
    }
}
```

❺ 整理脚本、样式等静态文件

JS 脚本、CSS 样式、图片等静态文件默认放置在应用 ch5_2 的 src/main/resources/static 目录下。

❻ View 视图页面

Thymeleaf 模板默认将视图页面放在 src/main/resources/templates 目录下，因此在 src/main/resources/templates 目录下新建 HTML 页面文件 index.html。在该页面中使用 Thymeleaf 模板显示控制器类 TestThymeleafController 中的 model 对象数据，具体代码如下：

```html
<!DOCTYPE html>
<html xmlns:th="http://www.thymeleaf.org">
<head>
<meta charset="UTF-8">
<title>Insert title here</title>
<link rel="stylesheet" th:href="@{css/bootstrap.min.css}"/>
</head>
<body>
    <!--面板-->
    <div class="panel panel-primary">
        <!--面板头信息-->
        <div class="panel-heading">
            <!--面板标题-->
            <h3 class="panel-title">第一个基于 Thymeleaf 与 BootStrap 的 Spring Boot Web 应用</h3>
        </div>
    </div>
    <!--容器-->
    <div class="container">
        <div>
            <h4>图书列表</h4>
```

```html
            </div>
            <div class="row">
                <!--col-md 针对桌面显示器，col-sm 针对平板-->
                <div class="col-md-4 col-sm-6">
                    <a href="">
                        <img th:src="'images/' + ${aBook.picture}" alt="图书封面"
                            style="height: 180px; width: 40%;"/>
                    </a>
                    <!--caption 容器中放置其他基本信息，比如标题、文本描述等-->
                    <div class="caption">
                        <h4 th:text="${aBook.bname}"></h4>
                        <p th:text="${aBook.author}"></p>
                        <p th:text="${aBook.isbn}"></p>
                        <p th:text="${aBook.price}"></p>
                        <p th:text="${aBook.publishing}"></p>
                    </div>
                </div>
                <!--循环取出集合数据-->
                <div class="col-md-4 col-sm-6" th:each="book:${books}">
                    <a href="">
                        <img th:src="'images/' + ${book.picture}" alt="图书封面"
                            style="height: 180px; width: 40%;"/>
                    </a>
                    <div class="caption">
                        <h4 th:text="${book.bname}"></h4>
                        <p th:text="${book.author}"></p>
                        <p th:text="${book.isbn}"></p>
                        <p th:text="${book.price}"></p>
                        <p th:text="${book.publishing}"></p>
                    </div>
                </div>
            </div>
        </div>
    </body>
</html>
```

❼ 运行

首先运行 Ch52Application 主类，然后访问"http://localhost:8080/ch5_2/"，运行结果如图 5.8 所示。

图 5.8 例 5-5 的运行结果

5.3 使用 Spring Boot 处理 JSON 数据

在 Spring Boot 的 Web 应用中内置了 JSON 数据的解析功能，默认使用 Jackson 自动完成解析（不需要加载 Jackson 依赖包），当控制器返回一个 Java 对象或集合数据时，Spring Boot 自动将其转换成 JSON 数据，使用起来方便、简洁。

在 Spring Boot 处理 JSON 数据时需要用到两个重要的 JSON 格式转换注解，分别是 @RequestBody 和@ResponseBody。

（1）@RequestBody：用于将请求体中的数据绑定到方法的形参中，该注解应用在方法的形参上。

（2）@ResponseBody：用于直接返回 JSON 对象，该注解应用在方法上。

下面通过一个实例讲解使用 Spring Boot 处理 JSON 数据的过程，该实例针对返回实体对象、ArrayList 集合、Map<String, Object>集合以及 List<Map<String, Object>>集合分别处理。

【例 5-6】使用 Spring Boot 处理 JSON 数据的过程。

其具体实现步骤如下：

❶ 创建实体类

在 ch5_2 应用的 com.ch.ch5_2.model 包中创建实体类 Person，具体代码如下：

```java
package com.ch.ch5_2.model;
import lombok.Data;
@Data
public class Person {
    private String pname;
    private String password;
    private Integer page;
}
```

❷ 创建视图页面

在 ch5_2 应用的 src/main/resources/templates 目录下创建视图页面 input.html，在 input.html 页面中引入 jQuery 框架，并使用它的 ajax 方法进行异步请求，具体代码如下：

```html
<!DOCTYPE html>
<html xmlns:th="http://www.thymeleaf.org">
<head>
<meta charset="UTF-8">
<title>Insert title here</title>
<link rel="stylesheet" th:href="@{css/bootstrap.min.css}" />
<!--默认访问 src/main/resources/static 下的 css 文件夹-->
<!--引入 jQuery-->
<script type="text/javascript" th:src="@{js/jquery-3.6.0.min.js}"></script>
<script type="text/javascript">
    function testJson() {
        //获取输入的值 pname 为 id
        var pname = $("#pname").val();
        var password = $("#password").val();
        var page = $("#page").val();
        alert(password);
        $.ajax({
            //发送请求的 URL 字符串
            url: "testJson",
            //定义回调响应的数据格式为 JSON 字符串，该属性可以省略
            dataType: "json",
            //请求类型
            type: "post",
```

```
                    //定义发送请求的数据格式为 JSON 字符串
                    contentType: "application/json",
                    //data 表示发送的数据
                    data: JSON.stringify({pname:pname,password:password,page:page}),
                    //成功响应的结果
                    success: function(data){
                        if(data != null){
                            //返回一个 Person 对象
                            //alert("输入的用户名:" + data.pname + ",密码: " + data.password +
                            //",年龄: " + data.page);
                            //ArrayList<Person>对象
                            /**for(var i = 0; i < data.length; i++){
                                alert(data[i].pname);
                            }**/
                            //返回一个 Map<String, Object>对象
                            //alert(data.pname);         //pname 为 key
                            //返回一个 List<Map<String, Object>>对象
                            for(var i = 0; i < data.length; i++){
                                alert(data[i].pname);
                            }
                        }
                    },
                    //请求出错
                    error:function(){
                        alert("数据发送失败");
                    }
                });
            }
    </script>
</head>
<body>
    <div class="panel panel-primary">
        <div class="panel-heading">
            <h3 class="panel-title">处理 JSON 数据</h3>
        </div>
    </div>
    <div class="container">
        <div>
            <h4>添加用户</h4>
        </div>
        <div class="row">
            <div class="col-md-6 col-sm-6">
                <form class="form-horizontal" action="">
                    <div class="form-group">
                        <div class="input-group col-md-6">
                            <span class="input-group-addon">
                                <i class="glyphicon glyphicon-pencil"></i>
                            </span>
                            <input class="form-control" type="text"
                              id="pname" th:placeholder="请输入用户名"/>
                        </div>
                    </div>
                    <div class="form-group">
                        <div class="input-group col-md-6">
                            <span class="input-group-addon">
                                <i class="glyphicon glyphicon-pencil"></i>
                            </span>
                            <input class="form-control" type="password"
                              id="password" th:placeholder="请输入密码"/>
```

```html
                        </div>
                    </div>
                    <div class="form-group">
                        <div class="input-group col-md-6">
                            <span class="input-group-addon">
                                <i class="glyphicon glyphicon-pencil"></i>
                            </span>
                            <input class="form-control" type="text"
                                id="page" th:placeholder="请输入年龄"/>
                        </div>
                    </div>
                    <div class="form-group">
                        <div class="col-md-6">
                            <div class="btn-group btn-group-justified">
                                <div class="btn-group">
                                    <button type="button" onclick="testJson()"
                                        class="btn btn-success">
                                        <span class="glyphicon glyphicon-
                                            share"></span>
                                         测试
                                    </button>
                                </div>
                            </div>
                        </div>
                    </div>
                </form>
            </div>
        </div>
    </div>
</body>
</html>
```

❸ 创建控制器

在 ch5_2 应用的 com.ch.ch5_2.controller 包中创建控制器类 TestJsonController，在该类中有两个处理方法，一个是界面导航方法 input，另一个是接收页面请求的方法，具体代码如下：

```java
package com.ch.ch5_2.controller;
import java.util.ArrayList;
import java.util.HashMap;
import java.util.List;
import java.util.Map;
import org.springframework.stereotype.Controller;
import org.springframework.web.bind.annotation.GetMapping;
import org.springframework.web.bind.annotation.RequestBody;
import org.springframework.web.bind.annotation.RequestMapping;
import org.springframework.web.bind.annotation.ResponseBody;
import com.ch.ch5_2.model.Person;
@Controller
public class TestJsonController {
    /**
     * 进入视图页面
     */
    @GetMapping("/input")
    public String input() {
        return "input";
    }
    /**
     * 接收页面请求的 JSON 数据
     */
    @RequestMapping("/testJson")
    @ResponseBody
```

第 5 章 Spring Boot 的 Web 开发

```
/*@RestController 注解相当于@ResponseBody+@Controller 的作用。
①如果只是使用@RestController注解Controller,则Controller中的方法无法返回JSP
页面或者 HTML,返回的内容就是 return 的内容。
②如果需要返回到指定页面,则需要用@Controller 注解。如果需要返回 JSON、XML 或自定
义 mediaType 内容到页面,则需要在对应的方法上添加@ResponseBody 注解。
*/
public List<Map<String, Object>> testJson (@RequestBody Person user) {
    //打印接收的 JSON 格式数据
    System.out.println("pname=" + user.getPname() +
            ", password=" + user.getPassword() + ",page=" + user.getPage());
    //返回 Person 对象
    //return user;
    /**ArrayList<Person> allp = new ArrayList<Person>();
    Person p1 = new Person();
    p1.setPname("陈恒1");
    p1.setPassword("123456");
    p1.setPage(80);
    allp.add(p1);
    Person p2 = new Person();
    p2.setPname("陈恒2");
    p2.setPassword("78910");
    p2.setPage(90);
    allp.add(p2);
    //返回 ArrayList<Person>对象
    return allp;
    **/
    Map<String, Object> map = new HashMap<String, Object>();
    map.put("pname", "陈恒2");
    map.put("password", "123456");
    map.put("page", 25);
    //返回一个 Map<String, Object>对象
    //return map;
    //返回一个 List<Map<String, Object>>对象
    List<Map<String, Object>> allp = new ArrayList<Map<String, Object>>();
    allp.add(map);
    Map<String, Object> map1 = new HashMap<String, Object>();
    map1.put("pname", "陈恒3");
    map1.put("password", "54321");
    map1.put("page", 55);
    allp.add(map1);
    return allp;
}
```

❹ 运行

首先运行 Ch52Application 主类,然后访问"http://localhost:8080/ch5_2/input",运行效果如图 5.9 所示。

图 5.9　input.html 的运行效果

5.4 Spring Boot 的文件上传与下载

文件的上传与下载是 Web 应用开发中常用的功能之一，本节将讲解如何在 Spring Boot 的 Web 应用开发中实现文件的上传与下载。

在实际的 Web 应用开发中，为了成功上传文件，必须将表单的 method 设置为 post，并将 enctype 设置为 multipart/form-data，只有这样设置，浏览器才能将所选文件的二进制数据发送给服务器。

Spring Boot 的 spring-boot-starter-web 已经集成了 Spring MVC，所以使用 Spring Boot 实现文件的上传更加便捷。

下面通过一个实例讲解 Spring Boot 的文件上传与下载的实现过程。

【例 5-7】Spring Boot 的文件上传与下载。

其具体实现步骤如下：

❶ 设置上传文件大小的限制

在 Web 应用 ch5_2 的配置文件 application.properties 中添加如下配置限制上传文件的大小。

```
#在上传文件时，默认单个上传文件的大小是 1MB，max-file-size 用于设置单个上传文件的大小
spring.servlet.multipart.max-file-size=50MB
#默认总文件大小是 10MB，max-request-size 用于设置总上传文件的大小
spring.servlet.multipart.max-request-size=500MB
```

❷ 创建选择文件视图页面

在 ch5_2 应用的 src/main/resources/templates 目录下创建选择文件视图页面 uploadFile.html，在该页面中有一个 enctype 属性值为 multipart/form-data 的 form 表单，具体代码如下：

```html
<!DOCTYPE html>
<html xmlns:th="http://www.thymeleaf.org">
<head>
<meta charset="UTF-8">
<title>Insert title here</title>
<link rel="stylesheet" th:href="@{css/bootstrap.min.css}" />
</head>
<body>
<div class="panel panel-primary">
    <div class="panel-heading">
        <h3 class="panel-title">文件上传示例</h3>
    </div>
</div>
<div class="container">
    <div class="row">
        <div class="col-md-6 col-sm-6">
            <form class="form-horizontal" action="upload"
                method="post" enctype="multipart/form-data">
                <div class="form-group">
                    <div class="input-group col-md-6">
                        <span class="input-group-addon">
                            <i class="glyphicon glyphicon-pencil"></i>
                        </span>
                        <input class="form-control" type="text"
                          name="description" th:placeholder="文件描述"/>
                    </div>
                </div>
                <div class="form-group">
                    <div class="input-group col-md-6">
```

```html
                    <span class="input-group-addon">
                        <i class="glyphicon glyphicon-search"></i>
                    </span>
                    <input class="form-control" type="file"
                        name="myfile" th:placeholder="请选择文件"/>
                </div>
            </div>
            <div class="form-group">
                <div class="col-md-6">
                    <div class="btn-group btn-group-justified">
                        <div class="btn-group">
                            <button type="submit" class="btn btn-success">
                                <span class="glyphicon glyphicon-
                                    share"></span>
                                 上传文件
                            </button>
                        </div>
                    </div>
                </div>
            </div>
        </form>
    </div>
</div>
</body>
</html>
```

❸ 创建实体类

在 ch5_2 应用的 com.ch.ch5_2.model 包中创建实体类 MyFile 封装文件对象,具体代码如下:

```java
package com.ch.ch5_2.model;
import lombok.Data;
@Data
public class MyFile {
    int fno;
    String fname;
}
```

❹ 创建控制器

在 ch5_2 应用的 com.ch.ch5_2.controller 包中创建控制器类 TestFileUpload,在该类中有 4 个处理方法,一个是界面导航方法 uploadFile,一个是实现文件上传的 upload 方法,一个是显示将要被下载文件的 showDownLoad 方法,一个是实现下载功能的 download 方法,具体代码如下:

```java
package com.ch.ch5_2.controller;
import java.io.File;
import java.io.FileInputStream;
import java.io.IOException;
import java.net.URLEncoder;
import java.util.ArrayList;
import com.ch.ch5_2.model.MyFile;
import jakarta.servlet.ServletOutputStream;
import jakarta.servlet.http.HttpServletRequest;
import jakarta.servlet.http.HttpServletResponse;
import org.springframework.stereotype.Controller;
import org.springframework.ui.Model;
import org.springframework.web.bind.annotation.RequestMapping;
import org.springframework.web.bind.annotation.RequestParam;
import org.springframework.web.multipart.MultipartFile;
```

```java
@Controller
public class TestFileUpload {
    /**
     * 进入文件选择页面
     */
    @RequestMapping("/uploadFile")
    public String uploadFile() {
        return "uploadFile";
    }
    /**
     * 上传文件自动绑定到MultipartFile对象中，在这里使用处理方法的形参接收请求参数
     */
    @RequestMapping("/upload")
    public String upload(HttpServletRequest request, @RequestParam("description") String description,@RequestParam("myfile") MultipartFile myfile)
    throws IllegalStateException, IOException {
        //如果选择了上传文件，将文件上传到指定的目录uploadFiles
        if(!myfile.isEmpty()) {
            //上传文件的路径
            String path = request.getServletContext().getRealPath("/uploadFiles/");
            //获得上传文件的原名
            String fileName = myfile.getOriginalFilename();
            File filePath = new File(path + File.separator + fileName);
            //如果文件目录不存在，创建目录
            if(!filePath.getParentFile().exists()) {
                filePath.getParentFile().mkdirs();
            }
            //将上传文件保存到一个目标文件中
            myfile.transferTo(filePath);
        }
        //转发到一个请求处理方法，查询将要下载的文件
        return "forward:/showDownLoad";
    }
    /**
     * 显示要下载的文件
     */
    @RequestMapping("/showDownLoad")
    public String showDownLoad(HttpServletRequest request, Model model) {
        String path = request.getServletContext().getRealPath("/uploadFiles/");
        File fileDir = new File(path);
        //从指定目录获得文件列表
        File filesList[] = fileDir.listFiles();
        ArrayList<MyFile> myList = new ArrayList<MyFile>();
        for (int i = 0; i < filesList.length; i++) {
            MyFile mf = new MyFile();
            mf.setFno(i + 1);
            mf.setFname(filesList[i].getName());
            myList.add(mf);
        }
        model.addAttribute("filesList", myList);
        return "showFile";
    }
    /**
     * 实现下载功能
     */
    @RequestMapping("/download")
    public void download(
        HttpServletRequest request, HttpServletResponse response,
        @RequestParam("filename") String filename) throws IOException {
```

```java
            //下载文件的路径
            String path = request.getServletContext().getRealPath("/uploadFiles/");
            //构建将要下载的文件对象
            File downFile = new File(path + File.separator + filename);
            //使用URLEncoder.encode对文件名进行编码
            filename = URLEncoder.encode(filename,"UTF-8");
            response.setHeader("Content-Type", "application/x-msdownload");
            response.setHeader("Content-Disposition", "attachment; filename=" + filename);
            FileInputStream in = null;                //输入流
            ServletOutputStream out = null;           //输出流
            //读入文件
            in = new FileInputStream(downFile);
            //得到响应对象的输出流，用于向客户端输出二进制数据
            out = response.getOutputStream();
            out.flush();
            int aRead = 0;
            byte b[] = new byte[1024];
            while ((aRead = in.read(b)) != -1 & in != null) {
                out.write(b, 0, aRead);
            }
            out.flush();
            in.close();
            out.close();
        }
}
```

❺ 创建文件下载视图页面

在 ch5_2 应用的 src/main/resources/templates 目录下创建文件下载视图页面 showFile.html，具体代码如下：

```html
<!DOCTYPE html>
<html xmlns:th="http://www.thymeleaf.org">
<head>
    <meta charset="UTF-8">
    <title>Insert title here</title>
    <link rel="stylesheet" th:href="@{css/bootstrap.min.css}" />
<body>
<div class="panel panel-primary">
    <div class="panel-heading">
        <h3 class="panel-title">文件下载示例</h3>
    </div>
</div>
<div class="container">
    <div class="panel panel-primary">
        <div class="panel-heading">
            <h3 class="panel-title">文件列表</h3>
        </div>
        <div class="panel-body">
            <div class="table table-responsive">
                <table class="table table-bordered table-hover">
                    <tbody class="text-center">
                    <tr th:each="myFile:${filesList}">
                        <td>
                            <span th:text="${myFile.fno}"></span>
                        </td>
                        <td>
                            <a th:href="@{download(filename=${myFile.fname})}">
                                <span th:text="${myFile.fname}"></span>
```

```
                        </a>
                    </td>
                </tr>
            </tbody>
        </table>
    </div>
  </div>
 </div>
</body>
</html>
```

❻ 运行

首先运行 Ch52Application 主类,然后访问"http://localhost:8080/ch5_2/uploadFile",运行效果如图 5.10 所示。

图 5.10　文件选择界面

在图 5.10 中输入文件描述并选择上传文件后,单击"上传文件"按钮,实现文件的上传。在文件上传成功后打开如图 5.11 所示的下载文件列表页面。

图 5.11　下载文件列表页面

单击图 5.11 中的文件名即可下载文件,至此文件的上传与下载示例演示完毕。

5.5　Spring Boot 的异常统一处理

在 Spring Boot 应用的开发中,不管是对底层数据库操作,还是对业务层、控制层操作,都不可避免地遇到各种可预知的、不可预知的异常需要处理。如果每个过程都单独处理异常,那么系统的代码耦合度高,工作量大且不好统一,以后维护的工作量也很大。

第 5 章　Spring Boot 的 Web 开发

如果能将所有类型的异常处理从各层中解耦出来，这样既保证了相关处理过程的功能较单一，也实现了异常信息的统一处理和维护。幸运的是，Spring 框架支持这样的实现。本节将以@ExceptionHandler 与@ControllerAdvice 注解两种方式讲解 Spring Boot 应用的异常统一处理。

▶ 5.5.1　自定义 error 页面

在 Spring Boot Web 应用的 src/main/resources/templates 目录下添加 error.html 页面，当访问发生错误或异常时，Spring Boot 将自动找到该页面作为错误页面。Spring Boot 为错误页面提供了以下属性。

（1）timestamp：错误发生的时间。
（2）status：HTTP 状态码。
（3）error：错误原因。
（4）exception：异常的类名。
（5）message：异常消息（如果这个错误是由异常引起的）。
（6）errors：BindingResult 异常中的各种错误（如果这个错误是由异常引起的）。
（7）trace：异常跟踪信息（如果这个错误是由异常引起的）。
（8）path：错误发生时请求的 URL 路径。

下面通过一个实例讲解在 Spring Boot 应用的开发中如何使用自定义 error 页面。

【例 5-8】自定义 error 页面。

其具体实现步骤如下：

❶ 创建基于 Thymeleaf 模板引擎的 Spring Boot Web 应用 ch5_3

参照 5.2.6 节的例 5-5，创建基于 Thymeleaf 模板引擎的 Spring Boot Web 应用 ch5_3。

❷ 设置 Web 应用 ch5_3 的上下文路径

在 ch5_3 的 application.properties 文件中配置如下内容：

```
server.servlet.context-path=/ch5_3
```

❸ 创建自定义异常类 MyException

在应用 ch5_3 的 src/main/java 目录下创建名为 com.ch.ch5_3.exception 的包，并在该包中创建名为 MyException 的异常类，具体代码如下：

```java
package com.ch.ch5_3.exception;
public class MyException extends Exception {
    private static final long serialVersionUID = 1L;
    public MyException() {
        super();
    }
    public MyException(String message) {
        super(message);
    }
}
```

❹ 创建控制器类 TestHandleExceptionController

在应用 ch5_3 的 src/main/java 目录下创建名为 com.ch.ch5_3.controller 的包，并在该包中创建名为 TestHandleExceptionController 的控制器类。在该控制器类中有 4 个请求处理方法，一个是导航到 index.html，另外 3 个分别抛出不同的异常（并没有处理异常）。其具体代码如下：

```java
package com.ch.ch5_3.controller;
import java.sql.SQLException;
import org.springframework.stereotype.Controller;
import org.springframework.web.bind.annotation.GetMapping;
import com.ch.ch5_3.exception.MyException;
@Controller
public class TestHandleExceptionController {
    @GetMapping("/")
    public String index() {
        return "index";
    }
    @GetMapping("/db")
    public void db() throws SQLException {
        throw new SQLException("数据库异常");
    }
    @GetMapping("/my")
    public void my() throws MyException {
        throw new MyException("自定义异常");
    }
    @GetMapping("/no")
    public void no() throws Exception {
        throw new Exception("未知异常");
    }
}
```

❺ 整理脚本、样式等静态文件

JS 脚本、CSS 样式、图片等静态文件默认放置在 src/main/resources/static 目录下，在 ch5_3 应用中引入了与 ch5_2 一样的 BootStrap 和 jQuery。

❻ View 视图页面

Thymeleaf 模板默认将视图页面放在 src/main/resources/templates 目录下，因此在 src/main/resources/templates 目录下新建 HTML index.html 和 error.html 页面文件。

在 index.html 页面中有 4 个超链接请求，3 个请求在控制器中有对应处理，另一个请求是 404 错误。其具体代码如下：

```html
<!DOCTYPE html>
<html xmlns:th="http://www.thymeleaf.org">
<head>
<meta charset="UTF-8">
<title>index</title>
<link rel="stylesheet" th:href="@{css/bootstrap.min.css}"/>
</head>
<body>
    <div class="panel panel-primary">
        <div class="panel-heading">
            <h3 class="panel-title">异常处理示例</h3>
        </div>
    </div>
    <div class="container">
        <div class="row">
            <div class="col-md-4 col-sm-6">
                <a th:href="@{db}">处理数据库异常</a><br>
                <a th:href="@{my}">处理自定义异常</a><br>
                <a th:href="@{no}">处理未知错误</a>
                <hr>
                <a th:href="@{nofound}">404 错误</a>
            </div>
        </div>
```

在 error.html 页面中使用 Spring Boot 为错误页面提供的属性显示错误消息，具体代码如下：

```html
<!DOCTYPE html>
<html xmlns:th="http://www.thymeleaf.org">
<head>
<meta charset="UTF-8">
<title>error</title>
<link rel="stylesheet" th:href="@{css/bootstrap.min.css}"/>
</head>
<body>
    <div class="panel-l container clearfix">
        <div class="error">
            <p class="title"><span class="code" th:text="${status}"></span>非常抱歉，没有找到您要查看的页面</p>
            <div class="common-hint-word">
                <div th:text="${#dates.format(timestamp,'yyyy-MM-dd HH:mm:ss')}"></div>
                <div th:text="${error}"></div>
            </div>
        </div>
    </div>
</body>
</html>
```

❼ 运行

首先运行 Ch53Application 主类，然后访问 "http://localhost:8080/ch5_3/" 打开 index.html 页面，运行效果如图 5.12 所示。

图 5.12　index.html 页面

当单击图 5.12 中的超链接时，Spring Boot 应用将根据链接请求到控制器中找对应的处理。例如，当单击图 5.12 中的 "处理数据库异常" 链接时，将执行控制器中的 public void db() throws SQLException 方法，而该方法仅抛出了 SQLException 异常，并没有处理异常。如果 Spring Boot 发现有异常抛出并没有处理，将自动在 src/main/resources/templates 目录下找到 error.html 页面显示异常信息，效果如图 5.13 所示。

图 5.13　error.html 页面

从例 5-8 的运行结果可以看出，使用自定义 error 页面并没有真正处理异常，只是将异常或错误信息显示给客户端，因为在服务器控制台上同样抛出了异常，如图 5.14 所示。

图 5.14 异常信息

5.5.2 @ExceptionHandler 注解

在 5.5.1 小节中使用自定义 error 页面并没有真正处理异常，在本小节可以使用 @ExceptionHandler 注解处理异常。如果在 Controller 中有一个使用@ExceptionHandler 注解修饰的方法，那么当 Controller 的任何方法抛出异常时都由该方法处理异常。

下面通过实例讲解如何使用@ExceptionHandler 注解处理异常。

【例 5-9】使用@ExceptionHandler 注解处理异常。

其具体实现步骤如下：

❶ 在控制器类中添加使用@ExceptionHandler 注解修饰的方法

在例 5-8 的控制器类 TestHandleExceptionController 中添加一个使用@ExceptionHandler 注解修饰的方法，具体代码如下：

```
@ExceptionHandler(value=Exception.class)
public String handlerException(Exception e) {
    //数据库异常
    if (e instanceof SQLException) {
        return "sqlError";
    } else if (e instanceof MyException) {     //自定义异常
        return "myError";
    } else {            //未知异常
        return "noError";
    }
}
```

❷ 创建 sqlError、myError 和 noError 页面

在 ch5_3 的 src/main/resources/templates 目录下创建 sqlError、myError 和 noError 页面。当发生 SQLException 异常时，Spring Boot 处理后显示 sqlError 页面；当发生 MyException 异常时，Spring Boot 处理后显示 myError 页面；当发生未知异常时，Spring Boot 处理后显示 noError 页面。这里省略它们的具体代码。

❸ 运行

再次运行 Ch53Application 主类后，访问"http://localhost:8080/ch5_3/"打开 index.html 页面，当单击"处理数据库异常"链接时，执行控制器中的 public void db() throws SQLException 方法，该方法抛出了 SQLException，这时 Spring Boot 会自动执行使用@ExceptionHandler 注解修饰的方法 public String handlerException(Exception e)进行异常处理并打开 sqlError.html 页面，同时观察控制台有没有抛出异常信息。注意，当单击"404 错误"链接时，仍然由自定义 error 页面显示错误信息，这是因为没有执行控制器中抛出异常的方法，进而不会执行使

用@ExceptionHandler 注解修饰的方法。

从例 5-9 可以看出，在控制器中添加使用@ExceptionHandler 注解修饰的方法才能处理异常。在一个 Spring Boot 应用中往往存在多个控制器，不太合适在每个控制器中添加使用@ExceptionHandler 注解修饰的方法进行异常处理。开发人员可以将使用@ExceptionHandler 注解修饰的方法放到一个父类中，然后让所有需要处理异常的控制器继承该类。例如，可以将例 5-9 中使用@ExceptionHandler 注解修饰的方法移到一个父类 BaseController 中，然后让控制器类 TestHandleExceptionController 继承该父类，这样即可处理异常。

▶ 5.5.3 @ControllerAdvice 注解

使用 5.5.2 小节中的 Controller 进行异常处理也有缺点，那就是代码耦合性太高，此时可以使用@ControllerAdvice 注解降低这种父子耦合关系。

@ControllerAdvice 注解是一个增强的 Controller，使用该 Controller 可以实现全局异常处理、全局数据绑定以及全局数据预处理 3 个方面的功能。本节将学习如何使用@ControllerAdvice 注解进行全局异常处理。

使用@ControllerAdvice 注解的类是当前 Spring Boot 应用中所有类的统一异常处理类，在该类中使用@ExceptionHandler 注解的方法来统一处理异常，不需要在每个 Controller 中逐一定义异常处理方法，这是因为它对所有注解了@RequestMapping 的控制器方法有效。

下面通过实例讲解如何使用@ControllerAdvice 注解进行全局异常处理。

【例 5-10】 使用@ControllerAdvice 注解进行全局异常处理。

其具体实现步骤如下：

❶ 创建使用@ControllerAdvice 注解的类

在 ch5_3 的 com.ch.ch5_3.controller 包中创建名为 GlobalExceptionHandlerController 的类。使用@ControllerAdvice 注解修饰该类，并将例 5-9 中使用@ExceptionHandler 注解修饰的方法移到该类中，具体代码如下：

```java
package com.ch.ch5_3.controller;
import java.sql.SQLException;
import org.springframework.web.bind.annotation.ControllerAdvice;
import org.springframework.web.bind.annotation.ExceptionHandler;
import com.ch.ch5_3.exception.MyException;
@ControllerAdvice
public class GlobalExceptionHandlerController {
    @ExceptionHandler(value=Exception.class)
    public String handlerException(Exception e) {
        //数据库异常
        if (e instanceof SQLException) {
            return "sqlError";
        } else if (e instanceof MyException) {   //自定义异常
            return "myError";
        } else {        //未知异常
            return "noError";
        }
    }
}
```

❷ 运行

再次运行 Ch53Application 主类后，访问"http://localhost:8080/ch5_3/"打开 index.html 页面测试即可。

5.6 Spring Boot 对 JSP 的支持

尽管 Spring Boot 建议使用 HTML 完成动态页面，但也有部分 Java Web 应用使用 JSP 完成动态页面。Spring Boot 官方不推荐使用 JSP 技术，但考虑到是常用的技术，本节将介绍 Spring Boot 如何集成 JSP 技术。

下面通过实例讲解 Spring Boot 如何集成 JSP 技术。

【例 5-11】Spring Boot 集成 JSP 技术。

其具体实现步骤如下：

❶ 创建 Spring Boot Web 应用 ch5_4

在 IDEA 中创建基于 Lombok 和 Spring Web 依赖的 Spring Boot Web 应用 ch5_4。

❷ 修改 pom.xml 文件，添加 Servlet、Tomcat 和 JSTL 依赖

因为在 JSP 页面中使用 EL 和 JSTL 标签显示数据，所以在 pom.xml 文件中除了添加 Servlet 和 Tomcat 依赖以外，还需要添加 JSTL 依赖，具体代码如下：

```xml
<!--添加Servlet依赖-->
<dependency>
    <groupId>jakarta.servlet</groupId>
    <artifactId>jakarta.servlet-api</artifactId>
    <version>6.0.0</version>
    <scope>provided</scope>
</dependency>
<!--添加Tomcat依赖-->
<dependency>
    <groupId>org.springframework.boot</groupId>
    <artifactId>spring-boot-starter-tomcat</artifactId>
    <scope>provided</scope>
</dependency>
<!--Jasper是Tomcat使用的引擎,使用tomcat-embed-jasper可以将Web应用在内嵌的Tomcat下运行-->
<dependency>
    <groupId>org.apache.tomcat.embed</groupId>
    <artifactId>tomcat-embed-jasper</artifactId>
    <scope>provided</scope>
</dependency>
<!--添加JSTL依赖-->
<dependency>
    <groupId>org.glassfish.web</groupId>
    <artifactId>jakarta.servlet.jsp.jstl</artifactId>
    <version>3.0.0</version>
</dependency>
```

❸ 设置 Web 应用 ch5_4 的上下文路径及页面配置信息

在 ch5_4 的 application.properties 文件中配置如下内容：

```
server.servlet.context-path=/ch5_4
#设置页面前缀目录
spring.mvc.view.prefix=/WEB-INF/jsp/
#设置页面后缀
spring.mvc.view.suffix=.jsp
```

❹ 创建实体类 Book

创建名为 com.ch.ch5_4.model 的包，并在该包中创建名为 Book 的实体类。此实体类用在模板页面展示数据，代码与例 5-5 中的 Book 一样，这里不再赘述。

第 5 章　Spring Boot 的 Web 开发

❺ 创建控制器类 ThymeleafController

创建名为 com.ch.ch5_4.controller 的包，并在该包中创建名为 ThymeleafController 的控制器类。在该控制器类中实例化 Book 类的多个对象，并保存到集合 ArrayList<Book>中。其代码与例 5-5 中的 ThymeleafController 一样，这里不再赘述。

❻ 整理脚本、样式等静态文件

JS 脚本、CSS 样式、图片等静态文件默认放置在 src/main/resources/static 目录下，ch5_4 应用引入的 BootStrap 和 jQuery 与例 5-5 中的一样，这里不再赘述。

❼ View 视图页面

从 application.properties 配置文件中可知将 JSP 文件路径指定到/WEB-INF/jsp/目录，因此需要在 src/main 目录下创建目录 webapp/WEB-INF/jsp/，并在该目录下创建 JSP 文件 index.jsp，具体代码如下：

```jsp
<%@ page language="java" contentType="text/html; charset=UTF-8" pageEncoding="UTF-8"%>
<!--引入JSTL标签-->
<%@ taglib prefix="c" uri="http://java.sun.com/jsp/jstl/core" %>
<%
    String path = request.getContextPath();
    String basePath = request.getScheme() + "://" + request.getServerName() + ":" + request.getServerPort() + path + "/";
%>
<!DOCTYPE html>
<html>
<head>
<base href="<%=basePath%>">
<meta charset="UTF-8">
<title>JSP 测试</title>
<link href="css/bootstrap.min.css" rel="stylesheet">
</head>
<body>
    <div class="panel panel-primary">
        <div class="panel-heading">
            <h3 class="panel-title">第一个基于 JSP 技术的 Spring Boot Web 应用</h3>
        </div>
    </div>
    <div class="container">
        <div>
        <h4>图书列表</h4>
    </div>
        <div class="row">
            <div class="col-md-4 col-sm-6">
                <!--使用EL表达式-->
                <a href="">
            <img src="images/${aBook.picture}" alt="图书封面" style="height:180px; width: 40%;"/>
                </a>
                <div class="caption">
                    <h4>${aBook.bname}</h4>
                    <p>${aBook.author}</p>
                    <p>${aBook.isbn}</p>
                    <p>${aBook.price}</p>
                    <p>${aBook.publishing}</p>
                </div>
            </div>
                <!--使用JSTL标签forEach循环取出集合数据-->
```

```
                <c:forEach var="book" items="${books}">
                    <div class="col-md-4 col-sm-6">
                        <a href="">
                            <img src="images/${book.picture}" alt="图书封面" style="height: 180px; width: 40%;"/>
                        </a>
                        <div class="caption">
                            <h4>${book.bname}</h4>
                            <p>${book.author}</p>
                            <p>${book.isbn}</p>
                            <p>${book.price}</p>
                            <p>${book.publishing}</p>
                        </div>
                    </div>
                </c:forEach>
            </div>
        </div>
    </body>
</html>
```

❽ 运行

首先运行 Ch54Application 主类，然后访问"http://localhost:8080/ch5_4/"，运行效果如图 5.15 所示。

图 5.15 例 5-11 的运行效果

本章小结

本章首先介绍了 Spring Boot 的 Web 开发支持，然后详细讲述了 Spring Boot 推荐使用的 Thymeleaf 模板引擎，包括 Thymeleaf 的基础语法、常用属性以及国际化。同时，本章还介绍了 Spring Boot 对 JSON 数据的处理、文件的上传与下载、异常的统一处理和对 JSP 的支持等 Web 应用开发的常用功能。

扫一扫

自测题

习题 5

使用 Hibernate Validator 验证如图 5.16 所示的表单信息，具体要求如下：

（1）用户名必须输入，并且长度为 5～20。

（2）年龄为 18～60。

（3）工作日期在系统时间之前。

图 5.16　输入页面

第 6 章　Spring Boot 的数据访问

学习目的与要求

本章将详细介绍 Spring Boot 访问数据库的解决方案。通过本章的学习，要求读者掌握 Spring Boot 访问关系数据库以及非关系数据库的解决方案。

本章主要内容

- Spring Data JPA
- Spring Boot 整合 REST
- Spring Boot 整合 MongoDB
- Spring Boot 整合 Redis
- 数据缓存 Cache

Spring Data 是 Spring 访问数据库的一整套解决方案，是一个伞形项目，包含大量关系数据库以及非关系数据库的数据访问解决方案。本章将详细介绍 Spring Data JPA、Spring Data REST、Spring Data MongoDB、Spring Data Redis 等 Spring Data 的子项目。

6.1　Spring Data JPA

Spring Data JPA 是 Spring Data 的子项目，在讲解 Spring Data JPA 之前先了解一下 Hibernate，这是因为 Spring Data JPA 是由 Hibernate 默认实现的。

Hibernate 是一个开源的对象关系映射框架，它对 JDBC 进行了非常轻量级的对象封装，它将 POJO（Plain Ordinary Java Object）简单的 Java 对象与数据库表建立映射关系，是一个全自动的 ORM（Object Relational Mapping）框架。Hibernate 可以自动生成 SQL 语句、自动执行，使得 Java 开发人员可以随心所欲地使用对象编程思维来操纵数据库。

JPA（Java Persistence API）是官方提出的 Java 持久化规范。JPA 通过注解或 XML 描述对象—关系（表）的映射关系，并将内存中的实体对象持久化到数据库。

Spring Data JPA 通过提供基于 JPA 的 Repository 极大地简化了 JPA 的写法，在几乎不写实现的情况下实现数据库的访问和操作。使用 Spring Data JPA 建立数据访问层十分方便，只需要定义一个继承 JpaRepository 接口的接口即可。

继承了 JpaRepository 接口的自定义数据访问接口具有 JpaRepository 接口的所有数据访问操作方法。JpaRepository 接口的核心源代码如下：

```
package org.springframework.data.jpa.repository;
…
@NoRepositoryBean
public interface JpaRepository<T, ID> extends ListCrudRepository<T, ID>,
ListPagingAndSortingRepository<T, ID>, QueryByExampleExecutor<T> {
    void flush();
    <S extends T> S saveAndFlush(S entity);
```

```
        <S extends T> List<S> saveAllAndFlush(Iterable<S> entities);
        void deleteAllInBatch(Iterable<T> entities);
        void deleteAllByIdInBatch(Iterable<ID> ids);
        void deleteAllInBatch();
        T getReferenceById(ID id);
        <S extends T> List<S> findAll(Example<S> example);
        <S extends T> List<S> findAll(Example<S> example, Sort sort);
}
```

JpaRepository 接口提供的常用方法如下。

void flush()：将缓存的对象数据操作更新到数据库。

<S extends T> S saveAndFlush(S entity)：在保存对象的同时立即更新到数据库。

<S extends T> List<S> saveAllAndFlush(Iterable<S> entities)：在保存多个对象的同时立即更新到数据库。

void deleteAllInBatch(Iterable<T> entities)：批量删除提供的实体对象。

void deleteAllByIdInBatch(Iterable<ID> ids)：根据 id 批量删除提供的实体对象。

void deleteAllInBatch()：批量删除所有的实体对象。

T getReferenceById(ID id)：根据 id 获得对应的实体对象。

<S extends T> List<S> findAll(Example<S> example)：根据提供的 example 实例查询实体对象数据。

<S extends T> List<S> findAll(Example<S> example, Sort sort)：根据提供的 example 实例并按照指定规则查询实体对象数据。

▶ 6.1.1　Spring Boot 的支持

在 Spring Boot 应用中如果需要使用 Spring Data JPA 访问数据库，那么可以在通过 IDEA 创建 Spring Boot 应用时选择 Spring Data JPA 模块依赖。

❶ JDBC 的自动配置

Spring Data JPA 模块的依赖关系如图 6.1 所示。从图 6.1 所示的依赖关系可知 spring-boot-starter-data-jpa 依赖于 spring-boot-starter-jdbc，而 Spring Boot 对 spring-boot-starter-jdbc 做了自动配置。JDBC 自动配置的源代码位于 org.springframework.boot.autoconfigure.jdbc 包下。从 DataSourceProperties 类可以看出能够使用 "spring.datasource" 为前缀的属性在 application.properties 配置文件中配置 datasource。

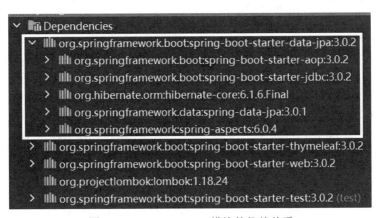

图 6.1　Spring Data JPA 模块的依赖关系

❷ JPA 的自动配置

Spring Boot 对 JPA 的自动配置位于 org.springframework.boot.autoconfigure.orm.jpa 包下。从 HibernateJpaAutoConfiguration 类可以看出 Spring Boot 对 JPA 的默认实现是 Hibernate。从 JpaProperties 类可以看出能够使用"spring.jpa"为前缀的属性在 application.properties 配置文件中配置 JPA。

❸ Spring Data JPA 的自动配置

Spring Boot 对 Spring Data JPA 的自动配置位于 org.springframework.boot.autoconfigure.data.jpa 包下。从 JpaRepositoriesAutoConfiguration 类可以看出 JpaRepositoriesAutoConfiguration 依赖于 HibernateJpaAutoConfiguration 配置；从 JpaRepositoriesRegistrar 类可以看出 Spring Boot 自动开启了对 Spring Data JPA 的支持，即开发人员无须在配置类中显式声明 @EnableJpaRepositories。

❹ Spring Boot 应用的 Spring Data JPA

从上述分析可知，在 Spring Boot 应用中使用 Spring Data JPA 访问数据库时，除了添加 spring-boot-starter-data-jpa 依赖以外，只需定义 DataSource、持久化实体类和数据访问层，并在需要使用数据访问的地方（例如 Service 层）依赖注入数据访问层即可。

6.1.2 简单条件查询

从前面的学习可知，只需定义一个继承 JpaRepository 接口的接口即可使用 Spring Data JPA 建立数据访问层，因此自定义的数据访问接口完全继承了 JpaRepository 的接口方法，更重要的是在自定义的数据访问接口中可以根据查询关键字定义查询方法，这些查询方法需要符合它的命名规则，一般是根据持久化实体类的属性名来确定。

❶ 查询关键字

目前，Spring Data JPA 支持的查询关键字如表 6.1 所示。

表 6.1 查询关键字

关 键 字	示 例	JPQL 代码段
And	findByLastnameAndFirstname	… where x.lastname = ?1 and x.firstname = ?2
Or	findByLastnameOrFirstname	… where x.lastname = ?1 or x.firstname = ?2
Is,Equals	findByFirstname,findByFirstnameIs,findByFirstnameEquals	… where x.firstname = ?1
Between	findByStartDateBetween	… where x.startDate between ?1 and ?2
LessThan	findByAgeLessThan	… where x.age < ?1
LessThanEqual	findByAgeLessThanEqual	… where x.age <= ?1
GreaterThan	findByAgeGreaterThan	… where x.age > ?1
GreaterThanEqual	findByAgeGreaterThanEqual	… where x.age >= ?1
After	findByStartDateAfter	… where x.startDate > ?1
Before	findByStartDateBefore	… where x.startDate < ?1
IsNull	findByAgeIsNull	… where x.age is null
IsNotNull、NotNull	findByAge(Is)NotNull	… where x.age not null
Like	findByFirstnameLike	… where x.firstname like ?1

续表

关 键 字	示 例	JPQL 代码段
NotLike	findByFirstnameNotLike	… where x.firstname not like ?1
StartingWith	findByFirstnameStartingWith	… where x.firstname like ?1 参数后加%，即以参数开头的模糊查询
EndingWith	findByFirstnameEndingWith	… where x.firstname like ?1 参数前加%，即以参数结尾的模糊查询
Containing	findByFirstnameContaining	… where x.firstname like ?1 参数两边加%，即包含参数的模糊查询
OrderBy	findByAgeOrderByLastnameDesc	… where x.age = ?1 order by x.lastname desc
Not	findByLastnameNot	… where x.lastname <> ?1
In	findByAgeIn(Collection\<Age> ages)	… where x.age in ?1
NotIn	findByAgeNotIn(Collection\<Age> ages)	… where x.age not in ?1
True	findByActiveTrue()	… where x.active = true
False	findByActiveFalse()	… where x.active = false
IgnoreCase	findByFirstnameIgnoreCase	… where UPPER(x.firstame) = UPPER(?1)

❷ 限制查询结果数量

在 Spring Data JPA 中使用 Top 和 First 关键字限制查询结果数量，示例如下：

```java
public interface UserRepository extends JpaRepository<MyUser, Integer>{
    /**
     * 获得符合查询条件的前 10 条
     */
    public List<MyUser> findTop10ByUnameLike(String uname);
    /**
     * 获得符合查询条件的前 15 条
     */
    public List<MyUser> findFirst15ByUnameLike(String uname);
}
```

❸ 简单条件查询示例

下面通过实例讲解在 Spring Boot Web 应用中如何使用 Spring Data JPA 进行简单条件查询。

【例 6-1】使用 Spring Data JPA 进行简单条件查询。

其具体实现步骤如下：

1）创建数据库

本书采用的关系数据库是 MySQL 8，为了演示本例，首先通过 "CREATE DATABASE springbootjpa;" 命令创建名为 springbootjpa 的数据库。

2）创建基于 Thymeleaf 和 Spring Data JPA 的 Spring Boot Web 应用 ch6_1

在 IDEA 中创建基于 Lombok、Thymeleaf 和 Spring Data JPA 的 Spring Boot Web 应用 ch6_1。

3）修改 pom.xml 文件，添加 MySQL 依赖

在 pom.xml 文件中添加如下依赖：

```xml
<dependency>
    <groupId>mysql</groupId>
    <artifactId>mysql-connector-java</artifactId>
    <version>8.0.29</version>
</dependency>
```

4）设置 Web 应用 ch6_1 的上下文路径及数据源配置信息

在 ch6_1 的 application.properties 文件中配置如下内容：

```
server.servlet.context-path=/ch6_1
#数据库地址
spring.datasource.url=jdbc:mysql://localhost:3306/springbootjpa?useUnicode=true&characterEncoding=UTF-8&allowMultiQueries=true&serverTimezone=GMT%2B8
#数据库用户名
spring.datasource.username=root
#数据库密码
spring.datasource.password=root
#数据库驱动
spring.datasource.driver-class-name=com.mysql.cj.jdbc.Driver
####
#JPA 持久化配置
####
#指定数据库类型
spring.jpa.database=MYSQL
#指定是否在日志中显示 SQL 语句
spring.jpa.show-sql=true
#指定自动创建、更新数据库表等配置，update 表示如果数据库中存在持久化类对应的表就不创建，
#不存在就创建对应的表
spring.jpa.hibernate.ddl-auto=update
#让控制器输出的 JSON 字符串的格式更美观
spring.jackson.serialization.indent-output=true
```

5）创建持久化实体类 MyUser

创建名为 com.ch.ch6_1.entity 的包，并在该包中创建名为 MyUser 的持久化实体类，具体代码如下：

```
package com.ch.ch6_1.entity;
import jakarta.persistence.*;
import lombok.Data;
import java.io.Serializable;
@Entity
@Table(name = "user_table")
@Data
public class MyUser implements Serializable{
    private static final long serialVersionUID = 1L;
    @Id
    @GeneratedValue(strategy = GenerationType.IDENTITY)
    private int id; //主键
    /**使用@Column 注解，可以配置列的相关属性（例如列名、长度等），
     * 可以省略，默认为属性名小写，如果属性名是词组，将在中间加上"_"
     */
    private String uname;
    private String usex;
    private int age;
}
```

在持久化类中，@Entity 注解表明该实体类是一个与数据库表映射的实体类。@Table 表示实体类与哪个数据库表映射，如果没有通过 name 属性指定表名，默认为小写的类名。如果类名为词组，将在中间加上"_"（例如 MyUser 类对应的表名为 my_user）。@Id 注解的属性表示该属性映射为数据库表的主键。@GeneratedValue 注解默认使用的主键生成方式为自增，如果是 MySQL、SQL Server 等关系数据库，可以映射成一个递增的主键；如果是 Oracle 等关系数据库，Hibernate 将自动生成一个名为 HIBERNATE_SEQUENCE 的序列。

第 6 章 Spring Boot 的数据访问

6) 创建数据访问层

创建名为 com.ch.ch6_1.repository 的包，并在该包中创建名为 UserRepository 的接口。该接口继承 JpaRepository 接口，具体代码如下：

```java
package com.ch.ch6_1.repository;
import java.util.List;
import org.springframework.data.jpa.repository.JpaRepository;
import com.ch.ch6_1.entity.MyUser;
/**
 * 这里不需要使用@Repository注解数据访问层，
 * 因为Spring Boot自动配置了JpaRepository
 */
public interface UserRepository extends JpaRepository<MyUser, Integer>{
    public MyUser findByUname(String uname);
    public List<MyUser> findByUnameLike(String uname);
}
```

由于 UserRepository 接口继承了 JpaRepository 接口，所以在 UserRepository 接口中除了上述自定义的两个接口方法以外（方法名的命名规范参照表 6.1），还有 JpaRepository 的接口方法。

7) 创建业务层

创建名为 com.ch.ch6_1.service 的包，并在该包中创建 UserService 接口和接口的实现类 UserServiceImpl。这里省略 UserService 接口的代码。

UserServiceImpl 实现类的核心代码如下：

```java
@Service
public class UserServiceImpl implements UserService{
    @Autowired   //依赖注入数据访问层
    private UserRepository userRepository;
    @Override
    public void saveAll() {
        MyUser mu1 = new MyUser();
        mu1.setUname("陈恒1");
        mu1.setUsex("男");
        mu1.setAge(88);
        MyUser mu2 = new MyUser();
        mu2.setUname("陈恒2");
        mu2.setUsex("女");
        mu2.setAge(18);
        MyUser mu3 = new MyUser();
        mu3.setUname("陈恒3");
        mu3.setUsex("男");
        mu3.setAge(99);
        List<MyUser> users = new ArrayList<MyUser>();
        users.add(mu1);
        users.add(mu2);
        users.add(mu3);
        //调用父接口中的saveAllAndFlush方法
        userRepository.saveAllAndFlush(users);
    }
    @Override
    public List<MyUser> findAll() {
        //调用父接口中的findAll方法
        return userRepository.findAll();
    }
    @Override
```

```java
    public MyUser findByUname(String uname) {
        return userRepository.findByUname(uname);
    }
    @Override
    public List<MyUser> findByUnameLike(String uname) {
        return userRepository.findByUnameLike("%" + uname + "%");
    }
    @Override
    public MyUser getOne(int id) {
        //调用父接口中的getReferenceById方法
        return userRepository.getReferenceById(id);
    }
}
```

8）创建控制器类 UserTestController

创建名为 com.ch.ch6_1.controller 的包，并在该包中创建名为 UserTestController 的控制器类。UserTestController 的核心代码如下：

```java
@Controller
public class UserTestController {
    @Autowired
    private UserService userService;
    @GetMapping("/save")
    @ResponseBody
    public String save() {
        userService.saveAll();
        return "保存用户成功！";
    }
    @GetMapping("/findByUname")
    public String findByUname(String uname, Model model) {
        model.addAttribute("title", "根据用户名查询一个用户");
        model.addAttribute("auser", userService.findByUname(uname));
        return "showAuser";
    }
    @GetMapping("/getOne")
    public String getOne(int id, Model model) {
        model.addAttribute("title", "根据用户id查询一个用户");
        model.addAttribute("auser",userService.getOne(id));
        return "showAuser";
    }
    @GetMapping("/findAll")
    public String findAll(Model model){
        model.addAttribute("title", "查询所有用户");
        model.addAttribute("allUsers",userService.findAll());
        return "showAll";
    }
    @GetMapping("/findByUnameLike")
    public String findByUnameLike(String uname, Model model){
        model.addAttribute("title", "根据用户名模糊查询所有用户");
        model.addAttribute("allUsers",userService.findByUnameLike(uname));
        return "showAll";
    }
}
```

9）整理脚本、样式等静态文件

JS 脚本、CSS 样式、图片等静态文件默认放置在 src/main/resources/static 目录下，ch6_1 应用引入的 BootStrap 和 jQuery 与例 5-5 中的一样，这里不再赘述。

10）创建 View 视图页面

在 src/main/resources/templates 目录下创建视图页面 showAll.html 和 showAuser.html。
showAll.html 的具体代码如下：

```html
<!DOCTYPE html>
<html xmlns:th="http://www.thymeleaf.org">
<head>
<meta charset="UTF-8">
<title>显示查询结果</title>
<link rel="stylesheet" th:href="@{css/bootstrap.min.css}" />
</head>
<body>
    <div class="panel panel-primary">
        <div class="panel-heading">
            <h3 class="panel-title">Spring Data JPA 简单查询</h3>
        </div>
    </div>
    <div class="container">
        <div class="panel panel-primary">
            <div class="panel-heading">
                <h3 class="panel-title"><span th:text="${title}"></span></h3>
            </div>
            <div class="panel-body">
                <div class="table table-responsive">
                    <table class="table table-bordered table-hover">
                        <tbody class="text-center">
                            <tr th:each="user:${allUsers}">
                                <td><span th:text="${user.id}"></span></td>
                                <td><span th:text="${user.uname}"></span></td>
                                <td><span th:text="${user.usex}"></span></td>
                                <td><span th:text="${user.age}"></span></td>
                            </tr>
                        </tbody>
                    </table>
                </div>
            </div>
        </div>
    </div>
</body>
</html>
```

showAuser.html 的具体代码如下：

```html
<!DOCTYPE html>
<html xmlns:th="http://www.thymeleaf.org">
<head>
<meta charset="UTF-8">
<title>显示查询结果</title>
<link rel="stylesheet" th:href="@{css/bootstrap.min.css}" />
</head>
<body>
    <div class="panel panel-primary">
        <div class="panel-heading">
            <h3 class="panel-title">Spring Data JPA 简单查询</h3>
        </div>
    </div>
    <div class="container">
        <div class="panel panel-primary">
            <div class="panel-heading">
                <h3 class="panel-title"><span th:text="${title}"></span></h3>
```

```
                </div>
                <div class="panel-body">
                    <div class="table table-responsive">
                        <table class="table table-bordered table-hover">
                            <tbody class="text-center">
                                <tr>
                                    <td><span th:text="${auser.id}"></span></td>
                                    <td><span th:text="${auser.uname}"></span></td>
                                    <td><span th:text="${auser.usex}"></span></td>
                                    <td><span th:text="${auser.age}"></span></td>
                                </tr>
                            </tbody>
                        </table>
                    </div>
                </div>
            </div>
    </body>
</html>
```

11）运行

首先运行 Ch61Application 主类，然后访问 "http://localhost:8080/ch6_1/save/"，运行效果如图 6.2 所示。

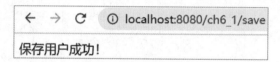

图 6.2　保存用户

"http://localhost:8080/ch6_1/save/" 成功运行后，在 MySQL 的 springbootjpa 数据库中将创建一张名为 user_table 的数据库表，并插入 3 条记录。

通过访问 "http://localhost:8080/ch6_1/findAll" 查询所有用户，运行效果如图 6.3 所示。

Spring Data JPA简单查询				
查询所有用户			
1	陈恒1	男	88
2	陈恒2	女	18
3	陈恒3	男	99

图 6.3　查询所有用户

通过访问 "http://localhost:8080/ch6_1/findByUnameLike?uname=陈" 模糊查询所有陈姓用户，运行效果图与图 6.3 一样。

通过访问 "http://localhost:8080/ch6_1/findByUname?uname=陈恒 2" 查询一个名为陈恒 2 的用户的信息，运行效果如图 6.4 所示。

图 6.4　查询一个名为陈恒 2 的用户的信息

通过访问"http://localhost:8080/ch6_1/getOne?id=1"查询一个 id 为 1 的用户的信息，运行效果如图 6.5 所示。

图 6.5　查询一个 id 为 1 的用户的信息

▶ 6.1.3　关联查询

在 Spring Data JPA 中有一对一、一对多、多对多等关系映射，本小节将对这些关系映射进行讲解。

❶ @OneToOne

一对一关系在现实生活中是十分常见的，比如一个大学生只有一张一卡通，一张一卡通只属于一个大学生。再比如人与身份证的关系也是一对一的关系。

在 Spring Data JPA 中可以用两种方式描述一对一关系映射，一种是通过外键的方式（一个实体通过外键关联到另一个实体的主键）；另一种是通过一张关联表来保存两个实体一对一的关系。下面通过外键的方式讲解一对一关系映射。

【例 6-2】使用 Spring Data JPA 实现人与身份证的一对一关系映射。

首先为例 6-2 创建基于 Lombok 和 Spring Data JPA 依赖的 Spring Boot Web 应用 ch6_2。ch6_2 应用的数据库、pom.xml 以及 application.properties 与 ch6_1 应用的基本一样，这里不再赘述。

其他内容的具体实现步骤如下：

1）创建持久化实体类

创建名为 com.ch.ch6_2.entity 的包，并在该包中创建名为 Person 和 IdCard 的持久化实体类。

Person 的代码如下：

```
package com.ch.ch6_2.entity;
import java.io.Serializable;
```

```java
import jakarta.persistence.*;
import com.fasterxml.jackson.annotation.JsonIgnore;
import com.fasterxml.jackson.annotation.JsonIgnoreProperties;
import lombok.Data;
@Entity
@Table(name="person_table")
/**解决No serializer found for class org.hibernate.proxy.pojo.bytebuddy.
ByteBuddyInterceptor 异常*/
@JsonIgnoreProperties(value = {"hibernateLazyInitializer"})
@Data
public class Person implements Serializable{
    private static final long serialVersionUID = 1L;
    @Id
    @GeneratedValue(strategy = GenerationType.IDENTITY)
    private int id;    //自动递增的主键
    private String pname;
    private String psex;
    private int page;
    @OneToOne(
        optional = true,
        fetch = FetchType.LAZY,
        targetEntity = IdCard.class,
        cascade = CascadeType.ALL
    )
    /*指明 Person 对应表的 id_Card_id 列作为外键与 IdCard 对应表的 id 列进行关联，
unique= true 指明id_Card_id列的值不可重复*/
    @JoinColumn(
        name = "id_Card_id",
        referencedColumnName = "id",
        unique= true
    )
    @JsonIgnore
    /*如果A对象持有B的引用,B对象持有A的引用,这样就形成了循环引用,如果直接使用JSON
转换会报错，使用@JsonIgnore解决该错误*/
    private IdCard idCard;
}
```

在上述实体类 Person 中，@OneToOne 注解有 5 个属性，即 targetEntity、cascade、fetch、optional 和 mappedBy。

targetEntity 属性：class 类型属性，定义关系类的类型，默认是该成员属性对应的类类型，所以通常不需要提供定义。

cascade 属性：CascadeType[]类型，该属性定义类和类之间的级联关系，定义的级联关系将被容器视为对当前类对象及其关联类对象采取相同的操作，而且这种关系是递归调用的。cascade 的值能从 CascadeType.PERSIST（级联保存）、CascadeType.REMOVE（级联删除）、CascadeType.REFRESH（级联刷新）、CascadeType.MERGE（级联更新）中选择一个或多个，另外还有一个选择是使用CascadeType.ALL，表示选择全部（以上4项）。

FetchType.LAZY：懒加载，在加载一个实体时，定义懒加载的属性不会马上从数据库中加载。**FetchType.EAGER**：急加载，在加载一个实体时，定义急加载的属性会立即从数据库中加载。

optional = true，表示 idCard 属性可以为 null，也就是允许没有身份证，例如未成年人没有身份证。

mappedBy 标签一定是定义在关系的被维护端，它指向关系的维护端；只有@OneToOne、

@OneToMany、@ManyToMany 上才有 mappedBy 属性,ManyToOne 不存在该属性;拥有 mappedBy 注解的实体类为关系的被维护端。

IdCard 的代码如下:

```java
package com.ch.ch6_2.entity;
import java.io.Serializable;
import java.util.Calendar;
import jakarta.persistence.*;
import com.fasterxml.jackson.annotation.JsonIgnoreProperties;
import lombok.Data;
@Entity
@Table(name = "idcard_table")
@JsonIgnoreProperties(value = { "hibernateLazyInitializer"})
@Data
public class IdCard implements Serializable{
    private static final long serialVersionUID = 1L;
    @Id
    @GeneratedValue(strategy = GenerationType.IDENTITY)
    private int id;    //自动递增的主键
    private String code;
    /**
     * @Temporal 主要是用来指明 java.util.Date 或 java.util.Calendar 类型的属性具体
     * 与数据库 3 个类型(date、time、timestamp)中的哪一个进行映射
     */
    @Temporal(value = TemporalType.DATE)
    private Calendar birthday;
    private String address;
    /**
     * optional = false 设置 person 属性值不能为 null,也就是身份证必须有对应的主人。
     * mappedBy = "idCard"与 Person 类中的 idCard 属性一致
     *拥有 mappedBy 注解的实体类为关系的被维护端。
     */
    @OneToOne(
        optional = false,
        fetch = FetchType.LAZY,
        targetEntity = Person.class,
        mappedBy = "idCard",
        cascade = CascadeType.ALL
    )
    private Person person;    //对应的人
}
```

2)创建数据访问层

创建名为 com.ch.ch6_2.repository 的包,并在该包中创建名为 IdCardRepository 和 PersonRepository 的接口。

IdCardRepository 的代码如下:

```java
package com.ch.ch6_2.repository;
import java.util.List;
import org.springframework.data.jpa.repository.JpaRepository;
import com.ch.ch6_2.entity.IdCard;
public interface IdCardRepository extends JpaRepository<IdCard, Integer>{
    /**
     * 根据人员 ID 查询身份信息(关联查询,根据 person 属性的 id)
     * 相当于 JPQL 语句: select ic from IdCard ic where ic.person.id = ?1
     */
    public IdCard findByPerson_id(Integer id);
```

```
    /**
     * 根据地址和身份证号查询身份信息
     * 相当于JPQL语句: select ic from IdCard ic where ic.address = ?1 and ic.code =?2
     */
    public List<IdCard> findByAddressAndCode(String address, String code);
}
```

按照 Spring Data JPA 的规则，查询两个有关联关系的对象可以通过方法名中的 "_" 下画线来标识，例如根据人员 ID 查询身份信息 findByPerson_id。JPQL（Java Persistence Query Language）是一种和 SQL 非常类似的中间性和对象化查询语言，它最终被编译成针对不同底层数据库的 SQL 查询，从而屏蔽不同数据库的差异。JPQL 语句可以是 select 语句、update 语句或 delete 语句，它们都通过 Query 接口封装执行。JPQL 的具体学习内容不是本书的重点，需要学习的读者请参考相关内容学习。

PersonRepository 的代码如下：

```
package com.ch.ch6_2.repository;
import java.util.List;
import org.springframework.data.jpa.repository.JpaRepository;
import com.ch.ch6_2.entity.Person;
public interface PersonRepository extends JpaRepository<Person, Integer>{
    /**
     * 根据身份ID查询人员信息（关联查询，根据idCard属性的id）
     * 相当于JPQL语句: select p from Person p where p.idCard.id = ?1
     */
    public Person findByIdCard_id(Integer id);
    /**
     * 根据人名和性别查询人员信息
     * 相当于JPQL语句: select p from Person p where p.pname = ?1 and p.psex = ?2
     */
    public List<Person> findByPnameAndPsex(String pname, String psex);
}
```

3）创建业务层

创建名为 com.ch.ch6_2.service 的包，并在该包中创建名为 PersonAndIdCardService 的接口和接口实现类 PersonAndIdCardServiceImpl。这里省略 PersonAndIdCardService 接口的代码。

PersonAndIdCardServiceImpl 的核心代码如下：

```
@Service
public class PersonAndIdCardServiceImpl implements PersonAndIdCardService{
    @Autowired
    private IdCardRepository idCardRepository;
    @Autowired
    private PersonRepository personRepository;
    @Override
    public void saveAll() {
        //保存身份证
        IdCard ic1 = new IdCard();
        ic1.setCode("123456789");
        ic1.setAddress("北京");
        Calendar c1 = Calendar.getInstance();
        c1.set(2023, 8, 13);
        ic1.setBirthday(c1);
        IdCard ic2 = new IdCard();
        ic2.setCode("000123456789");
        ic2.setAddress("上海");
        Calendar c2 = Calendar.getInstance();
```

```java
        c2.set(2023, 8, 14);
        ic2.setBirthday(c2);
        IdCard ic3 = new IdCard();
        ic3.setCode("1111123456789");
        ic3.setAddress("广州");
        Calendar c3 = Calendar.getInstance();
        c3.set(2023, 8, 15);
        ic3.setBirthday(c3);
        List<IdCard> idCards = new ArrayList<IdCard>();
        idCards.add(ic1);
        idCards.add(ic2);
        idCards.add(ic3);
        idCardRepository.saveAllAndFlush(idCards);
        //保存人员
        Person p1 = new Person();
        p1.setPname("陈恒1");
        p1.setPsex("男");
        p1.setPage(88);
        p1.setIdCard(ic1);
        Person p2 = new Person();
        p2.setPname("陈恒2");
        p2.setPsex("女");
        p2.setPage(99);
        p2.setIdCard(ic2);
        Person p3 = new Person();
        p3.setPname("陈恒3");
        p3.setPsex("女");
        p3.setPage(18);
        p3.setIdCard(ic3);
        List<Person> persons = new ArrayList<Person>();
        persons.add(p1);
        persons.add(p2);
        persons.add(p3);
        personRepository.saveAllAndFlush(persons);
}
@Override
public List<Person> findAllPerson() {
    return personRepository.findAll();
}
@Override
public List<IdCard> findAllIdCard() {
    return idCardRepository.findAll();
}
/**
 * 根据人员ID查询身份信息（关联查询）
 */
@Override
public IdCard findByPerson_id(Integer id) {
    return idCardRepository.findByPerson_id(id);
}
@Override
public List<IdCard> findByAddressAndCode(String address, String code) {
    return idCardRepository.findByAddressAndCode(address, code);
}
/**
 * 根据身份ID查询人员信息（关联查询）
 */
@Override
```

```
    public Person findByIdCard_id(Integer id) {
        return personRepository.findByIdCard_id(id);
    }
    @Override
    public List<Person> findByPnameAndPsex(String pname, String psex) {
        return personRepository.findByPnameAndPsex(pname, psex);
    }
    @Override
    public IdCard getOneIdCard(Integer id) {
        return idCardRepository.getReferenceById(id);
    }
    @Override
    public Person getOnePerson(Integer id) {
        return personRepository.getReferenceById(id);
    }
}
```

4）创建控制器类

创建名为 com.ch.ch6_2.controller 的包，并在该包中创建名为 TestOneToOneController 的控制器类。

TestOneToOneController 的核心代码如下：

```
@RestController
public class TestOneToOneController {
    @Autowired
    private PersonAndIdCardService personAndIdCardService;
    @GetMapping("/save")
    public String save() {
        personAndIdCardService.saveAll();
        return "人员和身份保存成功！";
    }
    @GetMapping("/findAllPerson")
    public List<Person> findAllPerson() {
        return personAndIdCardService.findAllPerson();
    }
    @GetMapping("/findAllIdCard")
    public List<IdCard> findAllIdCard() {
        return personAndIdCardService.findAllIdCard();
    }
    /**
     * 根据人员 ID 查询身份信息（关联查询）
     */
    @GetMapping("/findByPerson_id")
    public IdCard findByPerson_id(Integer id) {
        return personAndIdCardService.findByPerson_id(id);
    }
    @GetMapping("/findByAddressAndCode")
    public List<IdCard> findByAddressAndCode(String address, String code){
        return personAndIdCardService.findByAddressAndCode(address, code);
    }
    /**
     * 根据身份 ID 查询人员信息（关联查询）
     */
    @GetMapping("/findByIdCard_id")
    public Person findByIdCard_id(Integer id) {
        return personAndIdCardService.findByIdCard_id(id);
    }
    @GetMapping("/findByPnameAndPsex")
    public List<Person> findByPnameAndPsex(String pname, String psex) {
```

```
        return personAndIdCardService.findByPnameAndPsex(pname, psex);
    }
    @GetMapping("/getOneIdCard")
    public IdCard getOneIdCard(Integer id) {
        return personAndIdCardService.getOneIdCard(id);
    }
    @GetMapping("/getOnePerson")
    public Person getOnePerson(Integer id) {
        return personAndIdCardService.getOnePerson(id);
    }
}
```

5）运行

首先运行 Ch62Application 主类，然后访问"http://localhost:8080/ch6_2/save"，运行效果如图 6.6 所示。

图 6.6　保存数据

"http://localhost:8080/ch6_2/save"成功运行后，在 MySQL 的 springbootjpa 数据库中创建名为 idcard_table 和 person_table 的数据库表（实体类成功加载后就已经创建好数据表），并分别插入 3 条记录。

通过"http://localhost:8080/ch6_2/findByIdCard_id?id=1"查询身份证 id 为 1 的人员信息（关联查询），运行效果如图 6.7 所示。

```
{
    "id" : 1,
    "pname" : "陈恒1",
    "psex" : "男",
    "page" : 88
}
```

图 6.7　查询身份证 id 为 1 的人员信息

通过"http://localhost:8080/ch6_2/findByPerson_id?id=1"查询人员 id 为 1 的身份证信息（关联查询），运行效果如图 6.8 所示。

```
{
    "id" : 1,
    "code" : "123456789",
    "birthday" : "2023-09-12T16:00:00.000+00:00",
    "address" : "北京",
    "person" : {
        "id" : 1,
        "pname" : "陈恒1",
        "psex" : "男",
        "page" : 88
    }
}
```

图 6.8　查询人员 id 为 1 的身份证信息

❷ @OneToMany 和@ManyToOne

在实际生活中,作者和文章是一对多的双向关系,那么在 Spring Data JPA 中如何描述一对多的双向关系呢?

在 Spring Data JPA 中使用@OneToMany 和@ManyToOne 来表示一对多的双向关系。例如,一端(Author)使用@OneToMany,多端(Article)使用@ManyToOne。

在 JPA 规范中,一对多的双向关系由多端(例如 Article)来维护。也就是说多端为关系的维护端,负责关系的增、删、改、查;一端为关系的被维护端,不能维护关系。

一端(Author)使用@OneToMany 注解的 mappedBy="author"属性表明一端(Author)是关系的被维护端。多端(Article)使用@ManyToOne 和@JoinColumn 来注解属性 author,@ManyToOne 表明 Article 是多端,@JoinColumn 设置在 article 表的关联字段(外键)上。

【例 6-3】使用 Spring Data JPA 实现 Author 与 Article 的一对多关系映射。

在 ch6_2 应用中实现例 6-3,具体实现步骤如下:

1)创建持久化实体类

在 com.ch.ch6_2.entity 包中创建名为 Author 和 Article 的持久化实体类。

Author 的代码如下:

```
package com.ch.ch6_2.entity;
package com.ch.ch6_2.entity;
import java.io.Serializable;
import java.util.List;
import jakarta.persistence.*;
import com.fasterxml.jackson.annotation.JsonIgnoreProperties;
import lombok.Data;
@Entity
@Table(name = "author_table")
@JsonIgnoreProperties(value = { "hibernateLazyInitializer"})
@Data
public class Author implements Serializable{
    private static final long serialVersionUID = 1L;
    @Id
    @GeneratedValue(strategy = GenerationType.IDENTITY)
    private int id;
    //作者名
    private String aname;
    //文章列表,作者与文章是一对多的关系
    @OneToMany(
        mappedBy = "author",
        cascade=CascadeType.ALL,
        targetEntity = Article.class,
        fetch=FetchType.LAZY
    )
    private List<Article> articleList;
}
```

Article 的代码如下:

```
package com.ch.ch6_2.entity;
import java.io.Serializable;
import jakarta.persistence.*;
import com.fasterxml.jackson.annotation.JsonIgnore;
import com.fasterxml.jackson.annotation.JsonIgnoreProperties;
import lombok.Data;
@Entity
@Table(name = "article_table")
@JsonIgnoreProperties(value = { "hibernateLazyInitializer"})
```

```java
@Data
public class Article implements Serializable{
    private static final long serialVersionUID = 1L;
    @Id
    @GeneratedValue(strategy = GenerationType.IDENTITY)
    private int id;
    //标题
    @Column(nullable = false, length = 50)
    private String title;
    //文章内容
    @Lob    //大对象，映射为MySQL的Long文本类型
    @Basic(fetch = FetchType.LAZY)
    @Column(nullable = false)
    private String content;
    //所属作者，文章与作者是多对一的关系
    @ManyToOne(cascade={CascadeType.MERGE,CascadeType.REFRESH},optional=false)
    //可选属性optional=false，表示author不能为空。删除文章不影响用户
    @JoinColumn(name="id_author_id")        //设置在article表中的关联字段（外键）
    @JsonIgnore
    private Author author;
}
```

2）创建数据访问层

在 com.ch.ch6_2.repository 包中创建名为 AuthorRepository 和 ArticleRepository 的接口。

AuthorRepository 的代码如下：

```java
package com.ch.ch6_2.repository;
import org.springframework.data.jpa.repository.JpaRepository;
import com.ch.ch6_2.entity.Author;
public interface AuthorRepository extends JpaRepository<Author, Integer>{
    /**
     * 根据文章标题包含的内容查询作者（关联查询）
     * 相当于JPQL语句：select a from Author a inner join a.articleList t where t.title like %?1%
     */
    public Author findByArticleList_titleContaining(String title);
}
```

ArticleRepository 的代码如下：

```java
package com.ch.ch6_2.repository;
import java.util.List;
import org.springframework.data.jpa.repository.JpaRepository;
import com.ch.ch6_2.entity.Article;
public interface ArticleRepository extends JpaRepository<Article, Integer>{
    /**
     * 根据作者id查询文章信息（根据author属性的id关联查询）
     * 相当于JPQL语句：select a from Article a where a.author.id = ?1
     */
    public List<Article> findByAuthor_id(Integer id);
    /**
     * 根据作者名查询文章信息（根据author属性的aname关联查询）
     * 相当于JPQL语句：select a from Article a where a.author.aname = ?1
     */
    public List<Article> findByAuthor_aname(String aname);
}
```

3）创建业务层

在 com.ch.ch6_2.service 包中创建名为 AuthorAndArticleService 的接口和接口实现类 AuthorAndArticleServiceImpl。这里省略 AuthorAndArticleService 接口的代码。

AuthorAndArticleServiceImpl 的核心代码如下：

```java
@Service
public class AuthorAndArticleServiceImpl implements AuthorAndArticleService{
    @Autowired
    private AuthorRepository authorRepository;
    @Autowired
    private ArticleRepository articleRepository;
    @Override
    public void saveAll() {
        //保存作者（先保存一的一端）
        Author a1 = new Author();
        a1.setAname("陈恒1");
        Author a2 = new Author();
        a2.setAname("陈恒2");
        ArrayList<Author> allAuthor = new ArrayList<Author>();
        allAuthor.add(a1);
        allAuthor.add(a2);
        authorRepository.saveAll(allAuthor);
        //保存文章
        Article at1 = new Article();
        at1.setTitle("JPA的一对多111");
        at1.setContent("其实一对多映射关系很常见111。");
        //设置关系
        at1.setAuthor(a1);
        Article at2 = new Article();
        at2.setTitle("JPA的一对多222");
        at2.setContent("其实一对多映射关系很常见222。");
        //设置关系
        at2.setAuthor(a1);     //文章2与文章1的作者相同
        Article at3 = new Article();
        at3.setTitle("JPA的一对多333");
        at3.setContent("其实一对多映射关系很常见333。");
        //设置关系
        at3.setAuthor(a2);
        Article at4 = new Article();
        at4.setTitle("JPA的一对多444");
        at4.setContent("其实一对多映射关系很常见444。");
        //设置关系
        at4.setAuthor(a2);     //文章3与文章4的作者相同
        ArrayList<Article> allAt = new ArrayList<Article>();
        allAt.add(at1);
        allAt.add(at2);
        allAt.add(at3);
        allAt.add(at4);
        articleRepository.saveAll(allAt);
    }
    @Override
    public List<Article> findByAuthor_id(Integer id) {
        return articleRepository.findByAuthor_id(id);
    }
    @Override
    public List<Article> findByAuthor_aname(String aname) {
        return articleRepository.findByAuthor_aname(aname);
    }
    @Override
    public Author findByArticleList_titleContaining(String title) {
```

```
        return authorRepository.findByArticleList_titleContaining(title);
    }
}
```

4）创建控制器类

在 com.ch.ch6_2.controller 包中创建名为 TestOneToManyController 的控制器类。TestOneToManyController 的核心代码如下：

```
@RestController
public class TestOneToManyController {
    @Autowired
    private AuthorAndArticleService authorAndArticleService;
    @GetMapping("/saveOneToMany")
    public String save() {
        authorAndArticleService.saveAll();
        return "作者和文章保存成功！";
    }
    @GetMapping("/findArticleByAuthor_id")
    public List<Article> findByAuthor_id(Integer id) {
        return authorAndArticleService.findByAuthor_id(id);
    }
    @GetMapping("/findArticleByAuthor_aname")
    public List<Article> findByAuthor_aname(String aname){
        return authorAndArticleService.findByAuthor_aname(aname);
    }
    @GetMapping("/findByArticleList_titleContaining")
    public Author findByArticleList_titleContaining(String title) {
        return authorAndArticleService.findByArticleList_titleContaining(title);
    }
}
```

5）运行

首先运行 Ch62Application 主类，然后访问"http://localhost:8080/ch6_2/saveOneToMany"，运行效果如图 6.9 所示。

图 6.9　保存数据

"http://localhost:8080/ch6_2/saveOneToMany"成功运行后，在 MySQL 的 springbootjpa 数据库中创建名为 author_table 和 article_table 的数据库表，并在 author_table 表中插入两条记录，同时在 article_table 表中插入 4 条记录。

通过"http://localhost:8080/ch6_2/findArticleByAuthor_id?id=2"查询作者 id 为 2 的文章列表（关联查询），运行效果如图 6.10 所示。

图 6.10　查询作者 id 为 2 的文章列表

通过"http://localhost:8080/ch6_2/findArticleByAuthor_aname?aname=陈恒1"查询作者名为陈恒1的文章列表（关联查询），运行效果如图6.11所示。

```
[ {
  "id" : 1,
  "title" : "JPA的一对多111",
  "content" : "其实一对多映射关系很常见111。"
}, {
  "id" : 2,
  "title" : "JPA的一对多222",
  "content" : "其实一对多映射关系很常见222。"
} ]
```

图6.11　查询作者名为陈恒1的文章列表

通过"http://localhost:8080/ch6_2/findByArticleList_titleContaining?title=对多1"查询文章标题包含"对多1"的作者（关联查询），运行效果如图6.12所示。

```
{
  "id" : 1,
  "aname" : "陈恒1",
  "articleList" : [ {
    "id" : 1,
    "title" : "JPA的一对多111",
    "content" : "其实一对多映射关系很常见111。"
  }, {
    "id" : 2,
    "title" : "JPA的一对多222",
    "content" : "其实一对多映射关系很常见222。"
  } ]
}
```

图6.12　查询文章标题包含"对多1"的作者

❸ @ManyToMany

在实际生活中，用户和权限是多对多的关系。一个用户可以有多个权限，一个权限也可以被很多用户拥有。

在Spring Data JPA中使用@ManyToMany来注解多对多的映射关系，由一个关联表来维护。关联表的表名默认为主表名+下画线+从表名（主表是指关系维护端对应的表，从表是指关系被维护端对应的表）。关联表只有两个外键字段，分别指向主表ID和从表ID。字段的名称默认为主表名+下画线+主表中的主键列名，从表名+下画线+从表中的主键列名。需要注意的是，在多对多关系中一般不设置级联保存、级联删除、级联更新等操作。

【例6-4】使用Spring Data JPA实现用户（User）与权限（Authority）的多对多关系映射。

在ch6_2应用中实现例6-4，具体实现步骤如下：

1）创建持久化实体类

在com.ch.ch6_2.entity包中创建名为User和Authority的持久化实体类。

User的代码如下：

```
package com.ch.ch6_2.entity;
import java.io.Serializable;
```

```java
import java.util.List;
import jakarta.persistence.*;
import com.fasterxml.jackson.annotation.JsonIgnoreProperties;
import lombok.Data;
@Entity
@Table(name = "user")
@JsonIgnoreProperties(value = { "hibernateLazyInitializer"})
@Data
public class User implements Serializable{
    private static final long serialVersionUID = 1L;
    @Id
    @GeneratedValue(strategy = GenerationType.IDENTITY)
    private int id;
    private String username;
    private String password;
    @ManyToMany
    @JoinTable(name = "user_authority",joinColumns = @JoinColumn(name = "user_id"),
        inverseJoinColumns = @JoinColumn(name = "authority_id"))
    /**1. 关系维护端，负责多对多关系的绑定和解除
    2. @JoinTable 注解的 name 属性指定关联表的名字，joinColumns 指定外键的名字，关联
到关系维护端（User）
    3. inverseJoinColumns 指定外键的名字，需要关联的关系，称为被维护端（Authority）
    4. 其实可以不使用@JoinTable 注解，默认生成的关联表名称为主表表名+下画线+从表表名，
即表名为 user_authority。
    关联到主表的外键名：主表名+下画线+主表中的主键列名，即 user_id。
    关联到从表的外键名：主表中用于关联的属性名+下画线+从表的主键列名，即 authority_id。
主表就是关系维护端对应的表，从表就是关系被维护端对应的表
    */
    private List<Authority> authorityList;
}
```

Authority 的代码如下：

```java
package com.ch.ch6_2.entity;
import java.io.Serializable;
import java.util.List;
import jakarta.persistence.*;
import com.fasterxml.jackson.annotation.JsonIgnore;
import com.fasterxml.jackson.annotation.JsonIgnoreProperties;
import lombok.Data;
@Entity
@Table(name = "authority")
@JsonIgnoreProperties(value = { "hibernateLazyInitializer"})
@Data
public class Authority implements Serializable{
    private static final long serialVersionUID = 1L;
    @Id
    @GeneratedValue(strategy = GenerationType.IDENTITY)
    private int id;
    @Column(nullable = false)
    private String name;
    @ManyToMany(mappedBy = "authorityList")
    @JsonIgnore
    private List<User> userList;
}
```

2）创建数据访问层

在 com.ch.ch6_2.repository 包中创建名为 UserRepository 和 AuthorityRepository 的接口。
UserRepository 的代码如下：

```java
package com.ch.ch6_2.repository;
import java.util.List;
import org.springframework.data.jpa.repository.JpaRepository;
import com.ch.ch6_2.entity.User;
public interface UserRepository extends JpaRepository<User, Integer>{
    /**
     * 根据权限 id 查询拥有该权限的用户（关联查询）
     * 相当于 JPQL 语句：select u from User u inner join u.authorityList a where a.id = ?1
     */
    public List<User> findByAuthorityList_id(int id);
    /**
     * 根据权限名查询拥有该权限的用户（关联查询）
     * 相当于 JPQL 语句：select u from User u inner join u.authorityList a where a.name = ?1
     */
    public List<User> findByAuthorityList_name(String name);
}
```

AuthorityRepository 的代码如下：

```java
package com.ch.ch6_2.repository;
import java.util.List;
import org.springframework.data.jpa.repository.JpaRepository;
import com.ch.ch6_2.entity.Authority;
public interface AuthorityRepository extends JpaRepository<Authority, Integer>{
    /**
     * 根据用户 id 查询用户所拥有的权限（关联查询）
     * 相当于 JPQL 语句：select a from Authority a inner join a.userList u where u.id = ?1
     */
    public List<Authority> findByUserList_id(int id);
    /**
     * 根据用户名查询用户所拥有的权限（关联查询）
     * 相当于 JPQL 语句：select a from Authority a inner join a.userList u where u.username = ?1
     */
    public List<Authority> findByUserList_Username(String username);
}
```

3）创建业务层

在 com.ch.ch6_2.service 包中创建名为 UserAndAuthorityService 的接口和接口实现类 UserAndAuthorityServiceImpl。这里省略 UserAndAuthorityService 接口的代码。

UserAndAuthorityServiceImpl 的核心代码如下：

```java
@Service
public class UserAndAuthorityServiceImpl implements UserAndAuthorityService{
    @Autowired
    private AuthorityRepository authorityRepository;
    @Autowired
    private UserRepository userRepository;
    @Override
    public void saveAll() {
        //添加权限 1
        Authority at1 = new Authority();
        at1.setName("增加");
        authorityRepository.save(at1);
        //添加权限 2
        Authority at2 = new Authority();
```

```java
        at2.setName("修改");
        authorityRepository.save(at2);
        //添加权限 3
        Authority at3 = new Authority();
        at3.setName("删除");
        authorityRepository.save(at3);
        //添加权限 4
        Authority at4 = new Authority();
        at4.setName("查询");
        authorityRepository.save(at4);
        //添加用户 1
        User u1 = new User();
        u1.setUsername("陈恒1");
        u1.setPassword("123");
        ArrayList<Authority> authorityList1 = new ArrayList<Authority>();
        authorityList1.add(at1);
        authorityList1.add(at2);
        authorityList1.add(at3);
        u1.setAuthorityList(authorityList1);
        userRepository.save(u1);
        //添加用户 2
        User u2 = new User();
        u2.setUsername("陈恒2");
        u2.setPassword("234");
        ArrayList<Authority> authorityList2 = new ArrayList<Authority>();
        authorityList2.add(at2);
        authorityList2.add(at3);
        authorityList2.add(at4);
        u2.setAuthorityList(authorityList2);
        userRepository.save(u2);
    }
    @Override
    public List<User> findByAuthorityList_id(int id) {
        return userRepository.findByAuthorityList_id(id);
    }
    @Override
    public List<User> findByAuthorityList_name(String name) {
        return userRepository.findByAuthorityList_name(name);
    }
    @Override
    public List<Authority> findByUserList_id(int id) {
        return authorityRepository.findByUserList_id(id);
    }
    @Override
    public List<Authority> findByUserList_Username(String username) {
        return authorityRepository.findByUserList_Username(username);
    }
}
```

4）创建控制器类

在 com.ch.ch6_2.controller 包中创建名为 TestManyToManyController 的控制器类。TestManyToManyController 的核心代码如下：

```java
@RestController
public class TestManyToManyController {
    @Autowired
    private UserAndAuthorityService userAndAuthorityService;
```

```java
@GetMapping("/saveManyToMany")
public String save() {
    userAndAuthorityService.saveAll();
    return "权限和用户保存成功！";
}
@GetMapping("/findByAuthorityList_id")
public List<User> findByAuthorityList_id(int id) {
    return userAndAuthorityService.findByAuthorityList_id(id);
}
@GetMapping("/findByAuthorityList_name")
public List<User> findByAuthorityList_name(String name) {
    return userAndAuthorityService.findByAuthorityList_name(name);
}
@GetMapping("/findByUserList_id")
public List<Authority> findByUserList_id(int id) {
    return userAndAuthorityService.findByUserList_id(id);
}
@GetMapping("/findByUserList_Username")
public List<Authority> findByUserList_Username(String username) {
    return userAndAuthorityService.findByUserList_Username(username);
}
}
```

5）运行

首先运行 Ch62Application 主类，然后访问 "http://localhost:8080/ch6_2/saveManyToMany"，运行效果如图 6.13 所示。

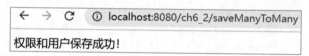

图 6.13　保存数据

通过 "http://localhost:8080/ch6_2/findByAuthorityList_id?id=1" 查询拥有 id 为 1 的权限的用户列表（关联查询），运行效果如图 6.14 所示。

图 6.14　查询拥有 id 为 1 的权限的用户列表

通过 "http://localhost:8080/ch6_2/findByAuthorityList_name?name=修改" 查询拥有 "修改" 权限的用户列表（关联查询），运行效果如图 6.15 所示。

第 6 章　Spring Boot 的数据访问

```
← → C  ① localhost:8080/ch6_2/findByAuthorityList_name?name=修改

[ {
  "id" : 1,
  "username" : "陈恒1",
  "password" : "123",
  "authorityList" : [ {
    "id" : 1,
    "name" : "增加"
  }, {
    "id" : 2,
    "name" : "修改"
  }, {
    "id" : 3,
    "name" : "删除"
  } ]
}, {
  "id" : 2,
  "username" : "陈恒2",
  "password" : "234",
  "authorityList" : [ {
    "id" : 2,
    "name" : "修改"
  }, {
```

图 6.15　查询拥有"修改"权限的用户列表

通过"http://localhost:8080/ch6_2/findByUserList_id?id=2"查询 id 为 2 的用户的权限列表（关联查询），运行效果如图 6.16 所示。

```
← → C  ① localhost:8080/ch6_2/findByUserList_id?id=2

[ {
  "id" : 2,
  "name" : "修改"
}, {
  "id" : 3,
  "name" : "删除"
}, {
  "id" : 4,
  "name" : "查询"
} ]
```

图 6.16　查询 id 为 2 的用户的权限列表

通过"http://localhost:8080/ch6_2/findByUserList_Username?username=陈恒 2"查询用户名为"陈恒 2"的用户的权限列表（关联查询），运行效果如图 6.17 所示。

```
← → C  ① localhost:8080/ch6_2/findByUserList_Username?username=陈恒2

[ {
  "id" : 2,
  "name" : "修改"
}, {
  "id" : 3,
  "name" : "删除"
}, {
  "id" : 4,
  "name" : "查询"
} ]
```

图 6.17　查询用户名为"陈恒 2"的用户的权限列表

在 6.1.2 小节和 6.1.3 小节中查询方法必须严格按照 Spring Data JPA 的查询关键字的命名规范进行命名。那么如何才能摆脱查询关键字和关联查询命名规范的约束呢？可以通过 @Query、@NamedQuery 直接定义 JPQL 语句进行数据的访问操作。

6.1.4 @Query 和 @Modifying 注解

❶ @Query 注解

使用 @Query 注解可以将 JPQL 语句直接定义在数据访问接口方法上，并且接口方法名不受查询关键字和关联查询命名规范的约束，示例代码如下：

```java
public interface AuthorityRepository extends JpaRepository<Authority, Integer>{
    /**
     * 根据用户名查询用户所拥有的权限（关联查询）
     */
    @Query("select a from Authority a inner join a.userList u where u.username = ?1")
    public List<Authority> findByUserListUsername(String username);
}
```

使用 @Query 注解定义 JPQL 语句，可以直接返回 List<Map<String, Object>>对象，示例代码如下：

```java
/**
 * 根据作者 id 查询文章信息（标题和内容）
 */
@Query("select new Map(a.title as title, a.content as content) from Article a where a.author.id = ?1 ")
public List<Map<String, Object>> findTitleAndContentByAuthorId(Integer id);
```

使用 @Query 注解定义 JPQL 语句，之前的方法是使用参数位置（"?1" 指获取方法形参列表中的第 1 个参数值，1 代表的是参数位置，以此类推）来获取参数值。除此之外，Spring Data JPA 还支持使用名称来获取参数值，使用格式为 ":参数名称"。示例代码如下：

```java
/**
 * 根据作者名和作者 id 查询文章信息
 */
@Query("select a from Article a where a.author.aname = :aname1 and a.author.id = :id1 ")
public List<Article> findArticleByAuthorAnameAndId(@Param("aname1") String aname, @Param("id1") Integer id);
```

❷ @Modifying 注解

用户可以使用 @Modifying 和 @Query 注解组合定义在数据访问接口方法上，进行更新查询操作，示例代码如下：

```java
/**
 * 根据作者 id 删除作者
 */
@Modifying
@Query("delete from Author a where a.id = ?1")
public int deleteAuthorByAuthorId(int id);
```

6.1.5 排序与分页查询

扫一扫

视频讲解

在实际应用开发中，排序与分页查询是必需的。Spring Data JPA 充分考虑了排序与分页查询的场景，为用户提供了 Sort 类、Page 接口以及 Pageable 接口。

例如以下数据访问接口：

```
public interface AuthorRepository extends JpaRepository<Author, Integer>{
    List<Author> findByAnameContaining(String aname, Sort sort);
}
```

那么在 Service 层可以这样使用排序：

```
public List<Author> findByAnameContaining(String aname, String sortColumn) {
    //按 sortColumn 降序排序
    return authorRepository.findByAnameContaining(aname, new Sort(Direction.DESC, sortColumn));
}
```

可以使用 Pageable 接口的实现类 PageRequest 的 of 方法构造分页查询对象，示例代码如下：

```
Page<Author> pageData = authorRepository.findAll(PageRequest.of(page-1, size, new Sort(Direction.DESC, "id")));
```

其中 Page 接口可以获得当前页面的记录、总页数、总记录数等信息，示例代码如下：

```
//获得当前页面的记录
List<Author> allAuthor = pageData.getContent();
model.addAttribute("allAuthor",allAuthor);
//获得总记录数
model.addAttribute("totalCount", pageData.getTotalElements());
//获得总页数
model.addAttribute("totalPage", pageData.getTotalPages());
```

下面通过实例讲解 Spring Data JPA 的排序与分页查询的使用方法。

【例 6-5】排序与分页查询的使用方法。

首先为例 6-5 创建基于 Thymeleaf、Lombok 和 Spring Data JPA 的 Spring Boot Web 应用 ch6_3。ch6_3 应用的数据库、pom.xml、application.properties 以及静态资源等内容与 ch6_1 应用的基本一样，这里不再赘述。

其他内容的具体实现步骤如下：

❶ 创建持久化实体类

在应用 ch6_3 的 src/main/java 目录下创建名为 com.ch.ch6_3.entity 的包，并在该包中创建名为 Article 和 Author 的持久化实体类，具体代码分别与 ch6_2 应用的 Article 和 Author 的代码一样，这里不再赘述。

❷ 创建数据访问层

在应用 ch6_3 的 src/main/java 目录下创建名为 com.ch.ch6_3.repository 的包，并在该包中创建名为 AuthorRepository 的接口。

AuthorRepository 的代码如下：

```
package com.ch.ch6_3.repository;
import java.util.List;
import org.springframework.data.domain.Sort;
import org.springframework.data.jpa.repository.JpaRepository;
import com.ch.ch6_3.entity.Author;
public interface AuthorRepository extends JpaRepository<Author, Integer>{
    /**
     * 查询作者名含有 name 的作者列表，并排序
     */
    List<Author> findByAnameContaining(String aname, Sort sort);
}
```

❸ 创建业务层

在应用 ch6_3 的 src/main/java 目录下创建名为 com.ch.ch6_3.service 的包，并在该包中创建名为 ArticleAndAuthorService 的接口和接口实现类 ArticleAndAuthorServiceImpl。

ArticleAndAuthorService 的代码如下：

```java
package com.ch.ch6_3.service;
import java.util.List;
import org.springframework.ui.Model;
import com.ch.ch6_3.entity.Author;
public interface ArticleAndAuthorService {
    /**
     * name 代表作者名的一部分（模糊查询），sortColumn 代表排序列
     */
    List<Author> findByAnameContaining(String aname, String sortColumn);
    /**
     * 分页查询作者，page 代表第几页
     */
    public String findAllAuthorByPage(Integer page, Model model);
}
```

ArticleAndAuthorServiceImpl 的核心代码如下：

```java
@Service
public class ArticleAndAuthorServiceImpl implements ArticleAndAuthorService{
    @Autowired
    private AuthorRepository authorRepository;
    @Override
    public List<Author> findByAnameContaining(String aname, String sortColumn) {
        //按 sortColumn 降序排序
        return authorRepository.findByAnameContaining(aname, Sort.by(Direction.DESC, sortColumn));
    }
    @Override
    public String findAllAuthorByPage(Integer page, Model model) {
        if(page == null) {    //第一次访问 findAllAuthorByPage 方法时
            page = 1;
        }
        int size = 2;         //每页显示两条
        //分页查询，of 方法的第一个参数代表第几页（比实际小 1），
        //第二个参数代表页面的大小，第三个参数代表排序规则
        Page<Author> pageData = authorRepository.findAll(PageRequest.of(page-1, size, Sort.by(Direction.DESC, "id")));
        //获得当前页面数据并转换成 List<Author>，转发到视图页面显示
        List<Author> allAuthor = pageData.getContent();
        model.addAttribute("allAuthor",allAuthor);
        //共多少条记录
        model.addAttribute("totalCount", pageData.getTotalElements());
        //共多少页
        model.addAttribute("totalPage", pageData.getTotalPages());
        //当前页
        model.addAttribute("page", page);
        return "index";
    }
}
```

❹ 创建控制器类

在应用 ch6_3 的 src/main/java 目录下创建名为 com.ch.ch6_3.controller 的包，并在该包中创建名为 TestSortAndPage 的控制器类。

TestSortAndPage 的核心代码如下：

```java
@Controller
public class TestSortAndPage {
    @Autowired
```

```
    private ArticleAndAuthorService articleAndAuthorService;
    @GetMapping("/findByAnameContaining")
    @ResponseBody
    public List<Author> findByAnameContaining(String aname, String sortColumn){
        return articleAndAuthorService.findByAnameContaining(aname, sortColumn);
    }
    @GetMapping("/findAllAuthorByPage")
    public String findAllAuthorByPage(Integer page, Model model){
        return articleAndAuthorService.findAllAuthorByPage(page, model);
    }
}
```

❺ 创建 View 视图页面

在 src/main/resources/templates 目录下创建视图页面 index.html。

index.html 的具体代码如下:

```html
<!DOCTYPE html>
<html xmlns:th="http://www.thymeleaf.org">
<head>
<meta charset="UTF-8">
<title>显示分页查询结果</title>
<link rel="stylesheet" th:href="@{css/bootstrap.min.css}" />
<link rel="stylesheet" th:href="@{css/bootstrap-theme.min.css}" />
</head>
<body>
    <div class="panel panel-primary">
        <div class="panel-heading">
            <h3 class="panel-title">Spring Data JPA 分页查询</h3>
        </div>
    </div>
    <div class="container">
        <div class="panel panel-primary">
            <div class="panel-body">
                <div class="table table-responsive">
                    <table class="table table-bordered table-hover">
                        <tbody class="text-center">
                            <tr th:each="author:${allAuthor}">
                                <td><span th:text="${author.id}"></span></td>
                                <td><span th:text="${author.aname}"></span></td>
                            </tr>
                            <tr>
                                <td colspan="2" align="right">
                                <ul class="pagination">
                                    <li><a>第<span th:text="${page}"></span>页</a></li>
                                    <li><a>共<span th:text="${totalPage}"></span>页</a></li>
                                    <li><a>共<span th:text="${totalCount}"></span>条</a></li>
                                    <li>
        <a th:href="@{findAllAuthorByPage(page=${page-1})}" th:if="${page != 1}">上一页</a>
                                    </li>
                                    <li>
        <a th:href="@{findAllAuthorByPage(page=${page+1})}" th:if="${page != totalPage}">下一页</a>
                                    </li>
                                </ul>
```

```
                            </td>
                        </tr>
                    </tbody>
                </table>
            </div>
        </div>
    </div>
</body>
</html>
```

❻ 运行

首先运行 Ch63Application 主类，然后通过"http://localhost:8080/ch6_3/findByAnameContaining?aname=陈&sortColumn=id"查询作者名含有"陈"的作者列表，并按照 id 降序。

通过"http://localhost:8080/ch6_3/findAllAuthorByPage"分页查询作者，并按照 id 降序，运行效果如图 6.18 所示。

图 6.18　分页查询作者

扫一扫

视频讲解

6.2　REST

本节将介绍 RESTful 风格接口，并通过 Spring Boot 来实现 RESTful。

▶ 6.2.1　REST 简介

REST 即表现层状态转化（英文为 Representational State Transfer，简称 REST），是 Roy Thomas Fielding 博士在 2000 年的一篇博士论文中提出来的一种软件架构风格。它是一种针对网络应用的设计和开发方式，可以降低开发的复杂性，提高系统的可伸缩性。在目前主流的 Web 服务实现方案中，因为 REST 模式的 Web 服务与复杂的 SOAP 和 XML-RPC 相比更加简洁，越来越多的 Web 服务开始采用 REST 风格设计和实现。

REST 是一组架构约束条件和原则。

（1）使用客户/服务器模型：客户和服务器之间通过一个统一的接口来互相通信。

（2）层次化的系统：在一个 REST 系统中，客户端并不会固定地与一个服务器"打交道"。

（3）无状态：在一个 REST 系统中，服务器端并不会保存有关客户的任何状态。也就是说，客户端自身负责用户状态的维持，并在每次发送请求时都需要提供足够的信息。

（4）可缓存：REST 系统需要能够恰当地缓存请求，以尽量减少服务器端和客户端之间的信息传输，从而提高性能。

（5）统一的接口：一个 REST 系统需要使用一个统一的接口来完成子系统之间以及服务与用户之间的交互，这使得 REST 系统中的各个子系统可以独自完成演化。

满足这些约束条件和原则的应用程序或设计就是 RESTful。需要注意的是，REST 是设计风格而不是标准。REST 通常基于 HTTP、URI、XML 以及 HTML 这些现有的广泛流行的协议和标准。

理解 RESTful 架构应该先理解 Representational State Transfer 这个词组到底是什么意思，它的每一个词表达了什么含义。

（1）资源（Resources）："表现层状态转化"中的"表现层"其实指的是"资源"的"表现层"。"资源"就是网络上的一个实体，或者说是网络上的一个具体信息。"资源"可以是一段文本、一张图片、一段视频，总之就是一个具体的实体。用户可以使用一个 URI（统一资源定位符）指向资源，每种资源对应一个特定的 URI。当需要获取资源时，访问它的 URI 即可，因此 URI 是每个资源的地址或独一无二的标识符。REST 风格的 Web 服务是通过一个简洁、清晰的 URI 来提供资源链接，客户端通过对 URI 发送 HTTP 请求获得这些资源，而获取和处理资源的过程让客户端应用的状态发生改变。

（2）表现层（Representation）："资源"是一种信息实体，可以有多种外在的表现形式，通常将"资源"呈现出来的形式称为它的"表现层"。例如，文本可以使用 TXT 格式表现，也可以使用 XML 格式、JSON 格式表现。

（3）状态转化（State Transfer）：客户端访问一个网站就代表了它和服务器的一个互动过程，在这个互动过程中将涉及数据和状态的变化。HTTP 是一个无状态的通信协议，这意味着所有状态都保存在服务器端。因此，如果客户端操作服务器，需要通过某种手段（例如 HTTP）让服务器端发生"状态变化"，而这种转化是建立在表现层之上的，所以就是"表现层状态转化"。

流行的各种 Web 框架（包括 Spring Boot）都支持 REST 开发。REST 并不是一种技术或者规范，而是一种架构风格，包括了如何标识资源、如何标识操作接口及操作的版本、如何标识操作的结果等，主要内容如下。

❶ 使用"api"作为上下文

在 REST 架构中建议使用"api"作为上下文，示例如下：

```
http://localhost:8080/api
```

❷ 增加一个版本标识

在 REST 架构中可以通过 URL 标识版本信息，示例如下：

```
http://localhost:8080/api/v1.0
```

❸ 标识资源

在 REST 架构中可以将资源名称放到 URL 中，示例如下：

```
http://localhost:8080/api/v1.0/user
```

❹ 确定 HTTP Method

HTTP 有 5 个常用的表示操作方式的动词，即 GET、POST、PUT、DELETE 和 PATCH。它们分别对应 5 种基本操作，其中 GET 用来获取资源，POST 用来增加资源（也可以用来更新资源），PUT 用来更新资源，DELETE 用来删除资源，PATCH 用来更新资源的部分属性。示例如下：

（1）新增用户。
```
POST http://localhost:8080/api/v1.0/user
```

（2）查询 id 为 123 的用户。
```
GET http://localhost:8080/api/v1.0/user/123
```

（3）更新 id 为 123 的用户。
```
PUT http://localhost:8080/api/v1.0/user/123
```

（4）删除 id 为 123 的用户。
```
DELETE http://localhost:8080/api/v1.0/user/123
```

❺ 确定 HTTP Status

常用的服务器向用户返回的状态码和提示信息如下。

（1）200 OK - [GET]：服务器成功返回用户请求的数据。

（2）201 CREATED - [POST/PUT/PATCH]：用户新建或修改数据成功。

（3）202 Accepted - [*]：一个请求已经进入后台排队（异步任务）。

（4）204 NO CONTENT - [DELETE]：用户删除数据成功。

（5）400 INVALID REQUEST - [POST/PUT/PATCH]：用户发出的请求有错误，服务器没有进行新建或修改数据的操作。

（6）401 Unauthorized - [*]：表示用户没有权限（令牌、用户名、密码错误）。

（7）403 Forbidden - [*]：表示用户得到授权（与 401 错误相对），但是访问是被禁止的。

（8）404 NOT FOUND - [*]：用户发出的请求针对的是不存在的记录，服务器没有进行操作。

（9）406 Not Acceptable - [GET]：用户请求的格式不可得（比如用户请求 JSON 格式，但是只有 XML 格式）。

（10）410 Gone -[GET]：用户请求的资源被永久删除，不会再得到。

（11）422 Unprocesable entity - [POST/PUT/PATCH]：当创建一个对象时发生一个验证错误。

（12）500 INTERNAL SERVER ERROR - [*]：服务器发生错误，用户将无法判断发出的请求是否成功。

▶ 6.2.2 Spring Boot 整合 REST

在 Spring Boot 的 Web 应用中自动支持 REST，也就是说只要 spring-boot-starter-web 依赖在 POM 中就支持 REST。

【例 6-6】一个 RESTful 应用示例。

假如在 ch6_2 应用的控制器类中有如下处理方法：

```
@RequestMapping("/findArticleByAuthor_id/{id}")
public List<Article> findByAuthor_id(@PathVariable("id") Integer id) {
    return authorAndArticleService.findByAuthor_id(id);
}
```

那么可以使用如下所示的 REST 风格的 URL 访问上述处理方法：

```
http://localhost:8080/ch6_2/findArticleByAuthor_id/2
```

在例 6-6 中使用了 URL 模板模式映射@RequestMapping("/findArticleByAuthor_id/{id}")，其中{XXX}为占位符，请求的 URL 可以是"/findArticleByAuthor_id/1"或"/findArticleByAuthor_id/2"。通过在处理方法中使用@PathVariable 获取{XXX}中的 XXX 变量值。@PathVariable 用于将请求 URL 中的模板变量映射到功能处理方法的参数上。如果{XXX}中的变量名 XXX 和形参名称一致，则@PathVariable 不用指定名称。

▶ 6.2.3 Spring Data REST

Spring Data JPA 基于 Spring Data 的 repository 之上，可以将 repository 自动输出为 REST 资源。目前，Spring Data REST 支持将 Spring Data JPA、Spring Data MongoDB、Spring Data Neo4j、Spring Data GemFire 以及 Spring Data Cassandra 的 repository 自动转换成 REST 服务。

Spring Boot 对 Spring Data REST 的自动配置存放在 org.springframework.boot.autoconfigure.data.rest 包中。通过 SpringBootRepositoryRestConfigurer 类的源代码可以得出，Spring Boot 已经自动配置了 RepositoryRestConfiguration，所以在 Spring Boot 应用中使用 Spring Data REST 只需引入 spring-boot-starter-data-rest 的依赖即可。

下面通过实例讲解 Spring Data REST 的构建过程。

【例 6-7】Spring Data REST 的构建过程。

其具体实现步骤如下：

❶ 创建 Spring Boot 应用 ch6_4

创建 Spring Boot 应用 ch6_4，依赖为 Lombok、Spring Data JPA 和 Rest Repositories，并在 pom.xml 文件中添加 MySQL 依赖。

❷ 设置应用 ch6_4 的上下文路径及数据源配置信息

在 ch6_4 的 application.properties 文件中配置如下内容：

```
server.servlet.context-path=/api
#数据库地址
spring.datasource.url=jdbc:mysql://localhost:3306/springbootjpa?characterEncoding=utf8
#数据库用户名
spring.datasource.username=root
#数据库密码
spring.datasource.password=root
#数据库驱动
spring.datasource.driver-class-name=com.mysql.jdbc.Driver
#指定数据库类型
spring.jpa.database=MYSQL
#指定是否在日志中显示 SQL 语句
spring.jpa.show-sql=true
#指定自动创建、更新数据库表等配置，update 表示如果数据库中存在持久化类对应的表就不创建，
#不存在就创建对应的表
spring.jpa.hibernate.ddl-auto=update
#让控制器输出的 JSON 字符串的格式更美观
spring.jackson.serialization.indent-output=true
```

❸ 创建持久化实体类 Student

在应用 ch6_4 的 src/main/java 目录下创建名为 com.ch.ch6_4.entity 的包，并在该包中创建名为 Student 的持久化实体类，具体代码如下：

```
package com.ch.ch6_4.entity;
```

```java
import java.io.Serializable;
import jakarta.persistence.*;
import lombok.Data;
@Entity
@Table(name = "student_table")
@Data
public class Student implements Serializable{
    private static final long serialVersionUID = 1L;
    @Id
    @GeneratedValue(strategy = GenerationType.IDENTITY)
    private int id;   //主键
    private String sno;
    private String sname;
    private String ssex;
    public Student() {
        super();
    }
    public Student(int id, String sno, String sname, String ssex) {
        super();
        this.id = id;
        this.sno = sno;
        this.sname = sname;
        this.ssex = ssex;
    }
}
```

❹ 创建数据访问层

在应用 ch6_4 的 src/main/java 目录下创建名为 com.ch.ch6_4.repository 的包，并在该包中创建名为 StudentRepository 的接口，该接口继承 JpaRepository 接口，具体代码如下：

```java
package com.ch.ch6_4.repository;
import java.util.List;
import org.springframework.data.jpa.repository.JpaRepository;
import org.springframework.data.repository.query.Param;
import org.springframework.data.rest.core.annotation.RestResource;
import com.ch.ch6_4.entity.Student;
public interface StudentRepository extends JpaRepository<Student, Integer>{
    /**
     * 自定义接口查询方法，暴露为 REST 资源
     */
    @RestResource(path = "snameStartsWith", rel = "snameStartsWith")
    List<Student> findBySnameStartsWith(@Param("sname") String sname);
}
```

在上述数据访问接口中定义了 findBySnameStartsWith，并使用@RestResource 注解将该方法暴露为 REST 资源。

至此，基于 Spring Data 的 REST 资源服务已经构建完毕，接下来使用 REST 客户端测试此服务。

▶ 6.2.4 REST 服务测试

在进行 Web 和移动端开发时经常会调用服务器端的 RESTful 接口进行数据请求，为了调试，一般会先用工具进行测试，通过测试后才开始在开发中使用。本小节将介绍如何使用 Postman 进行 6.2.3 小节的 RESTful 接口请求测试。

Postman 是一个接口测试工具，在做接口测试时，Postman 相当于一个客户端，它可以模

拟用户发起的各类HTTP请求，将请求数据发送至服务器端，获取对应的响应结果，从而验证响应中的结果数据是否和预期值相匹配。

Postman主要用来模拟各种HTTP请求（例如GET、POST、DELETE等），Postman与浏览器的区别在于有的浏览器不能输出JSON格式，而Postman能更直观地查看接口返回的结果。

用户可以从官网"https://www.postman.com/"下载对应的Postman安装程序，在安装成功后不需要创建账号即可使用。

❶ 获得列表数据

在RESTful架构中，每个网址代表一种资源（resource），所以网址中不能有动词，只能有名词，而且所用的名词往往与实体名对应。一般来说，数据库中的表都是同种记录的"集合（collection）"，所以API中的名词也应该使用复数，例如students。

运行ch6_4的主类Ch64Application，在student_table中添加几条学生信息后，在Postman中使用GET方式访问"http://localhost:8080/api/students"请求路径获得所有学生信息，如图6.19所示。

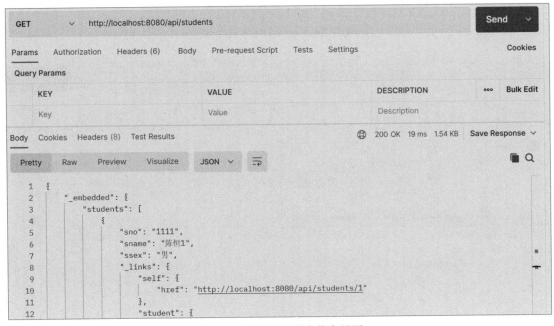

图6.19　获得所有学生信息界面

❷ 获得单一对象

在Postman中使用GET方式访问"http://localhost:8080/api/students/1"请求路径获得id为1的学生信息，如图6.20所示。

❸ 查询

在Postman中search调用自定义的接口查询方法,因此可以使用GET访问"http://localhost:8080/api/students/search/snameStartsWith?sname=陈"请求路径调用List<Student> findBySnameStartsWith(@Param("sname") String sname)接口方法,获得姓名前缀为"陈"的学生信息，如图6.21所示。

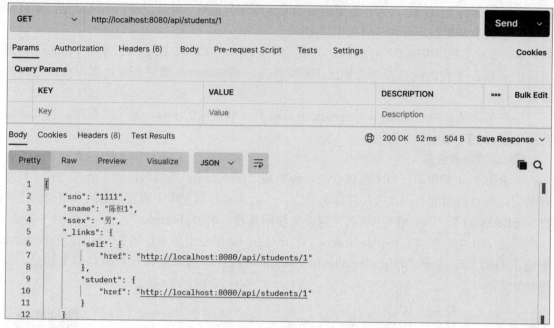

图 6.20　获得 id 为 1 的学生信息界面

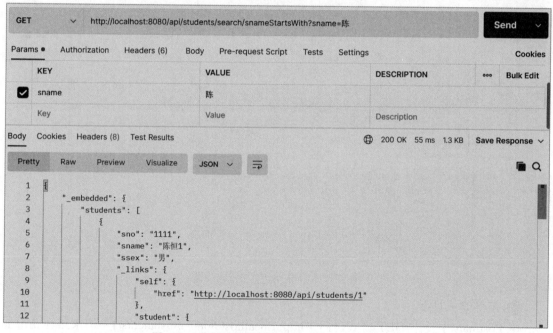

图 6.21　获得姓名前缀为"陈"的学生信息界面

❹ 分页查询

在 Postman 中使用 GET 方式访问"http://localhost:8080/api/students?page=0&size=2"请求路径获得第一页的学生信息（page=0 即第一页，size=2 即每页数量为 2），如图 6.22 所示。

从图 6.22 返回的结果可以看出，不仅获得了当前分页的数据，而且给出了上一页、下一页、第一页、最后一页的 REST 资源路径。

第 6 章　Spring Boot 的数据访问

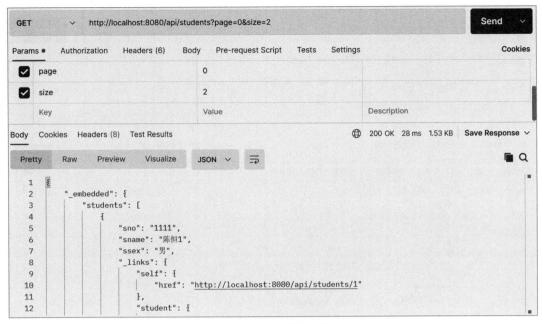

图 6.22　分页查询界面

❺ 排序

在 Postman 中使用 GET 方式访问 "http://localhost:8080/api/students?sort=sno,desc" 请求路径获得按照 sno 属性倒序的列表，如图 6.23 所示。

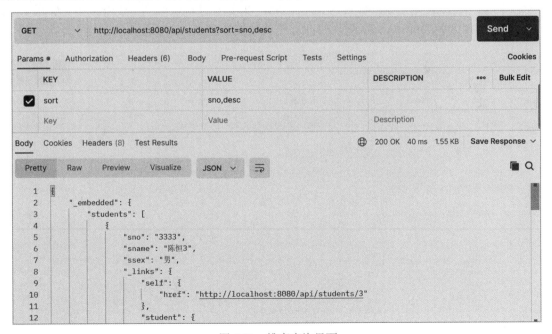

图 6.23　排序查询界面

❻ 保存

在 Postman 中发起 POST 方式请求 "http://localhost:8080/api/students" 实现新增功能，将要保存的数据放置在请求体中，数据类型为 JSON，如图 6.24 所示。

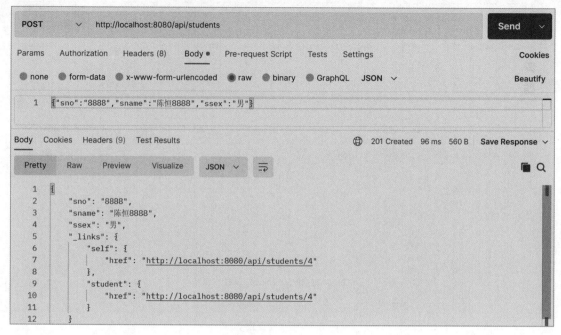

图 6.24 保存数据

从图 6.24 可以看出，保存成功后新数据的 id 为 4。

❼ 更新

假如需要更新新增的 id 为 4 的数据，可以在 Postman 中使用 PUT 方式访问 "http://localhost:8080/api/students/4"，修改提交的数据，如图 6.25 所示。

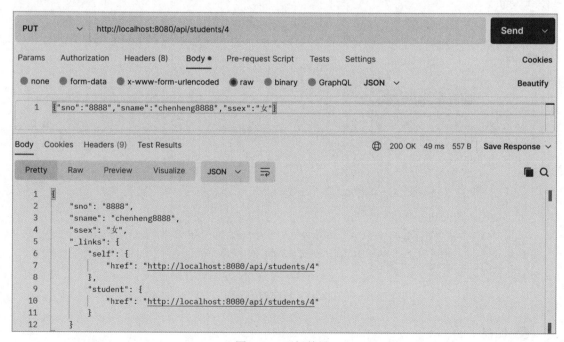

图 6.25 更新数据

❽ 删除

假如需要删除新增的 id 为 4 的数据，可以在 Postman 中使用 DELETE 方式访问 "http://localhost:8080/api/students/4"，删除数据，如图 6.26 所示。

图 6.26　删除成功

6.3　MongoDB

MongoDB 是一个基于分布式文件存储的 NoSQL 数据库，用 C++语言编写，旨在为 Web 应用提供可扩展的高性能数据存储解决方案。

MongoDB 是一个介于关系数据库和非关系数据库之间的产品，是非关系数据库中功能最丰富，最像关系数据库的。它支持的数据结构非常松散，是类似 JSON 的 BSON（Binary JSON，二进制 JSON）格式，因此可以存储比较复杂的数据类型。MongoDB 最大的特点是支持的查询语言非常强大，其语法有点类似于面向对象的查询语言，几乎可以实现类似关系数据库单表查询的绝大部分功能，而且还支持对数据建立索引。

本节不会介绍太多关于 MongoDB 数据库本身的知识，主要介绍 Spring Boot 对 MongoDB 的支持，以及基于 Spring Boot 和 MongoDB 的实例。

▶ 6.3.1　安装 MongoDB

从官方网站 "https://www.mongodb.com/download-center/community" 下载与自己的操作系统版本对应的 MongoDB，作者在编写本书时使用的 MongoDB 是 mongodb-win32-x86_64-2012plus-4.2.0-signed.msi。在成功下载以后，双击 mongodb-win32-x86_64-2012plus-4.2.0-signed.msi，按照默认设置安装即可。

用户可以使用 MongoDB 的图形界面管理工具 MongoDB Compass 可视化操作 MongoDB 数据库，可以使用 mongodb-win32-x86_64-2012plus-4.2.0-signed.msi 自带的 MongoDB Compass，也可以从官方网站 "https://www.mongodb.com/download-center/compass" 下载。

▶ 6.3.2　Spring Boot 整合 MongoDB

❶ Spring 对 MongoDB 的支持

Spring 对 MongoDB 的支持主要是通过 Spring Data MongoDB 实现的，Spring Data MongoDB 为用户提供了如下功能。

1）对象/文档映射注解

Spring Data MongoDB 提供了如表 6.2 所示的注解。

表 6.2　Spring Data MongoDB 提供的对象/文档映射注解

注　　解	含　　义
@Document	映射领域对象与 MongoDB 的一个文档
@Id	映射当前属性是文档对象 ID
@DBRef	当前属性将参考其他文档
@Field	为文档的属性定义名称
@Version	将当前属性作为版本

2）MongoTemplate

和 JdbcTemplate 一样，Spring Data MongoDB 为用户提供了一个 MongoTemplate，MongoTemplate 为用户提供了数据访问的方法。

3）Repository

类似于 Spring Data JPA，Spring Data MongoDB 也为用户提供了 Repository 的支持，使用方式和 Spring Data JPA 一样，示例如下：

```
public interface PersonRepository extends MongoRepository<Person, String>{
}
```

❷ Spring Boot 对 MongoDB 的支持

Spring Boot 对 MongoDB 的自动配置位于 org.springframework.boot.autoconfigure.mongo 包中，主要配置了数据库连接、MongoTemplate，用户可以在配置文件中使用以"spring.data.mongodb"为前缀的属性来配置 MongoDB 的相关信息。Spring Boot 对 MongoDB 提供了一些默认属性，例如默认端口号为 27017、默认服务器为 localhost、默认数据库为 test、默认无用户名和无密码访问方式，并默认开启了对 Repository 的支持，因此在 Spring Boot 应用中只需引入 spring-boot-starter-data-mongodb 依赖即可按照默认配置操作 MongoDB 数据库。

▶ 6.3.3　增、删、改、查

本节通过实例讲解如何在 Spring Boot 应用中对 MongoDB 数据库进行增、删、改、查。

【例 6-8】在 Spring Boot 应用中对 MongoDB 数据库进行增、删、改、查。

其具体实现步骤如下：

❶ 创建基于 spring-boot-starter-data-mongodb 依赖的 Spring Boot Web 应用 ch6_5

在 IDEA 中创建基于 Lombok、Spring Web 与 Spring Data MongoDB 依赖的 Spring Boot Web 应用 ch6_5。

❷ 配置 application.properties 文件

在 Spring Boot Web 应用 ch6_5 中使用 MongoDB 的默认数据库连接，因此不需要在 application.properties 文件中配置数据库连接信息。application.properties 文件的具体内容如下：

```
server.servlet.context-path=/ch6_5
#让控制器输出的JSON字符串的格式更美观
spring.jackson.serialization.indent-output=true
```

❸ 创建领域模型

在应用 ch6_5 的 src/main/java 目录中创建名为 com.ch.ch6_5.domain 的包，并在该包中创

第 6 章　Spring Boot 的数据访问

建领域模型 Person（人）以及 Person 去过的 Location（地点）。在 Person 类中使用@Document 注解将 Person 领域模型和 MongoDB 的文档进行映射。

Person 的代码如下：

```
package com.ch.ch6_5.domain;
import java.util.ArrayList;
import java.util.List;
import lombok.Data;
import org.springframework.data.annotation.Id;
import org.springframework.data.mongodb.core.mapping.Document;
import org.springframework.data.mongodb.core.mapping.Field;
@Document
@Data
public class Person {
    @Id
    private String pid;
    private String pname;
    private Integer page;
    private String psex;
    @Field("plocs")
    private List<Location> locations = new ArrayList<Location>();
    public Person() {
        super();
    }
    public Person(String pname, Integer page, String psex) {
        super();
        this.pname = pname;
        this.page = page;
        this.psex = psex;
    }
}
```

Location 的代码如下：

```
package com.ch.ch6_5.domain;
import lombok.Data;
@Data
public class Location {
    private String locName;
    private String year;
    public Location() {
        super();
    }
    public Location(String locName, String year) {
        super();
        this.locName = locName;
        this.year = year;
    }
}
```

❹ 创建数据访问接口

在应用 ch6_5 的 src/main/java 目录中创建名为 com.ch.ch6_5.repository 的包，并在该包中创建数据访问接口 PersonRepository，该接口继承 MongoRepository 接口。

PersonRepository 接口的代码如下：

```
package com.ch.ch6_5.repository;
import java.util.List;
import org.springframework.data.mongodb.repository.MongoRepository;
import org.springframework.data.mongodb.repository.Query;
```

163

```
import com.ch.ch6_5.domain.Person;
public interface PersonRepository extends MongoRepository<Person, String>{
    Person findByPname(String pname);    //支持方法名查询，方法名的命名规范参照表6.1
    @Query("{'psex':?0}")                //JSON字符串
    List<Person> selectPersonsByPsex(String psex);
}
```

❺ 创建控制器层

由于本实例业务简单，直接在控制器层调用数据访问层。创建名为 com.ch.ch6_5.controller 的包，并在该包中创建控制器类 TestMongoDBController。

TestMongoDBController 的核心代码如下：

```
@RestController
public class TestMongoDBController {
    @Autowired
    private PersonRepository personRepository;
    @RequestMapping("/save")
    public List<Person> save() {
        List<Location> locations1 = new ArrayList<Location>();
        Location loc1 = new Location("北京","2023");
        Location loc2 = new Location("上海","2024");
        locations1.add(loc1);
        locations1.add(loc2);
        List<Location> locations2 = new ArrayList<Location>();
        Location loc3 = new Location("广州","2025");
        Location loc4 = new Location("深圳","2026");
        locations2.add(loc3);
        locations2.add(loc4);
        List<Person> persons = new ArrayList<Person>();
        Person p1 = new Person("陈恒1", 88, "男");
        p1.setLocations(locations1);
        Person p2 = new Person("陈恒2", 99, "女");
        p2.setLocations(locations2);
        persons.add(p1);
        persons.add(p2);
        return personRepository.saveAll(persons);
    }
    @RequestMapping("/findByPname")
    public Person findByPname(String pname) {
        return personRepository.findByPname(pname);
    }
    @RequestMapping("/selectPersonsByPsex")
    public List<Person> selectPersonsByPsex(String psex) {
        return personRepository.selectPersonsByPsex(psex);
    }
    @RequestMapping("/updatePerson")
    public Person updatePerson(String oldPname, String newPname) {
        Person p1 = personRepository.findByPname(oldPname);
        if(p1 != null)
            p1.setPname(newPname);
        return personRepository.save(p1);
    }
    @RequestMapping("/deletePerson")
    public void updatePerson(String pname) {
        Person p1 = personRepository.findByPname(pname);
        personRepository.delete(p1);
    }
}
```

❻ 运行

首先运行 Ch65Application 主类，然后访问"http://localhost:8080/ch6_5/save"测试保存数据，运行效果如图 6.27 所示。

保存成功后，可以使用 MongoDB 的图形界面管理工具 MongoDB Compass 查看已保存的数据，如图 6.28 所示。

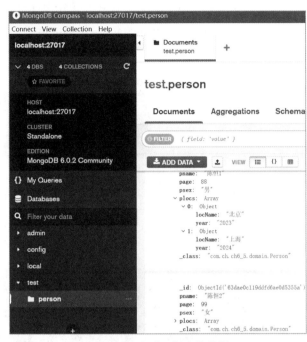

图 6.27　测试保存　　　　　　　　　　图 6.28　查看已保存的数据

通过"http://localhost:8080/ch6_5/findByPname?pname=陈恒 1"查询人名为"陈恒 1"的文档数据。

通过"http://localhost:8080/ch6_5/selectPersonsByPsex?psex=女"查询性别为"女"的文档数据。

通过"http://localhost:8080/ch6_5/updatePerson?oldPname=陈恒 1&newPname=陈恒 111"将人名为"陈恒 1"的数据修改成人名为"陈恒 111"。

通过"http://localhost:8080/ch6_5/deletePerson?pname=陈恒 111"将人名为"陈恒 111"的文档数据删除。

至此，通过 Spring Boot Web 应用对 MongoDB 数据库的操作演示完毕。

6.4　Redis

Redis 是一个开源的使用 ANSI C 语言编写、支持网络、可基于内存也可持久化的日志型 Key-Value 数据库，并且可以提供多种类型的 API。它支持字符串、哈希表、列表、集合、有序集合、位图、地理空间信息等数据类型，也可以作为高速缓存和消息队列代理，但是 Redis 在内存中存储数据，因此存放在 Redis 中的数据不应该超过内存容量，否则会导致操作系统的性能降低。

扫一扫

视频讲解

本节不会介绍太多关于 Redis 数据库本身的知识，主要介绍 Spring Boot 对 Redis 的支持，以及基于 Spring Boot 和 Redis 的实例。

▶ 6.4.1 安装 Redis

❶ 下载 Redis

在编写本书时，Redis 官方网站只提供了 Linux 版本的下载，因此用户只能通过 "https://github.com/MSOpenTech/redis/tags" 从 GitHub 上下载 Redis，本书下载的版本是 Redis-x64-3.2.100.zip。在运行中输入 cmd，然后把目录指向解压缩的 Redis 目录，如图 6.29 所示。

图 6.29　将目录指向解压缩的 Redis 目录

❷ 使用命令行启动 Redis 服务

使用 redis-server redis.windows.conf 命令行启动 Redis 服务，出现如图 6.30 所示的显示，表示成功启动 Redis 服务。

图 6.30　启动 Redis 服务

图 6.30 虽然启动了 Redis 服务，但是关闭 cmd 窗口 Redis 服务就会消失，所以需要把 Redis 设置成 Windows 下的服务。

关闭 cmd 后重新打开 cmd，进入 Redis 解压缩目录，执行设置服务命令 redis-server --service-install redis.windows-service.conf --loglevel verbose，如图 6.31 所示。

图 6.31　将 Redis 服务设置成 Windows 下的服务

第 6 章 Spring Boot 的数据访问

图 6.31 中没有报错，表示将 Redis 服务成功设置成 Windows 下的服务，刷新服务会看到 Redis 服务，如图 6.32 所示。

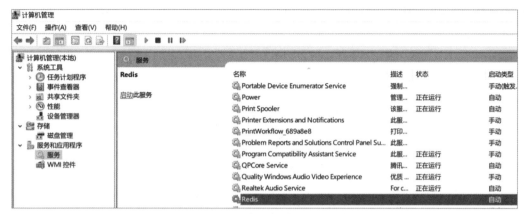

图 6.32 Windows 下的 Redis 服务

❸ 常用的 Redis 服务命令

卸载服务：redis-server --service-uninstall

开启服务：redis-server --service-start

停止服务：redis-server --service-stop

❹ 使用命令启动 Redis 服务

用户可以在 cmd 中使用 redis-server --service-start 命令启动 Redis 服务，如图 6.33 所示，同时可以在图 6.32 中启动 Redis 服务。

图 6.33 启动 Redis 服务

❺ 操作 Redis

如图 6.34 所示，在启动 Redis 服务以后，首先使用 redis-cli.exe -h 127.0.0.1 -p 6379 命令创建一个地址为 127.0.0.1、端口号为 6379 的 Redis 数据库服务，然后使用 set key value 和 get key 命令保存和获得数据。

图 6.34 测试 Redis

用户也可以使用 Redis 客户端查看数据（相当于 get key 命令）。本书提供一个使用 Java 开发的 Redis 客户端 RedisClient，下载地址是"https://github.com/caoxinyu/RedisClient"，解压缩后双击运行 C:\RedisClient-windows\release 目录下的 redisclient-win32.x86_64.2.0.jar，打开如图 6.35 所示的界面。

图 6.35 Redis 客户端 RedisClient 的界面

选择 Server→Add server 命令添加 Redis 服务，如图 6.36 所示。

这里添加一个地址为 127.0.0.1、端口号为 6379 的 Redis 数据库服务，如图 6.37 所示。

图 6.36 添加 Redis 服务

图 6.37 输入 Redis 服务信息界面

单击 OK 按钮，打开如图 6.38 所示的界面，可以看到 Redis 服务中共有 16 个数据库，其中 db0 是默认数据库，也就是说刚才存进去的 uname 就在该数据库中，因此展开 db0 数据库即可看到 uname 数据，如图 6.39 所示。

图 6.38 Redis 服务中的数据库

第 6 章　Spring Boot 的数据访问

图 6.39　查看数据

▶ 6.4.2　Spring Boot 整合 Redis

❶ Spring Data Redis

Spring 对 Redis 的支持是通过 Spring Data Redis 来实现的。Spring Data Redis 为用户提供了 RedisTemplate 和 StringRedisTemplate 两个模板来进行数据操作，其中 StringRedisTemplate 只针对键值都是字符串类型的数据进行操作。

RedisTemplate 和 StringRedisTemplate 模板提供的主要数据访问方法如表 6.3 所示。

表 6.3　RedisTemplate 和 StringRedisTemplate 的主要数据访问方法

方　　法	说　　明
opsForValue()	操作只有简单属性的数据
opsForList()	操作含有 List 的数据
opsForSet()	操作含有 Set 的数据
opsForZSet()	操作含有 Zset（有序的 Set）的数据
opsForHash()	操作含有 Hash 的数据

❷ Serializer

当数据存储到 Redis 中时，键和值都是通过 Spring 提供的 Serializer 序列化到数据的。RedisTemplate 默认使用 JdkSerializationRedisSerializer 序列化，StringRedisTemplate 默认使用 StringRedisSerializer 序列化。

❸ Spring Boot 的支持

Spring Boot 对 Redis 的支持位于 org.springframework.boot.autoconfigure.data.redis 包下，在 RedisAutoConfiguration 配置类中默认配置了 RedisTemplate 和 StringRedisTemplate，用户可以直接使用 Redis 存储数据。

在 RedisProperties 类中能够看到可以使用以 "spring.redis" 为前缀的属性在 application.properties 中配置 Redis，主要属性的默认配置如下：

```
spring.redis.database = 0          #数据库名 db0
```

```
spring.redis.host = localhost          #服务器地址
spring.redis.port = 6379               #连接端口号
spring.redis.max-idle = 8              #连接池的最大连接数
spring.redis.min-idle = 0              #连接池的最小连接数
spring.redis.max-active = 8            #在给定时间连接池可以分配的最大连接数
spring.redis.max-wait = -1             #当池被耗尽时，抛出异常之前连接分配应该阻塞的最大时
                                       #间量（以毫秒为单位），使用负值表示无限期地阻止
```

从上述默认属性值可以看出，默认为用户配置了数据库名为 db0、服务器地址为 localhost、端口号为 6379 的 Redis。

因此，在 Spring Boot 应用中只要引入 spring-boot-starter-data-redis 依赖就可以使用默认配置的 Redis 进行数据操作。

▶ 6.4.3 使用 StringRedisTemplate 和 RedisTemplate

本小节通过实例讲解如何在 Spring Boot 应用中使用 StringRedisTemplate 和 RedisTemplate 模板访问操作 Redis 数据库。

【例 6-9】在 Spring Boot 应用中使用 StringRedisTemplate 和 RedisTemplate 模板访问操作 Redis 数据库。

其具体实现步骤如下：

❶ 创建基于 spring-boot-starter-data-redis 依赖的 Spring Boot Web 应用 ch6_6

在 IDEA 中创建基于 Lombok、Spring Web 和 Spring Data Redis 依赖的 Spring Boot Web 应用 ch6_6。

❷ 配置 application.properties 文件

在 Spring Boot 应用 ch6_6 中使用 Redis 的默认数据库连接，所以不需要在 application.properties 文件中配置数据库连接信息。

❸ 创建实体类

在应用 ch6_6 的 src/main/java 目录中创建名为 com.ch.ch6_6.entity 的包，并在该包中创建名为 Student 的实体类。该类必须实现序列化接口，这是因为使用 Jackson 做序列化需要一个空构造。

Student 的代码如下：

```
package com.ch.ch6_6.entity;
import lombok.Data;
import java.io.Serializable;
@Data
public class Student implements Serializable{
    private static final long serialVersionUID = 1L;
    private String sno;
    private String sname;
    private Integer sage;
    public Student() {
        super();
    }
    public Student(String sno, String sname, Integer sage) {
        super();
        this.sno = sno;
        this.sname = sname;
        this.sage = sage;
    }
}
```

第 6 章 Spring Boot 的数据访问

❹ 创建数据访问层

在应用 ch6_6 的 src/main/java 目录中创建名为 com.ch.ch6_6.repository 的包，并在该包中创建名为 StudentRepository 的类，该类使用@Repository 注解标注为数据访问层。

StudentRepository 的代码如下：

```java
package com.ch.ch6_6.repository;
import java.util.List;
import jakarta.annotation.Resource;
import org.springframework.beans.factory.annotation.Autowired;
import org.springframework.data.redis.core.RedisTemplate;
import org.springframework.data.redis.core.StringRedisTemplate;
import org.springframework.data.redis.core.ValueOperations;
import org.springframework.stereotype.Repository;
import com.ch.ch6_6.entity.Student;
@Repository
public class StudentRepository{
    @SuppressWarnings("unused")
    @Autowired
    private StringRedisTemplate stringRedisTemplate;
    @SuppressWarnings("unused")
    @Autowired
    private RedisTemplate<Object, Object> redisTemplate;
    /**
     * 使用@Resource 注解指定 stringRedisTemplate，可注入基于字符串的简单属性操作方法
     * ValueOperations<String, String> valueOpsStr = stringRedisTemplate.opsForValue();
     */
    @Resource(name="stringRedisTemplate")
    ValueOperations<String, String> valueOpsStr;
    /**
     * 使用@Resource 注解指定 redisTemplate，可注入基于对象的简单属性操作方法
     * ValueOperations<Object, Object> valueOpsObject = redisTemplate.opsForValue();
     */
    @Resource(name="redisTemplate")
    ValueOperations<Object, Object> valueOpsObject;
    /**
     * 保存字符串到 Redis
     */
    public void saveString(String key, String value) {
        valueOpsStr.set(key, value);
    }
    /**
     * 保存对象到 Redis
     */
    public void saveStudent(Student stu) {
        valueOpsObject.set(stu.getSno(), stu);
    }
    /**
     * 保存 List 数据到 Redis
     */
    public void saveMultiStudents(Object key, List<Student> stus) {
        valueOpsObject.set(key, stus);
    }
    /**
     * 从 Redis 中获得字符串数据
     */
```

```
    public String getString(String key) {
        return valueOpsStr.get(key);
    }
    /**
     * 从Redis中获得对象数据
     */
    public Object getObject(Object key) {
        return valueOpsObject.get(key);
    }
}
```

❺ 创建控制器层

由于本实例业务简单,直接在控制器层调用数据访问层。在应用 ch6_6 的 src/main/java 目录中创建名为 com.ch.ch6_6.controller 的包,并在该包中创建控制器类 TestRedisController。TestRedisController 的核心代码如下:

```
@RestController
public class TestRedisController {
    @Autowired
    private StudentRepository studentRepository;
    @GetMapping("/save")
    public void save() {
        studentRepository.saveString("uname", "陈恒");
        Student s1 = new Student("111","陈恒1",77);
        studentRepository.saveStudent(s1);
        Student s2 = new Student("222","陈恒2",88);
        Student s3 = new Student("333","陈恒3",99);
        List<Student> stus = new ArrayList<Student>();
        stus.add(s2);
        stus.add(s3);
        studentRepository.saveMultiStudents("mutilStus",stus);
    }
    @GetMapping("/getUname")
    public String getUname(String key) {
        return studentRepository.getString(key);
    }
    @GetMapping("/getStudent")
    public Student getStudent(String key) {
        return (Student)studentRepository.getObject(key);
    }
    @GetMapping("/getMultiStus")
    public List<Student> getMultiStus(String key) {
        return (List<Student>)studentRepository.getObject(key);
    }
}
```

❻ 修改配置类 Ch66Application

大家知道 RedisTemplate 默认使用 JdkSerializationRedisSerializer 序列化数据,这使得使用 Redis Client 查看数据很不直观,因为 JdkSerializationRedisSerializer 使用二进制形式存储数据,所以在此自己配置 RedisTemplate,并定义 Serializer。

修改后的配置类 Ch66Application 的代码如下:

```
package com.ch.ch6_6;
import org.springframework.boot.SpringApplication;
import org.springframework.boot.autoconfigure.SpringBootApplication;
import org.springframework.context.annotation.Bean;
import org.springframework.data.redis.connection.RedisConnectionFactory;
```

```
import org.springframework.data.redis.core.RedisTemplate;
import org.springframework.data.redis.serializer.Jackson2JsonRedisSerializer;
import org.springframework.data.redis.serializer.StringRedisSerializer;
import com.fasterxml.jackson.annotation.JsonAutoDetect;
import com.fasterxml.jackson.annotation.PropertyAccessor;
import com.fasterxml.jackson.databind.ObjectMapper;
@SpringBootApplication
public class Ch66Application {
    public static void main(String[] args) {
        SpringApplication.run(Ch66Application.class, args);
    }
    @Bean
    public RedisTemplate<Object, Object> redisTemplate(RedisConnectionFactory redisConnectionFactory){
        RedisTemplate<Object, Object> rTemplate = new RedisTemplate<Object, Object>();
        rTemplate.setConnectionFactory(redisConnectionFactory);
        @SuppressWarnings({ "unchecked", "rawtypes" })
        Jackson2JsonRedisSerializer<Object>jackson2JsonRedisSerializer = new Jackson2JsonRedisSerializer(Object.class);
        ObjectMapper om = new ObjectMapper();
        om.setVisibility(PropertyAccessor.ALL,JsonAutoDetect.Visibility.ANY);
        om.enableDefaultTyping(ObjectMapper.DefaultTyping.NON_FINAL);
        jackson2JsonRedisSerializer.setObjectMapper(om);
        //设置值的序列化采用Jackson2JsonRedisSerializer
        rTemplate.setValueSerializer(jackson2JsonRedisSerializer);
        //设置键的序列化采用StringRedisSerializer
        rTemplate.setKeySerializer(new StringRedisSerializer());
        return rTemplate;
    }
}
```

❼ 测试运行

首先运行 Ch66Application 主类，然后通过 "http://localhost:8080/save" 测试存储数据，成功运行后通过 Redis Client 查看数据，如图 6.40 所示。

图 6.40　查看保存的数据

通过"http://localhost:8080/getUname?key=uname"查询 key 为 uname 的字符串值，如图 6.41 所示。

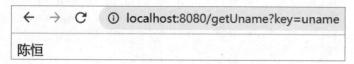

图 6.41　根据 key 查询字符串值

通过"http://localhost:8080/getStudent?key=111"查询 key 为 111 的 Student 对象值，如图 6.42 所示。

图 6.42　根据 key 查询简单对象

通过"http://localhost:8080/getMultiStus?key=mutilStus"查询 key 为 mutilStus 的 List 集合，如图 6.43 所示。

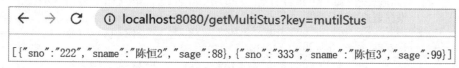

图 6.43　根据 key 查询 List 集合

6.5　数据缓存 Cache

大家知道内存的读取速度远大于硬盘的读取速度。当需要重复地获取相同数据时，一次一次地请求数据库或者远程服务，导致大量的时间消耗在数据库查询或者远程方法的调用上，最终导致程序的性能降低，这就是数据缓存要解决的问题。

本节将介绍 Spring Boot 应用中 Cache 的一般概念，Spring Cache 对 Cache 进行了抽象，提供了 CacheManager 和 Cache 接口，并提供了 @Cacheable、@CachePut、@CacheEvict、@Caching、@CacheConfig 等注解。Spring Boot 应用基于 Spring Cache，既提供了基于内存实现的缓存管理器用于单体应用系统，也集成了 Redis、EhCache 等缓存服务器用于大型系统或分布式系统。

▶ 6.5.1　Spring 缓存支持

Spring 框架定义了 org.springframework.cache.CacheManager 和 org.springframework.cache.Cache 接口来统一不同的缓存技术。针对不同的缓存技术，需要实现不同的 CacheManager。例如在使用 EhCache 作为缓存技术时需要注册实现 CacheManager 的 Bean，示例代码如下：

```
@Bean
public EhCacheCacheManager cacheManager(CacheManager ehCacheCacheManager) {
    return new EhCacheCacheManager(ehCacheCacheManager);
}
```

CacheManager 的常用实现如表 6.4 所示。

表 6.4　CacheManager 的常用实现

CacheManager	描述
SimpleCacheManager	使用简单的 Collection 来存储缓存，主要用于测试
NoOpCacheManager	仅用于测试，不会实际存储缓存
ConcurrentMapCacheManager	使用 ConcurrentMap 存储缓存，Spring 默认采用此技术存储缓存
EhCacheCacheManager	使用 EhCache 作为缓存技术
JCacheCacheManager	支持 JCache（JSR-107）标准实现作为缓存技术，例如 Apache Commons JCS
RedisCacheManager	使用 Redis 作为缓存技术
HazelcastCacheManager	使用 Hazelcast 作为缓存技术

一旦配置好 Spring 缓存支持，就可以在 Spring 容器管理的 Bean 中使用缓存注解（基于 AOP 原理），一般情况下都是在业务层（Service 类）使用这些注解。

❶ @Cacheable

@Cacheable 可以标记在一个方法上，也可以标记在一个类上。当标记在一个方法上时表示该方法是支持缓存的，当标记在一个类上时则表示该类的所有方法都是支持缓存的。对于一个支持缓存的方法，在方法执行前 Spring 先检查缓存中是否存在方法返回的数据，如果存在，则直接返回缓存数据；如果不存在，则调用方法并将方法返回值存入缓存。

@Cacheable 注解经常使用 value、key、condition 等属性。

value：缓存的名称，指定一个或多个缓存名称。例如@Cacheable(value="mycache")或者 @Cacheable(value={"cache1","cache2"})。该属性与 cacheNames 属性的含义相同。

key：缓存的 key，可以为空，如果指定需要按照 SpEL 表达式编写，如果不指定则默认按照方法的所有参数进行组合。例如@Cacheable(value="testcache",key="#student.id")。

condition：缓存的条件，可以为空，如果指定需要按照 SpEL 编写，返回 true 或者 false，只有为 true 才进行缓存。例如@Cacheable(value="testcache",condition="#student.id>2")。该属性与 unless 相反，当条件成立时不进行缓存。

❷ @CacheEvict

@CacheEvict 用来标注在需要清除缓存元素的方法或类上。当标注在一个类上时表示其中所有方法的执行都会触发缓存的清除操作。@CacheEvict 可以指定的属性有 value、key、condition、allEntries 和 beforeInvocation。其中 value、key 和 condition 的语义与@Cacheable 对应的属性类似。

allEntries：是否清空所有缓存内容，默认为 false，如果指定为 true，则方法调用后将立即清空所有缓存。例如@CacheEvict(value="testcache", allEntries=true)。

beforeInvocation：是否在方法执行前就清空，默认为 false，如果指定为 true，则在方法还没有执行时就清空缓存。在默认情况下，如果方法执行抛出异常，则不会清空缓存。

❸ @CachePut

@CachePut 也可以声明一个方法支持缓存功能，与@Cacheable 不同的是使用@CachePut 标注的方法在执行前不会检查缓存中是否存在之前执行过的结果，而是每次都会执行该方法，并将执行结果以键值对的形式存入指定的缓存中。

@CachePut 也可以标注在类或方法上。@CachePut 的属性与@Cacheable 的属性一样。

❹ @Caching

@Caching注解可以让用户在一个方法或者类上同时指定多个Spring Cache相关注解。其拥有3个属性，即cacheable、put和evict，分别用于指定@Cacheable、@CachePut和@CacheEvict。示例如下：

```
@Caching(
cacheable = @Cacheable("cache1"),
evict = { @CacheEvict("cache2"),@CacheEvict(value = "cache3", allEntries = true) }
)
```

❺ @CacheConfig

所有的Cache注解都需要提供Cache名称，如果每个Service方法上都包含相同的Cache名称，写起来可能会重复，此时可以使用@CacheConfig注解作用在类上，设置当前缓存的一些公共设置。

▶ 6.5.2 Spring Boot 缓存支持

在 Spring 中使用缓存技术的关键是配置缓存管理器 CacheManager，而 Spring Boot 为用户自动配置了多个 CacheManager 的实现。Spring Boot 的 CacheManager 的自动配置位于 org.springframework.boot.autoconfigure.cache 包中。Spring Boot 自动配置了 EhCacheCacheConfiguration、GenericCacheConfiguration、HazelcastCacheConfiguration、HazelcastJCacheCustomizationConfiguration、InfinispanCacheConfiguration、JcacheCacheConfiguration、NoOpCacheConfiguration、RedisCacheConfiguration 和 SimpleCacheConfiguration。在默认情况下，Spring Boot 使用的是 SimpleCacheConfiguration，即使用 ConcurrentMapCacheManager。Spring Boot 支持用以"spring.cache"为前缀的属性来进行缓存的相关配置。

在 Spring Boot 应用中，使用缓存技术只需在应用中引入相关缓存技术的依赖，并在配置类中使用@EnableCaching注解开启缓存支持即可。

下面通过实例讲解如何在 Spring Boot 应用中使用默认的缓存技术 ConcurrentMapCacheManager。

【例 6-10】在 Spring Boot 应用中使用默认的缓存技术 ConcurrentMapCacheManager。

其具体实现步骤如下：

❶ 创建基于 Lombok、spring-boot-starter-cache 和 spring-boot-starter-data-jpa 依赖的 Spring Boot Web 应用 ch6_7

在 IDEA 中创建基于 Lombok、Spring Web、Spring Cache Abstraction 和 Spring Data JPA 依赖的 Spring Boot Web 应用 ch6_7。

❷ 配置 application.properties 文件

在应用 ch6_7 中使用 Spring Data JPA 访问 MySQL 数据库，所以在 application.properties 文件中配置数据库连接信息，但因为使用默认的缓存技术 ConcurrentMapCacheManager，所以不需要缓存的相关配置。

应用 ch6_7 的 application.properties 文件的内容以及 pom.xml 中添加的 MySQL 连接依赖与例 6-1 基本一样，这里不再赘述。

❸ 创建持久化实体类

在应用 ch6_7 的 src/main/java 目录中创建名为 com.ch.ch6_7.entity 的包，并在该包中创建持久化实体类 Student。

第 6 章 Spring Boot 的数据访问

Student 的代码如下:

```java
package com.ch.ch6_7.entity;
import java.io.Serializable;
import jakarta.persistence.*;
import com.fasterxml.jackson.annotation.JsonIgnoreProperties;
import lombok.Data;
@Entity
@Table(name = "student_table")
@JsonIgnoreProperties(value = {"hibernateLazyInitializer"})
@Data
public class Student implements Serializable{
    private static final long serialVersionUID = 1L;
    @Id
    @GeneratedValue(strategy = GenerationType.IDENTITY)
    private int id;    //主键
    private String sno;
    private String sname;
    private String ssex;
    public Student() {
        super();
    }
    public Student(int id, String sno, String sname, String ssex) {
        super();
        this.id = id;
        this.sno = sno;
        this.sname = sname;
        this.ssex = ssex;
    }
}
```

❹ 创建数据访问接口

在应用 ch6_7 的 src/main/java 目录中创建名为 com.ch.ch6_7.repository 的包,并在该包中创建名为 StudentRepository 的数据访问接口。

StudentRepository 的代码如下:

```java
package com.ch.ch6_7.repository;
import org.springframework.data.jpa.repository.JpaRepository;
import com.ch.ch6_7.entity.Student;
public interface StudentRepository extends JpaRepository<Student, Integer>{}
```

❺ 创建业务层

在应用 ch6_7 的 src/main/java 目录中创建名为 com.ch.ch6_7.service 的包,并在该包中创建 StudentService 接口和该接口的实现类 StudentServiceImpl。这里省略 StudentService 接口的代码。

StudentServiceImpl 的核心代码如下:

```java
@Service
public class StudentServiceImpl implements StudentService{
    @Autowired
    private StudentRepository studentRepository;
    @Override
    @CachePut(value = "student", key="#student.id")
    public Student saveStudent(Student student) {
        Student s = studentRepository.save(student);
        System.out.println("为 key=" + student.getId() + "数据做了缓存");
        return s;
    }
```

```java
    @Override
    @CacheEvict(value = "student", key="#student.id")
    public void deleteCache(Student student) {
        System.out.println("删除了key=" + student.getId() + "的数据缓存");
    }
    @Override
    @Cacheable(value = "student")
    public Student selectOneStudent(Integer id) {
        Student s = studentRepository.getOne(id);
        System.out.println("为 key=" + id + "数据做了缓存");
        return s;
    }
}
```

在上述 Service 的实现类中，使用@CachePut 注解将新增的或更新的数据保存到缓存，其中缓存名为 student，数据的 key 是 student 的 id；使用@CacheEvict 注解从缓存 student 中删除 key 为 student 的 id 的数据；使用@Cacheable 注解将 key 为 student 的 id 的数据保存到名为 student 的缓存中。

❻ 创建控制器层

在应用 ch6_7 的 src/main/java 目录中创建名为 com.ch.ch6_7.controller 的包，并在该包中创建名为 TestCacheController 的控制器类。

TestCacheController 的核心代码如下：

```java
@RestController
public class TestCacheController {
    @Autowired
    private StudentService studentService;
    @GetMapping("/savePut")
    public Student save(Student student) {
        return studentService.saveStudent(student);
    }
    @GetMapping("/selectAble")
    public Student select(Integer id) {
        return studentService.selectOneStudent(id);
    }
    @GetMapping("/deleteEvict")
    public String deleteCache(Student student) {
        studentService.deleteCache(student);
        return "ok";
    }
}
```

❼ 开启缓存支持

在应用的主类 Ch67Application 中使用@EnableCaching 注解开启缓存支持，核心代码如下：

```java
@EnableCaching
@SpringBootApplication
public class Ch67Application {}
```

❽ 测试运行

1）测试@Cacheable

在启动应用程序的主类后，第一次访问"http://localhost:8080/ch6_7/selectAble?id=3"，第一次访问时将调用方法查询数据库，并将查询到的数据存储到缓存 student 中，此时控制台输出如图 6.44 所示。

图 6.44 第一次访问时的控制台输出结果

页面数据如图 6.45 所示。

图 6.45 第一次访问时的页面数据

再次访问 "http://localhost:8080/ch6_7/selectAble?id=3"，此时控制台没有输出 "为 key=3 数据做了缓存" 以及 Hibernate 的查询语句，这表明没有调用查询方法，页面数据直接从数据缓存中获得。

2）测试@CachePut

重启应用程序的主类，访问 "http://localhost:8080/ch6_7/savePut?sname=陈恒 5&sno=555&ssex=男"，此时控制台输出结果如图 6.46 所示，页面数据如图 6.47 所示。

图 6.46 测试@CachePut 控制台输出结果

图 6.47 测试@CachePut 页面数据

这时访问 "http://localhost:8080/ch6_7/selectAble?id=5"，控制台无输出，从缓存中直接获得数据，页面数据如图 6.47 所示。

3）测试@CacheEvict

重启应用程序的主类，首先访问 "http://localhost:8080/ch6_7/selectAble?id=1"，为 key 为 1

的数据做缓存，再次访问"http://localhost:8080/ch6_7/selectAble?id=1"，确认数据已从缓存中获取，然后访问"http://localhost:8080/ch6_7/deleteEvict?id=1"从缓存 student 中删除 key 为 1 的数据，此时控制台输出结果如图 6.48 所示。

图 6.48　测试@CacheEvict 删除缓存数据

最后再次访问"http://localhost:8080/ch6_7/selectAble?id=1"，此时重新做了缓存，控制台输出结果如图 6.49 所示。

图 6.49　测试@CacheEvict 重做缓存数据

▶ 6.5.3　使用 Redis Cache

在 Spring Boot 中使用 Redis Cache 只需添加 spring-boot-starter-data-redis 依赖即可，下面通过实例测试 Redis Cache。

【例 6-11】在例 6-9 的基础上测试 Redis Cache。

其具体实现步骤如下：

❶ 使用@Cacheable 注解修改控制器方法

将控制器中的 getUname 方法修改如下：

```
@GetMapping("/getUname")
@Cacheable(value = "myuname")
public String getUname(String key) {
    System.out.println("测试缓存");
    return personRepository.getString(key);
}
```

❷ 使用@EnableCaching 注解开启缓存支持

在应用程序的主类 Ch66Application 上使用@EnableCaching 注解开启缓存支持。

❸ 测试 Redis Cache

在启动应用程序的主类后多次访问"http://localhost:8080/getUname?key=uname"，但"测试缓存"字样在控制台中仅打印一次，页面查询结果不变，这说明只有在第一次访问时调用了查询方法，后面的多次访问都是从缓存直接获得数据。

本章小结

本章是本书的重点，讲解了 Spring Data JPA、Spring Data REST、Spring Boot 整合 MongoDB、Spring Boot 整合 Redis 以及数据缓存 Cache。通过本章的学习，读者不仅能够掌握 Spring Boot 访问关系数据库的解决方案，还能够掌握 Spring Boot 访问非关系数据库的解决方案。

扫一扫

自测题

习题 6

1. 在 Spring Boot 应用中数据缓存技术解决了什么问题？
2. 什么是 RESTful 架构？什么是 REST 服务？
3. Spring 框架提供了哪些缓存注解？这些注解如何使用？
4. 在 Spring Data JPA 中如何实现一对一、一对多、多对多关联查询？请举例说明。

第 7 章　MyBatis 与 MyBatis-Plus

学习目的与要求

本章将重点介绍 MyBatis 与 MyBatis-Plus 的基础知识，并详细介绍 Spring Boot 如何整合 MyBatis 与 MyBatis-Plus。通过本章的学习，要求读者掌握 Spring Boot 整合 MyBatis 与 MyBatis-Plus 的基本步骤。

本章主要内容

❖ Spring Boot 整合 MyBatis
❖ MyBatis 基础
❖ Spring Boot 整合 MyBatis-Plus
❖ MyBatis-Plus 基础

MyBatis-Plus 是增强版的 MyBatis，本章将学习 MyBatis 与 MyBatis-Plus 的基础知识以及在 Spring Boot 应用中如何整合 MyBatis 与 MyBatis-Plus。

扫一扫
视频讲解

7.1　Spring Boot 整合 MyBatis

MyBatis 本是 Apache Software Foundation 的一个开源项目 iBatis，2010 年这个项目由 Apache Software Foundation 迁移到 Google Code，并改名为 MyBatis。

MyBatis 是一个基于 Java 的持久层框架。MyBatis 提供的持久层框架包括 SQL Maps 和 Data Access Objects（DAO），它消除了几乎所有的 JDBC 代码和参数的手工设置以及结果集的检索。MyBatis 使用简单的 XML 或注解进行配置和原始映射，将接口和 Java 的 POJOs（Plain Old Java Objects，普通的 Java 对象）映射成数据库中的记录。

目前，Java 的持久层框架产品有许多，常见的有 Hibernate 和 MyBatis。MyBatis 是一个半自动映射的框架，因为 MyBatis 需要手动匹配 POJO、SQL 和映射关系；而 Hibernate 是一个全表映射的框架，只需提供 POJO 和映射关系即可。MyBatis 是一个小巧、方便、高效、简单、直接、半自动化的持久层框架；Hibernate 是一个强大、方便、高效、复杂、间接、全自动化的持久层框架。两个持久层框架各有优缺点，开发者应根据实际应用选择它们。

下面通过实例讲解如何在 Spring Boot 应用中使用 MyBatis 框架操作数据库（基于 XML 的映射配置）。

【例 7-1】在 Spring Boot 应用中使用 MyBatis 框架操作数据库（基于 XML 的映射配置）。其具体实现步骤如下：

❶ 创建 Spring Boot Web 应用

在创建 Spring Boot Web 应用 ch7_1 时选择 MyBatis Framework （mybatis-spring-boot-starter）依赖，如图 7.1 所示。在该应用中操作的数据库是 springtest，操作的数据表是 user 表。

图 7.1　应用 ch7_1 的依赖

第 7 章　MyBatis 与 MyBatis-Plus

❷ 修改 pom.xml 文件

在 pom.xml 文件中添加 MySQL 连接器依赖，具体如下：

```xml
<dependency>
    <groupId>mysql</groupId>
    <artifactId>mysql-connector-java</artifactId>
    <version>8.0.29</version>
</dependency>
```

❸ 设置 Web 应用 ch7_1 的上下文路径及数据源配置信息

在应用 ch7_1 的 application.properties 文件中配置如下内容：

```
server.servlet.context-path=/ch7_1
#数据库地址
spring.datasource.url=jdbc:mysql://localhost:3306/springtest?useUnicode=true&characterEncoding=UTF-8&allowMultiQueries=true&serverTimezone=GMT%2B8
#数据库用户名
spring.datasource.username=root
#数据库密码
spring.datasource.password=root
#数据库驱动
spring.datasource.driver-class-name=com.mysql.cj.jdbc.Driver
#设置包的别名（在 Mapper 映射文件中直接使用实体类名）
mybatis.type-aliases-package=com.ch.ch7_1.entity
#告诉系统到哪里去找 mapper.xml 文件（映射文件）
mybatis.mapperLocations=classpath:mappers/*.xml
#在控制台输出 SQL 语句日志
logging.level.com.ch.ch7_1.repository=debug
#让控制器输出的 JSON 字符串的格式更美观
spring.jackson.serialization.indent-output=true
```

❹ 创建实体类

创建名为 com.ch.ch7_1.entity 的包，并在该包中创建 MyUser 实体类，代码如下：

```java
package com.ch.ch7_1.entity;
import lombok.Data;
@Data
public class MyUser {
    private Integer uid;        //与数据表中的字段名相同
    private String uname;
    private String usex;
}
```

❺ 创建数据访问接口

创建名为 com.ch.ch7_1.repository 的包，并在该包中创建 MyUserRepository 接口。

MyUserRepository 的核心代码如下：

```java
/*
 * @Repository 可有可无，但有时提示依赖注入找不到（不影响运行），
 *加上后可以消除依赖注入的报错信息。
 * 这里不再需要@Mapper，是因为在启动类中使用@MapperScan 注解，
 * 将数据访问层的接口都注解为 Mapper 接口的实现类，
 * @Mapper 与@MapperScan 用一个即可
 */
@Repository
public interface MyUserRepository {
    List<MyUser> findAll();
}
```

❻ 创建 Mapper 映射文件

在 src/main/resources 目录下创建名为 mappers 的包，并在该包中创建 SQL 映射文件 MyUserMapper.xml，具体代码如下：

```xml
<?xml version="1.0" encoding="UTF-8" ?>
<!DOCTYPE mapper
PUBLIC "-//mybatis.org//DTD Mapper 3.0//EN"
"http://mybatis.org/dtd/mybatis-3-mapper.dtd">
<mapper namespace="com.ch.ch7_1.repository.MyUserRepository">
    <select id="findAll" resultType="MyUser">
        select * from user
    </select>
</mapper>
```

❼ 创建业务层

创建名为 com.ch.ch7_1.service 的包，并在该包中创建 MyUserService 接口和 MyUserServiceImpl 实现类。这里省略 MyUserService 的代码。

MyUserServiceImpl 的核心代码如下：

```java
@Service
public class MyUserServiceImpl implements MyUserService{
    @Autowired
    private MyUserRepository myUserRepository;
    @Override
    public List<MyUser> findAll() {
        return myUserRepository.findAll();
    }
}
```

❽ 创建控制器类 MyUserController

创建名为 com.ch.ch7_1.controller 的包，并在该包中创建控制器类 MyUserController。

MyUserController 的核心代码如下：

```java
@RestController
public class MyUserController {
    @Autowired
    private MyUserService myUserService;
    @GetMapping("/findAll")
    public List<MyUser> findAll(){
        return myUserService.findAll();
    }
}
```

❾ 在应用程序的主类中扫描 Mapper 接口

在应用程序的 Ch71Application 主类中使用@MapperScan 注解扫描 MyBatis 的 Mapper 接口，核心代码如下：

```java
@SpringBootApplication
//配置扫描 MyBatis 接口的包的路径
@MapperScan(basePackages={"com.ch.ch7_1.repository"})
public class Ch71Application {
    public static void main(String[] args) {
        SpringApplication.run(Ch71Application.class, args);
    }
}
```

❿ 运行

首先运行 Ch71Application 主类，然后访问 "http://localhost:8080/ch7_1/findAll"，运行效果如图 7.2 所示。

图 7.2 查询所有用户信息

7.2 MyBatis 基础

映射器是 MyBatis 最复杂且最重要的组件，由一个接口和一个 XML 文件（SQL 映射文件）组成。MyBatis 的映射器也可以使用注解完成，但在实际应用中使用不多，原因主要来自 3 个方面，其一，面对复杂的 SQL 会显得无力；其二，注解的可读性较差；其三，注解丢失了 XML 上下文相互引用的功能，因此推荐使用 XML 文件开发 MySQL 的映射器。

SQL 映射文件的常用配置元素如表 7.1 所示。

表 7.1 SQL 映射文件的常用配置元素

元素名称	描 述	备 注
select	查询语句，最常用、最复杂的元素之一	可以自定义参数，返回结果集等
insert	插入语句	执行后返回一个整数，代表插入的行数
update	更新语句	执行后返回一个整数，代表更新的行数
delete	删除语句	执行后返回一个整数，代表删除的行数
sql	定义一部分 SQL，在多个位置被引用	例如一张表的列名，一次定义，可以在多个 SQL 语句中使用
resultMap	用来描述从数据库结果集中来加载对象，是最复杂、最强大的元素之一	提供映射规则

▶ 7.2.1 <select>元素

在 SQL 映射文件中，<select>元素用于映射 SQL 的 select 语句，其示例代码如下：

```xml
<!--根据 uid 查询一个用户的信息-->
<select id="selectUserById" parameterType="Integer" resultType="MyUser">
    select * from user where uid = #{uid}
</select>
```

在上述示例代码中，id 的值是唯一标识符（对应 Mapper 接口的某个方法），它接收一个 Integer 类型的参数，返回一个 MyUser 类型的对象，结果集自动映射到 MyUser 的属性。需要注意的是，MyUser 的属性名称一定要与查询结果的列名相同。

<select>元素除了有上述示例代码中的几个属性以外，还有一些常用的属性，如表 7.2 所示。

扫一扫

视频讲解

表 7.2 <select>元素的常用属性

属 性 名 称	描 述
id	它和 Mapper 的命名空间组合起来使用（对应 Mapper 接口的某个方法），是唯一标识符，供 MyBatis 调用
parameterType	表示传入 SQL 语句的参数类型的全限定名或别名，是一个可选属性，MyBatis 能推断出具体传入语句的参数
resultType	SQL 语句执行后返回的类型（全限定名或者别名）。如果是集合类型，返回的是集合元素的类型，在返回时可以使用 resultType 或 resultMap
resultMap	它是映射集的引用，与<resultMap>元素一起使用，在返回时可以使用 resultType 或 resultMap
flushCache	设置在调用 SQL 语句后是否要求 MyBatis 清空之前的本地缓存和二级缓存，默认值为 false。如果设置为 true，则任何时候只要 SQL 语句被调用，都将清空本地缓存和二级缓存

续表

属性名称	描述
useCache	启动二级缓存的开关，默认值为 true，表示将查询结果存入二级缓存中
timeout	用于设置超时参数，单位是秒，超时将抛出异常
fetchSize	获取记录的总条数设定
statementType	告诉 MyBatis 使用哪个 JDBC 的 Statement 工作，取值为 STATEMENT（Statement）、PREPARED（PreparedStatement）、CALLABLE（CallableStatement），默认值为 PREPARED
resultSetType	这是针对 JDBC 的 ResultSet 接口而言的，其值可设置为 FORWARD_ONLY（只允许向前访问）、SCROLL_SENSITIVE（双向滚动，但不及时更新）、SCROLL_INSENSITIVE（双向滚动，及时更新）

❶ 使用 Map 接口传递参数

在实际开发中，查询 SQL 语句经常需要多个参数，比如多条件查询。当多个参数传递时，<select>元素的 parameterType 属性值的类型是什么呢？在 MyBatis 中允许 Map 接口通过键值对传递多个参数。

假设数据操作接口中有一个实现查询"陈"姓男性用户信息的方法：

```
List<MyUser> testMapSelect(Map<String, Object> param);
```

此时传递给 MyBatis 映射器的是一个 Map 对象，使用该 Map 对象在 SQL 中设置对应的参数，对应的 SQL 映射文件的代码如下：

```xml
<!--查询"陈"姓男性用户的信息-->
<select id="testMapSelect" resultType="MyUser" parameterType="map">
    select * from user
    where uname like concat('%',#{u_name},'%')
    and usex = #{u_sex}
</select>
```

上述 SQL 映射文件中的参数名 u_name 和 u_sex 是 Map 中的 key。

【例 7-2】 在例 7-1 的基础上实现使用 Map 接口传递参数。

为了节省篇幅，对于相同的实现不再赘述，其他的具体实现如下：

1）添加接口方法

在 Spring Boot Web 应用 ch7_1 的 MyUserRepository 接口中添加接口方法（见上述），实现查询"陈"姓男性用户的信息。

2）添加 SQL 映射

在应用 ch7_1 的 SQL 映射文件 MyUserMapper.xml 中添加 SQL 映射（见上述），实现查询"陈"姓男性用户的信息。

3）添加请求处理方法

在应用 ch7_1 的 MyUserController 控制器类中添加测试方法 testMapSelect，具体代码如下：

```java
@GetMapping("/testMapSelect")
public List<MyUser> testMapSelect() {
    //查询所有"陈"姓男性用户
    Map<String, Object> map = new HashMap<>();
    map.put("u_name", "陈");
    map.put("u_sex", "男");
    return myUserRepository.testMapSelect(map);
}
```

❷ 使用 Java Bean 传递参数

在 MyBatis 中，当需要将多个参数传递给映射器时，可以将它们封装在一个 Java Bean 中。下面通过具体实例讲解如何使用 Java Bean 传递参数。

【例 7-3】在例 7-2 的基础上实现使用 Java Bean 传递参数。

为了节省篇幅，对于相同的实现不再赘述，其他的具体实现如下：

1）添加接口方法

在 Spring Boot Web 应用 ch7_1 的 MyUserRepository 接口中添加接口方法 selectAllUserByJavaBean()，在该方法中使用 MyUser 类的对象将参数信息进行封装。接口方法 selectAllUserByJavaBean() 的定义如下：

```
List<MyUser> selectAllUserByJavaBean(MyUser user);
```

2）添加 SQL 映射

在应用 ch7_1 的 SQL 映射文件 MyUserMapper.xml 中添加接口方法对应的 SQL 映射，具体代码如下：

```xml
<!--通过Java Bean 传递参数查询"陈"姓男性用户的信息，#{uname}的 uname 为参数MyUser的属性-->
<select id="selectAllUserByJavaBean" resultType="MyUser" parameterType="MyUser">
    select * from user
    where uname like concat('%',#{uname},'%')
    and usex = #{usex}
</select>
```

3）添加请求处理方法

在应用 ch7_1 的 MyUserController 控制器类中添加请求处理方法 selectAllUserByJavaBean()，具体代码如下：

```java
@GetMapping("/selectAllUserByJavaBean")
public List<MyUser> selectAllUserByJavaBean() {
    //通过MyUser 封装参数，查询所有"陈"姓男性用户
    MyUser mu = new MyUser();
    mu.setUname("陈");
    mu.setUsex("男");
    return myUserRepository.selectAllUserByJavaBean(mu);
}
```

❸ 使用@Param 注解传递参数

不管是 Map 传参还是 Java Bean 传参，它们都是将多个参数封装在一个对象中，实际上传递的还是一个参数，而使用@Param 注解可以将多个参数依次传递给 MyBatis 映射器。接口方法的示例代码如下：

```
List<MyUser> selectAllUserByParam(@Param("puname") String uname, @Param("pusex") String usex);
```

在上述示例代码中，puname 和 pusex 是传递给 MyBatis 映射器的参数名。接口方法 selectAllUserByParam() 对应的 SQL 映射的具体代码如下：

```xml
<!--通过@Param注解传递参数查询"陈"姓男性用户的信息，这里不需要定义参数类型-->
<select id="selectAllUserByParam" resultType="MyUser">
    select * from user
    where uname like concat('%',#{puname},'%')
    and usex = #{pusex}
</select>
```

❹ 使用 POJO 存储结果集

在前面的例子中都是直接使用 Java Bean（MyUser）存储结果集，这是因为 MyUser 的属性名与查询结果集的列名相同。如果查询结果集的列名与 Java Bean 的属性名不同，那么可以结合<resultMap>元素将 Java Bean 的属性与查询结果集的列名一一对应，具体步骤如下。

首先创建一个名为 MapUser 的 POJO（Plain Ordinary Java Object，普通的 Java 类）类：

```
package com.ch.ch7_1.entity;
import lombok.Data;
@Data
public class MapUser {
    private Integer m_uid;
    private String m_uname;
    private String m_usex;
    @Override
    public String toString() {
        return "User [uid=" + m_uid +",uname=" + m_uname + ",usex=" + m_usex +"]";
    }
}
```

然后添加数据操作接口方法 selectAllUserPOJO()，该方法的返回值的类型是 List<MapUser>：

```
List<MapUser> selectAllUserPOJO();
```

最后在 SQL 映射文件中首先使用<resultMap>元素将 MapUser 类的属性与查询结果集的列名一一对应，然后添加接口方法 selectAllUserPOJO()对应的 SQL 映射：

```
<!--使用自定义结果集类型-->
<resultMap type="com.ch.ch7_1.entity.MapUser" id="myResult">
    <!-- property 是 MapUser 类中的属性-->
    <!-- column 是查询结果集的列名，可以来自不同的表-->
    <id property="m_uid" column="uid"/>
    <result property="m_uname" column="uname"/>
    <result property="m_usex" column="usex"/>
</resultMap>
<!--使用自定义结果集类型查询所有用户-->
<select id="selectAllUserPOJO" resultMap="myResult">
    select * from user
</select>
```

❺ 使用 Map 存储结果集

在 MyBatis 中任何查询结果都可以使用 Map 存储，具体示例如下。

首先添加数据操作接口方法 selectAllUserMap()，该方法的返回值的类型是 List<Map<String, Object>>：

```
List<Map<String, Object>> selectAllUserMap();
```

然后在 SQL 映射文件中添加接口方法 selectAllUserMap()对应的 SQL 映射：

```
<!--使用 Map 存储查询结果，查询结果的列名作为 Map 的 key，列值作为 Map 的 value -->
<select id="selectAllUserMap" resultType="map">
    select * from user
</select>
```

最后在控制器类中添加请求处理方法 selectAllUserMap()：

```
@GetMapping("/selectAllUserMap")
public List<Map<String, Object>> selectAllUserMap() {
    //使用 Map 存储查询结果
    return myUserRepository.selectAllUserMap();
}
```

▶ 7.2.2 \<insert\>、\<update\>以及\<delete\>元素

❶ \<insert\>元素

\<insert\>元素用于映射添加语句，MyBatis 在执行完一条添加语句后，将返回一个整数表示其影响的行数。它的属性与\<select\>元素的属性大部分相同，在本节讲解它的几个特有属性。

keyProperty：在添加时将自动生成的主键值回填给 PO（Persistant Object）类的某个属性，通常会设置为主键对应的属性。如果是联合主键，可以在多个值之间用逗号隔开。

keyColumn：设置第几列是主键，当主键列不是表中的第一列时需要设置。如果是联合主键，可以在多个值之间用逗号隔开。

useGeneratedKeys：该属性将使 MyBatis 使用 JDBC 的 getGeneratedKeys()方法获取由数据库内部产生的主键，例如 MySQL、SQL Server 等自动递增的字段，其默认值为 false。

MySQL、SQL Server 等数据库的表格可以采用自动递增的字段作为主键，有时可能需要使用这个刚产生的主键，用于关联其他业务。因为本书采用的数据库是 MySQL 数据库，所以可以直接使用自动递增主键回填的方法，具体步骤如下：

首先添加数据操作接口方法 addUserBack()，该方法的返回值的类型是 int：

```
int addUserBack(MyUser mu);
```

然后在 SQL 映射文件中添加接口方法 addUserBack()对应的 SQL 映射：

```xml
<!--添加一个用户,成功后将主键值回填给 uid（PO 类的属性）-->
<insert id="addUserBack" parameterType="MyUser" keyProperty=
        "uid" useGeneratedKeys="true">
    insert into user (uname,usex) values(#{uname},#{usex})
</insert>
```

❷ \<update\>与\<delete\>元素

\<update\>和\<delete\>元素比较简单，它们的属性和\<insert\>元素的属性基本一样，在执行后也返回一个整数，表示影响数据库的记录的行数。SQL 映射文件的示例代码如下：

```xml
<!--修改一个用户-->
<update id="updateUser" parameterType="MyUser">
    update user set uname = #{uname},usex = #{usex} where uid = #{uid}
</update>
<!--删除一个用户-->
<delete id="deleteUser" parameterType="Integer">
    delete from user where uid = #{uid}
</delete>
```

▶ 7.2.3 动态 SQL

开发人员通常根据需求手动拼接 SQL 语句，这是一个极其麻烦的工作，而 MyBatis 提供了对 SQL 语句动态组装的功能，恰好能解决这一问题。MyBatis 的动态 SQL 和使用 JSTL 或其他类似基于 XML 的文本处理器相似，常用元素有\<if\>、\<choose\>、\<when\>、\<otherwise\>、\<trim\>、\<where\>、\<foreach\>和\<bind\>等。

❶ \<if\>元素

动态 SQL 通常要做的事情是有条件地包含 where 子句的一部分，所以在 MyBatis 中\<if\>元素是最常用的元素，它类似于 Java 中的 if 语句，示例代码如下：

```xml
<!--使用 if 元素,根据条件动态查询用户信息-->
<select id="selectAllUserByIf" resultType="MyUser" parameterType="MyUser">
    select * from user where 1=1
```

```xml
        <if test="uname !=null and uname!=''">
            and uname like concat('%',#{uname},'%')
        </if>
        <if test="usex !=null and usex!=''">
            and usex = #{usex}
        </if>
</select>
```

❷ `<choose>`、`<when>`、`<otherwise>`元素

有时不需要用到所有的条件语句，而只需要从中择其一二。针对这种情况，MyBatis 提供了`<choose>`元素，它类似 Java 中的 switch 语句，示例代码如下：

```xml
<!--使用 choose、when、otherwise 元素，根据条件动态查询用户信息-->
<select id="selectUserByChoose" resultType="MyUser" parameterType="MyUser">
    select * from user where 1=1
    <choose>
        <when test="uname !=null and uname!=''">
            and uname like concat('%',#{uname},'%')
        </when>
        <when test="usex !=null and usex!=''">
            and usex = #{usex}
        </when>
        <otherwise>
            and uid > 3
        </otherwise>
    </choose>
</select>
```

❸ `<trim>`元素

`<trim>`元素可以在自己包含的内容前加上某些前缀，也可以在其后加上某些后缀，与之对应的属性是 prefix 和 suffix；可以把包含内容的首部的某些内容覆盖，即忽略，也可以把尾部的某些内容覆盖，对应的属性是 prefixOverrides 和 suffixOverrides。因为`<trim>`元素有这样的功能，所以可以非常简单地利用`<trim>`来代替`<where>`元素。示例代码如下：

```xml
<!--使用 trim 元素，根据条件动态查询用户信息-->
<select id="selectUserByTrim" resultType="MyUser" parameterType="MyUser">
    select * from user
    <trim prefix="where" prefixOverrides="and|or">
        <if test="uname !=null and uname!=''">
            and uname like concat('%',#{uname},'%')
        </if>
        <if test="usex !=null and usex!=''">
            and usex = #{usex}
        </if>
    </trim>
</select>
```

❹ `<where>`元素

`<where>`元素的作用是输出一个 where 语句，优点是不考虑`<where>`元素的条件输出，MyBatis 将智能处理。如果所有的条件都不满足，那么 MyBatis 将会查出所有记录，如果输出是以 and 开头，MyBatis 将把第一个 and 忽略；如果是以 or 开头，MyBatis 也将把它忽略。此外，在`<where>`元素中不考虑空格的问题，MyBatis 将智能加上。

```xml
<!--使用 where 元素，根据条件动态查询用户信息-->
<select id="selectUserByWhere" resultType="MyUser" parameterType="MyUser">
    select * from user
    <where>
        <if test="uname !=null and uname!=''">
```

第7章　MyBatis 与 MyBatis-Plus

```
            and uname like concat('%',#{uname},'%')
        </if>
        <if test="usex !=null and usex!=''">
            and usex = #{usex}
        </if>
    </where>
</select>
```

❺ `<foreach>` 元素

`<foreach>` 元素主要用于构建 in 条件，它可以在 SQL 语句中迭代一个集合。`<foreach>` 元素的属性主要有 item、index、collection、open、separator、close。item 表示集合中每一个元素进行迭代时的别名，index 指定一个名字，用于表示在迭代过程中每次迭代到的位置，open 表示该语句以什么开始，separator 表示在每次迭代之间以什么符号作为分隔符，close 表示以什么结束。在使用`<foreach>`时，最关键的也是最容易出错的是 collection 属性，该属性是必选的，但在不同情况下该属性的值是不一样的，主要有以下 3 种情况：

（1）如果传入的是单参数且参数类型是一个 List，collection 属性的值为 list。

（2）如果传入的是单参数且参数类型是一个 array 数组，collection 属性的值为 array。

（3）如果传入的参数是多个，需要把它们封装成一个 Map，当然单参数也可以封装成 Map。Map 的 key 是参数名，所以 collection 属性值是传入的 List 或 array 对象在自己封装的 Map 中的 key。

示例代码如下：

```
<!--使用 foreach 元素，查询用户信息-->
<select id="selectUserByForeach" resultType="MyUser" parameterType="List">
    select * from user where uid in
    <foreach item="item" index="index" collection="list"
        open="(" separator="," close=")">
        #{item}
    </foreach>
</select>
```

❻ `<bind>` 元素

在进行模糊查询时，如果使用"${}"拼接字符串，则无法防止 SQL 注入问题。如果使用字符串拼接函数或连接符号，不同数据库的拼接函数或连接符号不同，例如 MySQL 的 concat 函数、Oracle 的连接符号"||"，这样 SQL 映射文件就需要根据不同的数据库提供不同的实现，显然比较麻烦，且不利于代码的移植，幸运的是 MyBatis 提供了`<bind>`元素来解决这一问题。示例代码如下：

```
<!--使用 bind 元素进行模糊查询-->
<select id="selectUserByBind" resultType="MyUser" parameterType="MyUser">
    <!-- bind 中的 uname 是 com.po.MyUser 的属性名 -->
    <bind name="paran_uname" value="'%' + uname + '%'"/>
    select * from user where uname like #{paran_uname}
</select>
```

7.3　MyBatis-Plus 快速入门

▶ 7.3.1　MyBatis-Plus 简介

MyBatis-Plus 是 MyBatis 的一个增强工具，在 MyBatis 的基础上只做增强不做改变，为简化开发、提高效率而生。MyBatis-Plus 的特性具体如下。

扫一扫

视频讲解

（1）无侵入：只做增强不做改变，引入它不会对现有工程产生影响。
（2）损耗小：启动即会自动注入基本 CRUD，性能基本无损耗，直接面向对象操作。
（3）强大的 CRUD 操作：内置通用 Mapper、通用 Service，仅通过少量配置即可实现单表的大部分 CRUD 操作，更有强大的条件构造器，满足各类使用需求。
（4）支持 Lambda 形式调用：通过 Lambda 表达式方便编写各类查询条件，无须再担心写错字段。
（5）支持主键自动生成：支持多种主键策略，可自由配置，完美解决主键问题。
（6）支持 ActiveRecord 模式：支持 ActiveRecord 形式调用，实体类只需继承 Model 类即可进行强大的 CRUD 操作。
（7）支持自定义全局通用操作：支持全局通用方法注入。
（8）内置代码生成器：采用代码或者 Maven 插件可快速生成 Mapper、Model、Service、Controller 层代码，支持模板引擎，提供更多自定义配置。
（9）内置分页插件：基于 MyBatis 物理分页，开发者无须关心具体操作，配置好插件之后，实现分页等同于普通 List 遍历。

▶ 7.3.2 Spring Boot 整合 MyBatis-Plus

在 Spring Boot 应用中，添加 mybatis-plus-boot-starter 依赖即可整合 MyBatis-Plus，具体如下：

```xml
<dependency>
    <groupId>com.baomidou</groupId>
    <artifactId>mybatis-plus-boot-starter</artifactId>
    <version>3.x.y.z</version>
</dependency>
```

在 Spring Boot 应用中，通过 mybatis-plus-boot-starter 引入 MyBatis-Plus 依赖后，将自动引入 MyBatis、MyBatis-Spring 等相关依赖，所以不再需要引入这些依赖，避免了因版本差异而导致问题。

下面通过实例讲解如何在 Spring Boot 应用中使用 MyBatis-Plus 框架操作数据库。

【例 7-4】在 Spring Boot 应用中使用 MyBatis-Plus 框架操作数据库。

其具体实现步骤如下：

❶ 创建 Spring Boot Web 应用

创建基于 Lombok 依赖的 Spring Boot Web 应用 ch7_2，在该应用中操作的数据库与 7.1 节一样，都是 springtest，操作的数据表是 user 表。

❷ 修改 pom.xml 文件

在 pom.xml 文件中添加 MySQL 连接器与 MyBatis-Plus 依赖，具体如下：

```xml
<dependency>
    <groupId>mysql</groupId>
    <artifactId>mysql-connector-java</artifactId>
    <version>8.0.29</version>
</dependency>
<dependency>
    <groupId>com.baomidou</groupId>
    <artifactId>mybatis-plus-boot-starter</artifactId>
    <version>3.5.3.1</version>
</dependency>
```

第 7 章　MyBatis 与 MyBatis-Plus

❸ 设置 Web 应用 ch7_2 的上下文路径及数据源配置信息

在应用 ch7_2 的 application.properties 文件中配置如下内容：

```
server.servlet.context-path=/ch7_2
#数据库地址
spring.datasource.url=jdbc:mysql://localhost:3306/springtest?useUnicode=true&characterEncoding=UTF-8&allowMultiQueries=true&serverTimezone=GMT%2B8
#数据库用户名
spring.datasource.username=root
#数据库密码
spring.datasource.password=root
#数据库驱动
spring.datasource.driver-class-name=com.mysql.cj.jdbc.Driver
#设置包的别名（在 Mapper 映射文件中直接使用实体类名）
mybatis-plus.type-aliases-package=com.ch.ch7_2.entity
#告诉系统到哪里去找 mapper.xml 文件（映射文件）
mybatis-plus.mapper-locations=classpath:mappers/*.xml
#在控制台中输出 SQL 语句日志
logging.level.com.ch.ch7_2.mapper=debug
#让控制台输出的 JSON 字符串的格式更美观
spring.jackson.serialization.indent-output=true
```

❹ 创建实体类

创建名为 com.ch.ch7_2.entity 的包，并在该包中创建 MyUser 实体类，具体代码如下：

```java
package com.ch.ch7_2.entity;
import lombok.Data;
import com.baomidou.mybatisplus.annotation.TableName;
@Data
@TableName("user")
public class MyUser {
    private Integer uid;        //与数据表中的字段名相同
    private String uname;
    private String usex;
}
```

❺ 创建数据访问接口

创建名为 com.ch.ch7_2.mapper 的包，并在该包中创建 UserMapper 接口。UserMapper 接口通过继承 BaseMapper<MyUser> 接口（在 7.4 节将讲解该接口）对实体类 MyUser 对应的数据表 user 进行 CRUD 操作。UserMapper 接口的代码如下：

```java
package com.ch.ch7_2.mapper;
import com.baomidou.mybatisplus.core.mapper.BaseMapper;
import com.ch.ch7_2.entity.MyUser;
import org.springframework.stereotype.Repository;
import java.util.List;
@Repository
public interface UserMapper extends BaseMapper<MyUser> {
    List<MyUser> myFindAll();
}
```

❻ 创建 Mapper 映射文件

在 src/main/resources 目录下创建名为 mappers 的包，并在该包中创建 SQL 映射文件 MyUserMapper.xml（当 Mapper 接口中没有自定义方法时可以不创建此文件），具体代码如下：

```xml
<?xml version="1.0" encoding="UTF-8" ?>
<!DOCTYPE mapper
PUBLIC "-//mybatis.org//DTD Mapper 3.0//EN"
```

```
            "http://mybatis.org/dtd/mybatis-3-mapper.dtd">
<mapper namespace="com.ch.ch7_2.mapper.UserMapper">
    <select id="myFindAll" resultType="MyUser">
        select * from user
    </select>
</mapper>
```

❼ 创建控制器类 MyUserController

创建名为 com.ch.ch7_2.controller 的包,并在该包中创建控制器类 MyUserController。MyUserController 的核心代码如下:

```
@RestController
public class MyUserController {
    @Autowired
    private UserMapper userMapper;
    @GetMapping("/findAll")
    public List<MyUser> findAll(){
        //通过 BaseMapper 接口方法 selectList 查询
        return userMapper.selectList(null);
    }
    @GetMapping("/myFindAll")
    public List<MyUser> myFindAll(){
        //通过自定义方法 myFindAll 查询
        return userMapper.myFindAll();
    }
}
```

❽ 在应用程序的主类中扫描 Mapper 接口

在应用程序的 Ch72Application 主类中使用@MapperScan 注解扫描 MyBatis 的 Mapper 接口,核心代码如下:

```
@SpringBootApplication
@MapperScan(basePackages={"com.ch.ch7_2.mapper"})
public class Ch72Application {
    public static void main(String[] args) {
        SpringApplication.run(Ch72Application.class, args);
    }
}
```

❾ 运行

首先运行 Ch72Application 主类,然后访问"http://localhost:8080/ch7_2/findAll"和"http://localhost:8080/ch7_2/myFindAll"进行测试。

扫一扫

视频讲解

7.4 MyBatis-Plus 基础

▶ 7.4.1 MyBatis-Plus 注解

本小节将详细介绍 MyBatis-Plus 的相关注解类,具体如下。

❶ @TableName

当实体类的类名与要操作表的表名不一致时,需要使用@TableName 注解标识实体类对应的表。示例代码如下:

```
@TableName("user")
public class MyUser {}
```

@TableName 注解的所有属性都不是必须指定的,如表 7.3 所示。

表 7.3　@TableName 的属性

属性	类型	默认值	描述
value	String	""	表名
schema	String	""	指定模式名称。如果是 MySQL 数据库，则指定数据库名称；如果是 Oracle，则为 schema，例如 schema="scott"，scott 就是 Oracle 中的 schema
keepGlobalPrefix	boolean	false	是否保持使用全局的 tablePrefix 值（当设置全局 tablePrefix 时）
resultMap	String	""	XML 中 resultMap 的 ID（用于满足特定类型的实体类对象的绑定）
autoResultMap	boolean	false	是否自动构建 resultMap 并使用（如果设置 resultMap，则不会进行 resultMap 的自动构建与注入）
excludeProperty	String[]	{}	需要排除的属性名

❷ @TableId

@TableId 注解为主键注解，指定实体类中的某属性为主键字段。示例代码如下：

```
@TableName("user")
public class MyUser {
    @TableId(type=IdType.AUTO)
    private Integer uid;
}
```

@TableId 注解有 value 和 type 两个属性，其中 value 属性表示主键字段名，默认值为""；type 属性表示主键类型，默认值为 IdType.NONE。对于 type 属性值，IdType.AUTO 表示数据表 ID 自增；IdType.NONE 表示无状态，该类型为未设置主键类型；IdType.INPUT 表示插入前自行设置主键值；IdType.ASSIGN_ID 表示分配 ID（类型为 Long、Integer 或 String），使用 IdentifierGenerator 接口的 nextId 方法（默认实现类为 DefaultIdentifierGenerator 雪花算法）；IdType.ASSIGN_UUID 表示分配 UUID，类型为 String，使用 IdentifierGenerator 接口的 nextUUID 方法。

❸ @TableField

@TableField 注解为非主键字段注解。若实体类中的属性使用的是驼峰命名风格，而表中的字段使用的是下画线命名风格，例如实体类属性为 userName，表中字段为 user_name，此时 MyBatis-Plus 会自动将下画线命名风格转化为驼峰命名风格。若实体类中的属性和表中的字段不满足上述条件，例如实体类属性为 name，表中字段为 username，此时需要在实体类属性上使用@TableField("username")设置属性所对应的字段名。

@TableField 注解的所有属性都不是必须指定的，如表 7.4 所示。

表 7.4　@TableField 的属性

属性	类型	默认值	描述
value	String	""	数据库字段名
exist	boolean	true	是否为数据库表字段
condition	String	""	字段 where 实体查询比较条件，有设置值则以设置的值为准，没有则为默认全局的%s=#{%s}
update	String	""	字段 update set 部分注入，例如当 version 字段上注解 update="%s+1" 表示更新时会设置 version=version+1

续表

属性	类型	默认值	描述
insertStrategy	Enum	FieldStrategy.DEFAULT	IGNORED 为忽略判断；NOT_NULL 为非 NULL 判断；NOT_EMPTY 为非空判断（只针对字符串类型字段，其他类型字段仍然为非 NULL 判断）；DEFAULT 为追随全局配置；NEVER 为不加入 SQL。举例：NOT_NULL insert into table_a(\<if test="columnProperty != null"\>column\</if\>) values (\<if test="columnProperty != null"\>#{columnProperty}\</if\>)
updateStrategy	Enum	FieldStrategy.DEFAULT	举例：IGNORED update table_a set column=#{columnProperty}
whereStrategy	Enum	FieldStrategy.DEFAULT	举例：NOT_EMPTY where\<if test="columnProperty != null and columnProperty!=''"\>column=#{columnProperty}\</if\>
fill	Enum	FieldFill.DEFAULT	字段自动填充策略。DEFAULT 为默认不处理；INSERT 为插入时填充字段；UPDATE 为更新时填充字段；INSERT_UPDATE 为插入和更新时填充字段
select	boolean	true	是否进行 select 查询
keepGlobalFormat	boolean	false	是否保持使用全局的 format 进行处理
jdbcType	JdbcType	JdbcType.UNDEFINED	JDBC 类型（该默认值不代表会按照该值生效）
typeHandler	Class\<? extends TypeHandler\>	UnknownTypeHandler.class	类型处理器（该默认值不代表会按照该值生效）
numericScale	String	""	指定小数点后保留的位数

❹ @Version

乐观锁注解，@Version 标记在字段上。乐观锁指当要更新一条记录时，希望这条记录没有被别人更新。乐观锁的实现方式为取出记录时获取当前 version；更新时带上这个 version；执行更新时 set version = newVersion where version = oldVersion，如果 version 不对，则更新失败。具体示例如下：

```
@Data
@TableName("t_product")
public class Product {
    private Long id;
    private String name;
    private Integer price;
    @Version
    private Integer version;
}
@Configuration
public class MybatisPlusConfig {
    @Bean
    public MybatisPlusInterceptor mybatisPlusInterceptor() {
        MybatisPlusInterceptor interceptor =
            new MybatisPlusInterceptor();
        //乐观锁插件
        interceptor.addInnerInterceptor(new OptimisticLockerInnerInterceptor());
        return interceptor;
    }
}
```

这个注解@Version 就是实现乐观锁的重要注解，当更新数据库中的数据时，例如价格，version 就会加 1，如果 where 语句中的 version 版本不对，则更新失败。

❺ @EnumValue

普通枚举类注解，注解在枚举字段上。示例代码如下：

```
@Getter              //类中属性都生成getter方法
public enum SexEnum {
    MALE(1, "男"),
    FEMALE(2, "女");
    @EnumValue    //标记数据库中存储的值是sex
    private Integer sex;
    private String sexName;
    SexEnum(Integer sex, String sexName) {
        this.sex = sex;
        this.sexName = sexName;
    }
}
```

❻ @TableLogic

使用@TableLogic 表示实体类中的属性是逻辑删除的属性。逻辑删除即假删除，将对应数据中代表是否被删除字段的状态修改为"被删除状态"，之后在数据库中仍然能看到此条数据记录。

▶ 7.4.2 CRUD 接口

MyBatis-Plus 利用 MyBatis 接口编程实现机制，默认提供了一系列增、删、改、查基础方法，并且开发人员对于这些基础方法不需要编写 SQL 语句即可进行操作。

❶ Mapper CRUD 接口

MyBatis-Plus 内置了可以实现对单表 CRUD 的 BaseMapper<T>接口，泛型 T 为任意实体对象。BaseMapper<T>接口是针对 Dao 层的 CRUD 方法进行封装。

在自定义数据访问接口时，继承 BaseMapper<T>接口，即可使用 BaseMapper<T>接口方法进行单表的 CRUD，例如 public interface UserMapper extends BaseMapper<MyUser> {}。

BaseMapper<T>接口方法具体如下。

1) insert

BaseMapper<T>接口提供了实现插入一条记录的方法，具体如下：

```
//插入一条记录
int insert(T entity);
```

2) delete

BaseMapper<T>接口提供了许多删除方法，具体如下：

```
//根据 entity 条件删除记录
int delete(@Param(Constants.WRAPPER) Wrapper<T> wrapper);
//删除（根据ID 批量删除）
int deleteBatchIds(@Param(Constants.COLLECTION) Collection<? extends Serializable> idList);
//根据 ID 删除
int deleteById(Serializable id);
//根据 columnMap 条件删除记录
int deleteByMap(@Param(Constants.COLUMN_MAP) Map<String, Object> columnMap);
```

在上述方法中，wrapper（Wrapper<T>）表示实体对象封装操作类（即条件构造器，可以为 null）；idList（Collection<? extends Serializable>）表示主键 ID 列表（不能为 null 以及 empty）；id（Serializable）表示主键 ID；columnMap（Map<String, Object>）表示表字段 map 对象。

3）update

BaseMapper<T>接口提供了两个更新方法，具体如下：

```
//根据 whereWrapper 条件更新记录
int update(@Param(Constants.ENTITY) T updateEntity, @Param(Constants.WRAPPER) Wrapper<T> whereWrapper);
//根据 ID 修改
int updateById(@Param(Constants.ENTITY) T entity);
```

4）select

BaseMapper<T>接口提供了许多查询方法，具体如下：

```
//根据 ID 查询
T selectById(Serializable id);
//根据 entity 条件查询一条记录
T selectOne(@Param(Constants.WRAPPER) Wrapper<T> queryWrapper);
//查询（根据 ID 批量查询）
List<T> selectBatchIds(@Param(Constants.COLLECTION) Collection<? extends Serializable> idList);
//根据 entity 条件查询全部记录
List<T> selectList(@Param(Constants.WRAPPER) Wrapper<T> queryWrapper);
//查询（根据 columnMap 条件）
List<T> selectByMap(@Param(Constants.COLUMN_MAP) Map<String, Object> columnMap);
//根据 Wrapper 条件查询全部记录
List<Map<String, Object>> selectMaps(@Param(Constants.WRAPPER) Wrapper<T> queryWrapper);
//根据 Wrapper 条件查询全部记录。注意只返回第一个字段的值
List<Object> selectObjs(@Param(Constants.WRAPPER) Wrapper<T> queryWrapper);
//根据 entity 条件查询全部记录（并翻页）
IPage<T> selectPage(IPage<T> page, @Param(Constants.WRAPPER) Wrapper<T> queryWrapper);
//根据 Wrapper 条件查询全部记录（并翻页）
IPage<Map<String, Object>> selectMapsPage(IPage<T> page, @Param(Constants.WRAPPER) Wrapper<T> queryWrapper);
//根据 Wrapper 条件查询总记录数
Integer selectCount(@Param(Constants.WRAPPER) Wrapper<T> queryWrapper);
```

5）ActiveRecord 模式

所谓 ActiveRecord 模式，在 Spring Boot 应用中，如果已注入对应实体的 BaseMapper，例如 public interface UserMapper extends BaseMapper<MyUser>{}，那么实体类 MyUser 只需要继承 Model 类即可进行强大的 CRUD 操作，例如 public class MyUser extends Model<MyUser>{}。

❷ Service CRUD 接口

通用 Service CRUD 封装 IService<T>接口，进一步封装 CRUD，采用 get 查询单行、remove 删除、list 查询集合、page 分页等前缀命名方式区分 Mapper 层，避免混淆。创建 Service 接口及其实现类，示例代码如下：

```
public interface UserService extends IService<MyUser> { }
/* ServiceImpl 实现了 IService，提供了 IService 中基础功能的实现。若 ServiceImpl 无法满足业务需求，可以使用自定义的 UserService 方法，并在实现类中实现*/
@Service
public class UserServiceImpl extends ServiceImpl<UserMapper, MyUser> implements UserService { }
```

第 7 章　MyBatis 与 MyBatis-Plus

IService<T>接口针对业务逻辑层的封装，需要指定 Dao 层接口和对应的实体类，是在 BaseMapper<T>基础上的加强，ServiceImpl<M, T>是针对业务逻辑层的实现。

1）save

通用 Service CRUD 接口提供了以下 save 方法：

```
//插入一条记录（选择字段，策略插入）
boolean save(T entity);
//插入（批量）
boolean saveBatch(Collection<T> entityList);
//插入（批量）
boolean saveBatch(Collection<T> entityList, int batchSize);
//TableId注解存在更新记录，否则插入一条记录
boolean saveOrUpdate(T entity);
//根据updateWrapper尝试更新，否则继续执行saveOrUpdate(T)方法
boolean saveOrUpdate(T entity, Wrapper<T> updateWrapper);
//批量修改插入
boolean saveOrUpdateBatch(Collection<T> entityList);
//批量修改插入
boolean saveOrUpdateBatch(Collection<T> entityList, int batchSize);
```

2）remove

通用 Service CRUD 接口提供了以下 remove 方法：

```
//根据entity条件删除记录
boolean remove(Wrapper<T> queryWrapper);
//根据ID删除
boolean removeById(Serializable id);
//根据columnMap条件删除记录
boolean removeByMap(Map<String, Object> columnMap);
//删除（根据ID批量删除）
boolean removeByIds(Collection<? extends Serializable> idList);
```

3）update

通用 Service CRUD 接口提供了以下 update 方法：

```
//根据UpdateWrapper条件更新记录，需要设置sqlset
boolean update(Wrapper<T> updateWrapper);
//根据whereWrapper条件更新记录
boolean update(T updateEntity, Wrapper<T> whereWrapper);
//根据ID选择修改
boolean updateById(T entity);
//根据ID批量更新
boolean updateBatchById(Collection<T> entityList);
//根据ID批量更新
boolean updateBatchById(Collection<T> entityList, int batchSize);
```

4）get、list、page 及 count

通用 Service CRUD 接口提供了以下 get、list、page 及 count 查询方法：

```
//根据ID查询
T getById(Serializable id);
//根据 Wrapper 查询一条记录。结果集如果是多个会抛出异常，随机取一条加上限制条件wrapper.last("LIMIT 1")
T getOne(Wrapper<T> queryWrapper);
//根据Wrapper查询一条记录
T getOne(Wrapper<T> queryWrapper, boolean throwEx);
//根据Wrapper查询一条记录
Map<String, Object> getMap(Wrapper<T> queryWrapper);
```

```java
//根据Wrapper查询一条记录
<V> V getObj(Wrapper<T> queryWrapper, Function<? super Object, V> mapper);
//查询所有
List<T> list();
//查询列表
List<T> list(Wrapper<T> queryWrapper);
//查询(根据ID批量查询)
Collection<T> listByIds(Collection<? extends Serializable> idList);
//查询(根据columnMap条件)
Collection<T> listByMap(Map<String, Object> columnMap);
//查询所有列表
List<Map<String, Object>> listMaps();
//查询列表
List<Map<String, Object>> listMaps(Wrapper<T> queryWrapper);
//查询全部记录
List<Object> listObjs();
//查询全部记录
<V> List<V> listObjs(Function<? super Object, V> mapper);
//根据Wrapper条件查询全部记录
List<Object> listObjs(Wrapper<T> queryWrapper);
//根据Wrapper条件查询全部记录
<V> List<V> listObjs(Wrapper<T> queryWrapper, Function<? super Object, V> mapper);
//无条件分页查询
IPage<T> page(IPage<T> page);
//条件分页查询
IPage<T> page(IPage<T> page, Wrapper<T> queryWrapper);
//无条件分页查询
IPage<Map<String, Object>> pageMaps(IPage<T> page);
//条件分页查询
IPage<Map<String, Object>> pageMaps(IPage<T> page, Wrapper<T> queryWrapper);
//查询总记录数
int count();
//根据Wrapper条件查询总记录数
int count(Wrapper<T> queryWrapper);
```

5）链式query及update

通用Service CRUD接口提供了以下query及update链式方法：

```java
//链式查询，普通
QueryChainWrapper<T> query();
//链式查询，Lambda式。注意不支持Kotlin
LambdaQueryChainWrapper<T> lambdaQuery();
//示例
query().eq("column", value).one();
lambdaQuery().eq(Entity::getId, value).list();
//链式更改，普通
UpdateChainWrapper<T> update();
//链式更改，Lambda式。注意不支持Kotlin
LambdaUpdateChainWrapper<T> lambdaUpdate();
//示例
update().eq("column", value).remove();
lambdaUpdate().eq(Entity::getId, value).update(entity);
```

❸ Mapper CRUD接口和Service CRUD接口的使用示例

下面通过一个实例演示Mapper CRUD接口和Service CRUD接口的使用方法。

【例7-5】演示Mapper CRUD接口和Service CRUD接口的使用方法。

其具体实现步骤如下。

第 7 章 MyBatis 与 MyBatis-Plus

1）创建 Spring Boot Web 应用

创建基于 Lombok 依赖的 Spring Boot Web 应用 ch7_3，在该应用中操作的数据库与 7.1 节一样，都是 springtest，操作的数据表是 user 表。

2）修改 pom.xml 文件

在 pom.xml 文件中添加 MySQL 连接器与 MyBatis-Plus 依赖，因为与例 7-4 相同，这里不再赘述。

3）设置 Web 应用 ch7_3 的上下文路径及数据源配置信息

在应用 ch7_3 的 application.properties 文件中设置上下文路径及数据源配置信息，配置内容与例 7-4 基本相同，这里不再赘述。

4）创建实体类

创建名为 com.ch.ch7_3.entity 的包，并在该包中创建 MyUser 实体类，具体代码如下：

```java
package com.ch.ch7_3.entity;
import com.baomidou.mybatisplus.annotation.IdType;
import com.baomidou.mybatisplus.annotation.TableId;
import com.baomidou.mybatisplus.extension.activerecord.Model;
import lombok.Data;
import com.baomidou.mybatisplus.annotation.TableName;
@Data
@TableName("user")
public class MyUser extends Model<MyUser> {
    @TableId(value = "uid", type = IdType.AUTO)
    private Integer uid;
    private String uname;
    private String usex;
}
```

5）创建数据访问接口

创建名为 com.ch.ch7_3.mapper 的包，并在该包中创建 UserMapper 接口，UserMapper 接口通过继承 BaseMapper<MyUser>接口对实体类 MyUser 对应的数据表 user 进行 CRUD 操作。UserMapper 接口的代码如下：

```java
package com.ch.ch7_3.mapper;
import com.baomidou.mybatisplus.core.mapper.BaseMapper;
import com.ch.ch7_3.entity.MyUser;
import org.springframework.stereotype.Repository;
@Repository
public interface UserMapper extends BaseMapper<MyUser> {
}
```

6）创建 Service 接口及实现类

创建名为 com.ch.ch7_3.service 的包，并在该包中创建 UserService 接口及实现类 UserServiceImpl。

UserService 接口继承 IService<MyUser>接口，具体代码如下：

```java
package com.ch.ch7_3.service;
import com.baomidou.mybatisplus.extension.service.IService;
import com.ch.ch7_3.entity.MyUser;
public interface UserService extends IService<MyUser> {
}
```

实现类 UserServiceImpl 继承 ServiceImpl<UserMapper, MyUser>类，具体代码如下：

```java
package com.ch.ch7_3.service;
import com.baomidou.mybatisplus.extension.service.impl.ServiceImpl;
```

```
import com.ch.ch7_3.entity.MyUser;
import com.ch.ch7_3.mapper.UserMapper;
import org.springframework.stereotype.Service;
@Service
public class UserServiceImpl extends ServiceImpl<UserMapper, MyUser> implements UserService {}
```

7）配置分页插件 PaginationInnerInterceptor

MyBatis-Plus 基于 PaginationInnerInterceptor 拦截器，实现分页查询功能，所以需要事先配置该拦截器才能实现分页查询功能。

创建名为 com.ch.ch7_3.config 的包，并在该包中创建 MybatisPlusConfig 配置类，具体代码如下：

```
package com.ch.ch7_3.config;
import com.baomidou.mybatisplus.annotation.DbType;
import com.baomidou.mybatisplus.extension.plugins.MybatisPlusInterceptor;
import com.baomidou.mybatisplus.extension.plugins.inner.PaginationInnerInterceptor;
import org.springframework.context.annotation.Bean;
import org.springframework.context.annotation.Configuration;
@Configuration
public class MybatisPlusConfig{
    @Bean
    public MybatisPlusInterceptor mybatisPlusInterceptor() {
        MybatisPlusInterceptor interceptor = new MybatisPlusInterceptor();
        interceptor.addInnerInterceptor(new PaginationInnerInterceptor(DbType.MYSQL));
        return interceptor;
    }
}
```

8）创建控制器类 MyUserController

创建名为 com.ch.ch7_3.controller 的包，并在该包中创建控制器类 MyUserController。MyUserController 的代码如下：

```
package com.ch.ch7_3.controller;
import com.baomidou.mybatisplus.core.metadata.IPage;
import com.baomidou.mybatisplus.extension.plugins.pagination.Page;
import com.ch.ch7_3.entity.MyUser;
import com.ch.ch7_3.mapper.UserMapper;
import com.ch.ch7_3.service.UserService;
import org.springframework.beans.factory.annotation.Autowired;
import org.springframework.web.bind.annotation.GetMapping;
import org.springframework.web.bind.annotation.RestController;
import java.util.Arrays;
import java.util.List;
@RestController
public class MyUserController {
    @Autowired
    private UserMapper userMapper;
    @Autowired
    private UserService userService;
    @GetMapping("/testMapperSave")
    public MyUser testMapperSave(){
        MyUser mu = new MyUser();
        mu.setUname("testMapperSave 陈恒1");
        mu.setUsex("女");
        int result = userMapper.insert(mu);
        //在实体类主键属性使用@TableId注解后，主键自动回填
```

```java
        return mu;
    }
    @GetMapping("/testMapperDelete")
    public int testMapperDelete(){
        List<Long> list = Arrays.asList(17L, 7L, 19L);
        int result = userMapper.deleteBatchIds(list);
        return result;
    }
    @GetMapping("/testMapperUpdate")
    public MyUser testMapperUpdate(){
        MyUser mu = new MyUser();
        mu.setUid(1);
        mu.setUname("李四");
        mu.setUsex("男");
        int result = userMapper.updateById(mu);
        return mu;
    }
    @GetMapping("/testMapperSelect")
    public List<MyUser> testMapperSelect(){
        return userMapper.selectList(null);
    }
    @GetMapping("/testModelSave")
    public MyUser testModelSave(){
    MyUser mu = new MyUser();
        mu.setUname("testModelSave 陈恒 2");
        mu.setUsex("男");
        mu.insert();
        return mu;
    }
    @GetMapping("/testServiceSave")
    public List<MyUser> testServiceSave(){
        MyUser mu1 = new MyUser();
        mu1.setUname("testServiceSave 陈恒 1");
        mu1.setUsex("女");
        MyUser mu2 = new MyUser();
        mu2.setUname("testServiceSave 陈恒 2");
        mu2.setUsex("男");
        List<MyUser> list = Arrays.asList(mu1, mu2);
        boolean result = userService.saveBatch(list);
        return list;
    }
    @GetMapping("/testServiceUpdate")
    public List<MyUser> testServiceUpdate(){
        MyUser mu1 = new MyUser();
        mu1.setUid(23);
        mu1.setUname("testServiceSave 陈恒 11");
        mu1.setUsex("女");
        MyUser mu2 = new MyUser();
        mu2.setUid(24);
        mu2.setUname("testServiceSave 陈恒 22");
        mu2.setUsex("男");
        List<MyUser> list = Arrays.asList(mu1, mu2);
        boolean result = userService.updateBatchById(list);
        return list;
    }
    @GetMapping("/testServicePage")
    public List<MyUser> testServicePage(){
```

```java
            //1为当前页，5为页面大小
            IPage<MyUser> iPage = new Page<>(1, 5);
            IPage<MyUser> page = userService.page(iPage);
            //条件构造器
            //QueryWrapper<MyUser> wrapper = new QueryWrapper<>();
            //wrapper.like("uname", "陈");
            //IPage<MyUser> page = userService.page(iPage, wrapper);
            System.out.println(page.getPages());
            //返回当前页的记录
            return page.getRecords();
    }
    @GetMapping("/testServiceQuery")
    public List<MyUser> testServiceQuery(){
        //链式查询
        return userService.query().like("uname", "陈").list();
    }
}
```

9）在应用程序的主类中扫描 Mapper 接口

在应用程序的 Ch73Application 主类中使用@MapperScan 注解扫描 MyBatis 的 Mapper 接口，核心代码如下：

```java
@SpringBootApplication
@MapperScan(basePackages={"com.ch.ch7_3.mapper"})
public class Ch73Application {
    public static void main(String[] args) {
        SpringApplication.run(Ch73Application.class, args);
    }
}
```

10）运行

首先运行 Ch73Application 主类，然后访问 "http://localhost:8080/ch7_3/testMapperSave" 进行测试。

▶ 7.4.3 条件构造器

MyBatis-Plus 提供了构造条件的类 Wrapper，方便用户根据自己的意图定义所需要的条件。Wrapper 是一个抽象类，在一般情况下用它的子类 QueryWrapper 来实现自定义条件查询。在查询前首先创建条件构造器（QueryWrapper wrapper = new QueryWrapper< >()），然后调用构造器中的方法实现按条件查询。

AbstractWrapper 是 Wrapper 的子类，也是 QueryWrapper(LambdaQueryWrapper) 和 UpdateWrapper(LambdaUpdateWrapper) 的父类，用于生成 SQL 的 where 条件，entity 属性用于生成 SQL 的 where 条件。条件构造器中的方法具体如下。

❶ allEq：全部 eq（或个别 isNull）

```
allEq(Map<R, V> params)
allEq(Map<R, V> params, boolean null2IsNull)
allEq(boolean condition, Map<R, V> params, boolean null2IsNull)
```

对于 params，key 为数据库字段名，value 为字段值；null2IsNull 如果为 true，则在 map 的 value 为 null 时调用 isNull 方法，如果为 false，则忽略 value 为 null 的条件。

示例 1：map.put("id", 1); map.put("name", "老王"); map.put("age", null); wrapper.allEq(map)，等价 SQL 为 id = 1 and name = '老王' and age is null。

示例 2：wrapper.allEq(map, false)，等价 SQL 为 id = 1 and name = '老王'。

```
allEq(BiPredicate<R, V> filter, Map<R, V> params)
allEq(BiPredicate<R, V> filter, Map<R, V> params, boolean null2IsNull)
allEq(boolean condition, BiPredicate<R, V> filter, Map<R, V> params, boolean null2IsNull)
```

其中，filter 为过滤函数，判断是否允许字段传入比对条件中。

示例 1：wrapper.allEq((k,v) -> k.contains("a"), map)，等价于 SQL 为 name = '老王' and age is null。

示例 2：wrapper.allEq((k,v) -> k.contains("a"), map, false)，等价于 SQL 为 name = '老王'。

❷ eq：等于，=

```
eq(R column, Object val)
eq(boolean condition, R column, Object val)
```

示例：eq("name", "老王")，等价于 name = '老王'。

❸ ne：不等于，<>

```
ne(R column, Object val)
ne(boolean condition, R column, Object val)
```

示例：ne("name", "老王")，等价于 name <> '老王'。

❹ gt：大于，>

```
gt(R column, Object val)
gt(boolean condition, R column, Object val)
```

示例：gt("age", 7)，等价于 age > 7。

❺ ge：大于或等于，>=

```
ge(R column, Object val)
ge(boolean condition, R column, Object val)
```

示例：ge("age", 7)，等价于 age >= 7。

❻ lt：小于，<

```
lt(R column, Object val)
lt(boolean condition, R column, Object val)
```

示例：lt("age", 7)，等价于 age < 7。

❼ le：小于或等于，<=

```
le(R column, Object val)
le(boolean condition, R column, Object val)
```

示例：le("age", 7)，等价于 age <= 7。

❽ between 值 1 and 值 2

```
between(R column, Object val1, Object val2)
between(boolean condition, R column, Object val1, Object val2)
```

示例：between("age", 7, 30)，等价于 age between 7 and 30。

❾ not between 值 1 and 值 2

```
notBetween(R column, Object val1, Object val2)
notBetween(boolean condition, R column, Object val1, Object val2)
```

示例：notBetween("age", 7, 30)，等价于 age not between 7 and 30。

❿ like '%值%'

```
like(R column, Object val)
like(boolean condition, R column, Object val)
```

示例：like("name", "王")，等价于 name like '%王%'。

⑪ not like '%值%'

```
notLike(R column, Object val)
notLike(boolean condition, R column, Object val)
```

示例：notLike("name", "王")，等价于 name not like '%王%'。

⑫ likeLeft like '%值'

```
likeLeft(R column, Object val)
likeLeft(boolean condition, R column, Object val)
```

示例：likeLeft("name", "王")，等价于 name like '%王'。

⑬ likeRight like '值%'

```
likeRight(R column, Object val)
likeRight(boolean condition, R column, Object val)
```

示例：likeRight("name", "王")，等价于 name like '王%'。

⑭ isNull 及 isNotNull

```
isNull(R column)
isNull(boolean condition, R column)
isNotNull(R column)
isNotNull(boolean condition, R column)
```

示例 1：isNull("name")，等价于 name is null。

示例 2：isNotNull("name")，等价于 name is not null。

⑮ in 及 notIn

```
in(R column, Collection<?> value)
in(boolean condition, R column, Collection<?> value)
notIn(R column, Collection<?> value)
notIn(boolean condition, R column, Collection<?> value)
```

示例 1：in("age",{1,2,3})，等价于 age in (1,2,3)。

示例 2：notIn("age",{1,2,3})，等价于 age not in (1,2,3)。

⑯ inSql 及 notInSql

```
inSql(R column, String inValue)
inSql(boolean condition, R column, String inValue)
notInSql(R column, String inValue)
notInSql(boolean condition, R column, String inValue)
```

示例 1：inSql("age", "1,2,3,4,5,6")，等价于 age in (1,2,3,4,5,6)。

示例 2：inSql("id", "select id from table where id < 3")，等价于 id in (select id from table where id < 3)。

示例 3：notInSql("age", "1,2,3,4,5,6")，等价于 age not in (1,2,3,4,5,6)。

示例 4：notInSql("id", "select id from table where id < 3")，等价于 id not in (select id from table where id < 3)。

⑰ groupBy

```
groupBy(R... columns)
groupBy(boolean condition, R... columns)
```

示例：groupBy("id", "name")，等价于 group by id,name。

⑱ orderByAsc、orderByDesc 及 orderBy

```
orderByAsc(R... columns)
orderByAsc(boolean condition, R... columns)
```

```
orderByDesc(R... columns)
orderByDesc(boolean condition, R... columns)
orderBy(boolean condition, boolean isAsc, R... columns)
```

示例 1：orderByAsc("id", "name")，等价于 order by id ASC,name ASC。

示例 2：orderByDesc("id", "name")，等价于 order by id DESC,name DESC。

示例 3：orderBy(true, true, "id", "name")，等价于 order by id ASC,name ASC。

⑲ having

```
having(String sqlHaving, Object... params)
having(boolean condition, String sqlHaving, Object... params)
```

示例 1：having("sum(age) > 10")，等价于 having sum(age) > 10。

示例 2：having("sum(age) > {0}", 11)，等价于 having sum(age) > 11。

⑳ func

```
func(Consumer<Children> consumer)
func(boolean condition, Consumer<Children> consumer)
```

示例：func(i -> if(true) {i.eq("id", 1)} else {i.ne("id", 1)})。

㉑ or 及 and

```
or()
or(boolean condition)
or(Consumer<Param> consumer)
or(boolean condition, Consumer<Param> consumer)
and(Consumer<Param> consumer)
and(boolean condition, Consumer<Param> consumer)
```

示例 1：eq("id",1).or().eq("name","老王")，等价于 id = 1 or name = '老王'。

示例 2：or(i -> i.eq("name", "李白").ne("status", "活着"))，等价于 or (name = '李白' and status <> '活着')。

示例 3：and(i -> i.eq("name", "李白").ne("status", "活着"))，等价于 and (name = '李白' and status <> '活着')。

㉒ nested：正常嵌套，不带 and 或者 or

```
nested(Consumer<Param> consumer)
nested(boolean condition, Consumer<Param> consumer)
```

示例：nested(i -> i.eq("name", "李白").ne("status", "活着"))，等价于(name = '李白' and status <> '活着')。

㉓ apply：拼接 SQL

```
apply(String applySql, Object... params)
apply(boolean condition, String applySql, Object... params)
```

示例 1：apply("id = 1")，等价于 id = 1。

示例 2：apply("date_format(dateColumn,'%Y-%m-%d') = '2008-08-08'")，等价于 date_format(dateColumn,'%Y-%m-%d') = '2008-08-08'。

示例 3：apply("date_format(dateColumn,'%Y-%m-%d') = {0}", "2008-08-08")，等价于 date_format(dateColumn,'%Y-%m-%d') = '2008-08-08'。

㉔ last：无视优化规则直接拼接到 SQL 的最后

```
last(String lastSql)
last(boolean condition, String lastSql)
```

示例：last("limit 1")。

㉕ exists 及 notExists

```
exists(String existsSql)
exists(boolean condition, String existsSql)
notExists(String notExistsSql)
notExists(boolean condition, String notExistsSql)
```

示例 1：exists("select id from table where age = 1")，等价于 exists (select id from table where age = 1)。

示例 2：notExists("select id from table where age = 1")，等价于 not exists (select id from table where age = 1)。

㉖ QueryWrapper 之 select：设置查询字段

```
select(String... sqlSelect)
select(Predicate<TableFieldInfo> predicate)
select(Class<T> entityClass, Predicate<TableFieldInfo> predicate)
```

示例 1：select("id", "name", "age")。

示例 2：select(i -> i.getProperty().startsWith("test"))。

㉗ UpdateWrapper 之 set 及 setSql

```
set(String column, Object val)
set(boolean condition, String column, Object val)
setSql(String sql)
```

示例 1：set("name", "老李头")。

示例 2：set("name", "")，数据库字段值变为空字符串。

示例 3：set("name", null)，数据库字段值变为 null。

示例 4：setSql("name = '老李头'")。

本章小结

MyBatis-Plus 是增强版的 MyBatis，对 MyBatis 只做增强不做改变，因此灵活使用 MyBatis-Plus 的前提是掌握 MyBatis 基础知识。

本章详细介绍了 Spring Boot 如何整合 MyBatis 及 MyBatis-Plus，希望读者掌握 MyBatis 及 MyBatis-Plus 在 Spring Boot 应用中的整合开发过程。

习题 7

1. 简述 MyBatis 与 MyBatis-Plus 的关系。

2. 简述 MyBatis-Plus 的特性。

3. 在 MyBatis-Plus 中，当实体类的类名与要操作表的表名不一致时，需要使用（　　）注解标识实体类对应的表。

　　A. @TableName　　B. @TableId　　C. @TableField　　D. @TableLogic

第 8 章　Spring Boot 的安全控制

学习目的与要求

本章首先重点讲解 Spring Security 安全控制机制，然后介绍 Spring Security 操作实例。通过本章的学习，要求读者掌握如何使用 Spring Security 安全控制机制解决企业应用程序的安全问题。

本章主要内容

- Spring Security 快速入门
- Spring Security 操作实例

在 Web 应用开发中，安全毋庸置疑是十分重要的，使用 Spring Security 来保护 Web 应用是一个非常好的选择。Spring Security 是 Spring 框架的一个安全模块，可以非常方便地与 Spring Boot 应用无缝集成。

8.1 Spring Security 快速入门

8.1.1 什么是 Spring Security

Spring Security 是一个专门针对 Spring 应用系统的安全框架，充分利用了 Spring 框架的依赖注入和 AOP 功能，为 Spring 应用系统提供安全访问控制解决方案。

在 Spring Security 安全框架中有两个重要概念，即授权（Authorization）和认证（Authentication）。授权即确定用户在当前应用系统下所拥有的功能权限；认证即确认用户访问当前系统的身份。

8.1.2 Spring Security 的用户认证

验证用户的最常用方法之一是验证用户名和密码。Spring Security 为使用用户名和密码进行身份验证提供了全面的支持。

在 Spring Security 安全框架中，可以通过配置 AuthenticationManager 认证管理器完成用户认证。示例代码如下：

```
@Bean
public AuthenticationManager authenticationManager(AuthenticationConfiguration authenticationConfiguration) throws Exception{
    return authenticationConfiguration.getAuthenticationManager();
}
```

❶ 内存中的用户认证

在 Spring Security 安全框架中，InMemoryUserDetailsManager 实现了 UserDetailsService，以支持存储在内存中的用户名/密码的身份验证。InMemoryUserDetailsManager 通过实现

UserDetailsManager 接口提供对 UserDetails 的管理。当 Spring Security 配置为接受用户名和密码进行身份验证时会使用基于 UserDetails 的身份验证，示例代码如下：

```java
@Bean
public UserDetailsService users() {
    UserDetails user = User.builder()
        .username("chenheng")
        .password("xxxx")
        .roles("USER")
        .build();
    UserDetails admin = User.builder()
        .username("admin")
        .password("yyyy")
        .roles("DBA", "ADMIN")
        .build();
    return new InMemoryUserDetailsManager(user, admin);
}
```

在上述示例代码中添加了两个用户，一个用户名为"admin"，密码为"yyyy"，用户权限为"ROLE_ADMIN"和"ROLE_DBA"；另一个用户名为"chenheng"，密码为"xxxx"，用户权限为"ROLE_USER"。"ROLE_"在 Spring Security 保存用户权限的时候默认加上。

❷ 通用的用户认证

在实际应用中可以查询数据库获取用户和权限，这时需要自定义实现 org.springframework.security.core.userdetails.UserDetailsService 接口的类，并重写 public UserDetails loadUserByUsername(String username)方法查询对应的用户和权限，示例代码如下：

```java
@Service
public class MyUserSecurityService implements UserDetailsService{
    @Autowired
    private MyUserRepository myUserRepository;
    /**
     * 通过重写 loadUserByUsername 方法查询对应的用户
     * UserDetails 是 Spring Security 的一个核心接口
     * UserDetails 定义了可以获取用户名、密码、权限等与认证相关信息的方法
     */
    @Override
    public UserDetails loadUserByUsername(String username) throws UsernameNotFoundException {
        //根据用户名（页面接收的用户名）查询当前用户
        MyUser myUser = myUserRepository.findByUsername(username);
        if(myUser == null) {
            throw new UsernameNotFoundException("用户名不存在");
        }
        //GrantedAuthority 代表赋予当前用户的权限（认证权限）
        List<GrantedAuthority> authorities = new ArrayList<GrantedAuthority>();
        //获得当前用户权限集合
        List<Authority> roles = myUser.getAuthorityList();
        //将当前用户的权限保存为用户的认证权限
        for (Authority authority: roles) {
            GrantedAuthority sg=new SimpleGrantedAuthority(authority.getName());
            authorities.add(sg);
        }
        //org.springframework.security.core.userdetails.User 是 Spring Security
        //内部的实现，专门用于保存用户名、密码、权限等与认证相关的信息
```

```
            User su = new User(myUser.getUsername(), myUser.getPassword(),
authorities);
        return su;
    }
}
```

8.1.3 Spring Security 的请求授权

在 Spring Security 安全框架中可以通过配置 SecurityFilterChain 完成用户授权，示例代码如下：

```
@Bean
public SecurityFilterChain filterChain(HttpSecurity http) throws Exception {
    http
        .authorizeHttpRequests(authorize -> authorize
            .requestMatchers("/xxx").permitAll()
            .requestMatchers("/user/**").hasRole("USER")
            //其他所有请求认证授权后才能访问
            .anyRequest().authenticated()
        );
    return http.build();
}
```

在 filterChain(HttpSecurity http) 方法中，使用 HttpSecurity 的 authorizeHttpRequests() 方法的子节点给指定用户授权访问 URL 模式。大家可以通过 requestMatchers() 方法匹配 URL 路径，在匹配请求路径后可以针对当前用户对请求进行安全处理。Spring Security 提供了许多安全处理方法，部分方法如表 8.1 所示。

表 8.1 Spring Security 的部分安全处理方法

方　　法	用　　途
anyRequest()	匹配所有请求路径
access(String attribute)	当 Spring EL 表达式的结果为 true 时可以访问
anonymous()	匿名可以访问
authenticated()	用户登录后可以访问
denyAll()	用户不能访问
fullyAuthenticated()	用户完全认证可以访问（非 remember-me 下自动登录）
hasAnyAuthority(String...)	参数表示权限，用户权限与其中任一权限相同就可以访问
hasAnyRole(String...)	参数表示角色，用户角色与其中任一角色相同就可以访问
hasAuthority(String authority)	参数表示权限，用户权限与参数相同才可以访问
hasRole(String role)	参数表示角色，用户角色与参数相同才可以访问
permitAll()	任何用户都可以访问
rememberMe()	允许通过 remember-me 登录的用户访问

8.1.4 Spring Security 的核心类

Spring Security 的核心类包括 Authentication、SecurityContextHolder、UserDetails、UserDetailsService、GrantedAuthority、DaoAuthenticationProvider 和 PasswordEncoder。

❶ Authentication

Authentication 用来封装用户认证信息的接口，在用户登录认证之前，Spring Security 将相关信息封装为一个 Authentication 具体实现类的对象，在登录认证成功后将生成一个信息更全面、包含用户权限等信息的 Authentication 对象，然后将该对象保存在 SecurityContextHolder 所持有的 SecurityContext 中，以方便后续程序进行调用，例如当前用户名、访问权限等。

❷ SecurityContextHolder

SecurityContextHolder 顾名思义是用来持有 SecurityContext 的类。在 SecurityContext 中包含当前认证用户的详细信息。Spring Security 使用一个 Authentication 对象描述当前用户的相关信息。例如，最常见的是获得当前登录用户的用户名和权限，示例代码如下：

```java
/**
 * 获得当前用户名
 */
private String getUname() {
    return SecurityContextHolder.getContext().getAuthentication().getName();
}
/**
 * 获得当前用户权限
 */
private String getAuthorities() {
    Authentication authentication = SecurityContextHolder.getContext().getAuthentication();
    List<String> roles = new ArrayList<String>();
    for (GrantedAuthority ga: authentication.getAuthorities()) {
        roles.add(ga.getAuthority());
    }
    return roles.toString();
}
```

❸ UserDetails

UserDetails 是 Spring Security 的一个核心接口。该接口定义了一些可以获取用户名、密码、权限等与认证相关的信息的方法。通常需要在应用中获取当前用户的其他信息，例如 Email、电话等。如果这时只包含认证相关的 UserDetails 对象可能不能满足需要，在该情况下可以实现自己的 UserDetails，在该实现类中定义一些获取用户其他信息的方法，这样就可以直接从当前 SecurityContext 的 Authentication 的 principal 中获取用户的其他信息。

Authentication.getPrincipal()的返回类型是 Object，但通常返回的是一个 UserDetails 的实例，通过强制类型转换可以将 Object 转换为 UserDetails 类型。

❹ UserDetailsService

UserDetails 是通过 UserDetailsService 的 loadUserByUsername(String username)方法加载的。UserDetailsService 也是一个接口，也需要实现自己的 UserDetailsService 来加载自定义的 UserDetails 信息。

在登录认证时，Spring Security 将通过 UserDetailsService 的 loadUserByUsername(String username)方法获取对应的 UserDetails 进行认证，认证通过后将该 UserDetails 赋给认证通过的 Authentication 的 principal，然后再将该 Authentication 保存到 SecurityContext 中。在应用中，如果需要使用用户信息，可以通过 SecurityContextHolder 获取存放在 SecurityContext 中的 Authentication 的 principal，即 UserDetails 实例。

❺ GrantedAuthority

通过 Authentication 的 getAuthorities()方法可以返回当前 Authentication 对象拥有的权限

（一个 GrantedAuthority 类型的数组），即当前用户拥有的权限。GrantedAuthority 是一个接口，通常是通过 UserDetailsService 进行加载，然后赋给 UserDetails。

❻ DaoAuthenticationProvider

在 Spring Security 安全框架中，默认使用 DaoAuthenticationProvider 实现 AuthenticationProvider 接口进行用户认证的处理。在 DaoAuthenticationProvider 进行认证时，需要一个 UserDetailsService 来获取用户信息 UserDetails。当然用户可以实现自己的 AuthenticationProvider，进而改变认证方式。

❼ PasswordEncoder

在 Spring Security 安全框架中，通过 PasswordEncoder 接口完成对密码的加密。Spring Security 对 PasswordEncoder 有多种实现，包括 MD5 加密、SHA-256 加密等，开发者只需要直接使用即可。在 Spring Boot 应用中，使用 BCryptPasswordEncoder 加密是较好的选择。BCryptPasswordEncoder 使用 BCrypt 的强哈希加密实现，并可以由客户端指定加密强度，强度越高安全性越高。

▶ 8.1.5 Spring Security 的验证机制

Spring Security 的验证机制是由许多 Filter 实现的，Filter 将在 Spring MVC 前拦截请求，主要包括注销 Filter（LogoutFilter）、用户名密码验证 Filter（UsernamePasswordAuthenticationFilter）等内容。Filter 再交由其他组件完成细分的功能，最常用的 UsernamePasswordAuthenticationFilter 会持有一个 AuthenticationManager 引用，AuthenticationManager 是一个验证管理器，专门负责验证。AuthenticationManager 持有一个 AuthenticationProvider 集合，AuthenticationProvider 是做验证工作的组件，在验证成功或失败之后调用对应的 Hanlder（处理）。

8.2 Spring Boot 的支持

在 Spring Boot 应用中，用户只需要引入 spring-boot-starter-security 依赖即可使用 Spring Security 安全框架，这是因为 Spring Boot 对 Spring Security 提供了自动配置功能。从 org.springframework.boot.autoconfigure.security.SecurityProperties 类中可以看到使用以"spring.security"为前缀的属性设置了 Spring Security 的相关默认配置。

8.3 实际开发中的 Spring Security 操作实例

本节将讲解一个基于 Spring Data JPA 的 Spring Security 操作实例。

下面通过一个简单实例演示在 Spring Boot 应用中如何使用基于 Spring Data JPA 的 Spring Security 安全框架。

【例 8-1】在 Spring Boot 应用中使用基于 Spring Data JPA 的 Spring Security 安全框架。

其具体实现步骤如下：

❶ 创建 Spring Boot Web 应用 ch8

创建基于 Lombok、Spring Data JPA、Thymeleaf 及 Spring Security 的 Web 应用 ch8，如图 8.1 所示。

图 8.1 Web 应用 ch8 的依赖

❷ 修改 pom.xml 文件，添加 MySQL 依赖

在 pom.xml 文件中添加 MySQL 依赖，配置内容与例 6-1 中的相同，这里不再赘述。

❸ 设置 Web 应用 ch8 的上下文路径及数据源配置信息

应用 ch8 的 application.properties 文件中的配置内容与例 6-1 中的相同，这里不再赘述。

❹ 整理脚本、样式等静态文件

JS 脚本、CSS 样式、图片等静态文件默认放置在 src/main/resources/static 目录下，ch8 应用引入的 BootStrap 和 jQuery 与例 5-5 中的一样，这里不再赘述。

❺ 创建用户和权限持久化实体类

在应用 ch8 的 src/main/java 目录下创建名为 com.ch.ch8.entity 的包，并在该包中创建持久化实体类 MyUser 和 Authority。MyUser 类用来保存用户数据，用户名唯一；Authority 类用来保存权限信息。用户和权限是多对多的关系。

MyUser 的核心代码如下：

```
@Entity
@Table(name = "user")
@JsonIgnoreProperties(value = { "hibernateLazyInitializer"})
@Data
public class MyUser implements Serializable{
    private static final long serialVersionUID = 1L;
    @Id
    @GeneratedValue(strategy = GenerationType.IDENTITY)
    private int id;
    private String username;
    private String password;
    //这里不能是懒加载 lazy，否则在 MyUserSecurityService 的 loadUserByUsername 方法中获取不到权限
    @ManyToMany(cascade = {CascadeType.REFRESH}, fetch = FetchType.EAGER)
    @JoinTable(name = "user_authority",joinColumns = @JoinColumn(name = "user_id"),
    inverseJoinColumns = @JoinColumn(name = "authority_id"))
    private List<Authority> authorityList;
    //repassword 不映射到数据表
    @Transient
    private String repassword;
}
```

注意：在实际开发中 MyUser 还可实现 org.springframework.security.core.userdetails.UserDetails 接口，实现该接口后即可成为 Spring Security 所使用的用户。本例为了区分 Spring Data JPA 的 POJO 和 Spring Security 的用户对象，并没有实现 UserDetails 接口，而是在实现 UserDetailsService 接口的类中进行绑定。

Authority 的核心代码如下：

```
@Entity
@Table(name = "authority")
@JsonIgnoreProperties(value = { "hibernateLazyInitializer"})
@Data
public class Authority implements Serializable{
    private static final long serialVersionUID = 1L;
    @Id
    @GeneratedValue(strategy = GenerationType.IDENTITY)
    private int id;
    @Column(nullable = false)
```

```
    private String name;
    @ManyToMany(mappedBy = "authorityList")
    @JsonIgnore
    private List<MyUser> userList;
}
```

❻ 创建数据访问层接口

在应用 ch8 的 src/main/java 目录下创建名为 com.ch.ch8.repository 的包，并在该包中创建名为 MyUserRepository 的接口，该接口继承了 JpaRepository，核心代码如下：

```
public interface MyUserRepository extends JpaRepository<MyUser, Integer>{
    //根据用户名查询用户，方法名的命名符合 Spring Data JPA 规范
    MyUser findByUsername(String username);
}
```

在 com.ch.ch8.repository 包中创建名为 AuthorityRepository 的接口，该接口继承了 JpaRepository，具体代码如下：

```
public interface AuthorityRepository extends JpaRepository<Authority, Integer>{}
```

❼ 创建业务层

在应用 ch8 的 src/main/java 目录下创建名为 com.ch.ch8.service 的包，并在该包中创建 UserService 接口和 UserServiceImpl 实现类。这里省略 UserService 接口的代码。

UserServiceImpl 的核心代码如下：

```
@Service
public class UserServiceImpl implements UserService{
    @Autowired
    private MyUserRepository myUserRepository;
    @Autowired
    private AuthorityRepository authorityRepository;
    /**
     * 实现注册
     */
    @Override
    public String register(MyUser userDomain) {
        String username = userDomain.getUsername();
        List<Authority> authorityList = new ArrayList<Authority>();
        //管理员权限
        if("admin".equals(username)) {
            Authority a1 = new Authority();
            Authority a2 = new Authority();
            a1.setName("ROLE_ADMIN");
            a2.setName("ROLE_DBA");
            authorityList.add(a1);
            authorityList.add(a2);
        }else {           //用户权限
            Authority a1 = new Authority();
            a1.setName("ROLE_USER");
            authorityList.add(a1);
        }
        //注册权限
        authorityRepository.saveAll(authorityList);
        userDomain.setAuthorityList(authorityList);
        //加密密码
        String secret = new BCryptPasswordEncoder().encode(userDomain.getPassword());
        userDomain.setPassword(secret);
        //注册用户
        MyUser mu = myUserRepository.save(userDomain);
```

```java
            if(mu != null)      //注册成功
                return "/login";
            return "/register";          //注册失败
        }
        /**
         * 用户登录成功
         */
        @Override
        public String loginSuccess(Model model) {
            model.addAttribute("user", getUname());
            model.addAttribute("role", getAuthorities());
            return "/user/loginSuccess";
        }
        /**
         * 管理员登录成功
         */
        @Override
        public String main(Model model) {
            model.addAttribute("user", getUname());
            model.addAttribute("role", getAuthorities());
            return "/admin/main";
        }
        /**
         * 注销用户
         */
        @Override
        public String logout(HttpServletRequest request, HttpServletResponse response) {
            //获得用户认证信息
            Authentication authentication = SecurityContextHolder.getContext().getAuthentication();
            if(authentication != null) {
                //注销
                new SecurityContextLogoutHandler().logout(request, response, authentication);
            }
            return "redirect:/login?logout";
        }
        /**
         * 没有权限拒绝访问
         */
        @Override
        public String deniedAccess(Model model) {
            model.addAttribute("user", getUname());
            model.addAttribute("role", getAuthorities());
            return "deniedAccess";
        }
        /**
         * 获得当前用户名
         */
        private String getUname() {
            return SecurityContextHolder.getContext().getAuthentication().getName();
        }
        /**
         * 获得当前用户权限
         */
        private String getAuthorities() {
            Authentication authentication = SecurityContextHolder.getContext().getAuthentication();
            List<String> roles = new ArrayList<String>();
```

```
        for (GrantedAuthority ga: authentication.getAuthorities()) {
            roles.add(ga.getAuthority());
        }
        return roles.toString();
    }
}
```

⑧ 创建控制器类

在应用 ch8 的 src/main/java 目录下创建名为 com.ch.ch8.controller 的包，并在该包中创建控制器类 TestSecurityController，核心代码如下：

```
@Controller
public class TestSecurityController {
    @Autowired
    private UserService userService;
    @RequestMapping("/")
    public String index() {
        return "/index";
    }
    @RequestMapping("/toLogin")
    public String toLogin() {
        return "/login";
    }
    @RequestMapping("/toRegister")
    public String toRegister(@ModelAttribute("userDomain") MyUser userDomain) {
        return "/register";
    }
    @RequestMapping("/register")
    public String register(@ModelAttribute("userDomain") MyUser userDomain) {
        return userService.register(userDomain);
    }
    @RequestMapping("/login")
    public String login() {
        //这里什么都不做，由Spring Security负责登录验证
        return "/login";
    }
    @RequestMapping("/user/loginSuccess")
    public String loginSuccess(Model model) {
        return userService.loginSuccess(model);
    }
    @RequestMapping("/admin/main")
    public String main(Model model) {
        return userService.main(model);
    }
    @RequestMapping("/logout")
    public String logout(HttpServletRequest request, HttpServletResponse response) {
        return userService.logout(request, response);
    }
    @RequestMapping("/deniedAccess")
    public String deniedAccess(Model model) {
        return userService.deniedAccess(model);
    }
}
```

⑨ 创建应用的安全控制相关实现

在应用 ch8 的 src/main/java 目录下创建名为 com.ch.ch8.security 的包，并在该包中创建 MyUserSecurityService、MyAuthenticationSuccessHandler 和 SpringSecurityConfig 类。

MyUserSecurityService 实现了 UserDetailsService 接口，通过重写 loadUserByUsername (String username)方法查询对应的用户，并将用户名、密码、权限等与认证相关的信息封装在 UserDetails 对象中。

MyUserSecurityService 的代码如下：

```java
package com.ch.ch8.security;
import com.ch.ch8.entity.Authority;
import com.ch.ch8.entity.MyUser;
import com.ch.ch8.repository.MyUserRepository;
import org.springframework.beans.factory.annotation.Autowired;
import org.springframework.security.core.GrantedAuthority;
import org.springframework.security.core.authority.SimpleGrantedAuthority;
import org.springframework.security.core.userdetails.User;
import org.springframework.security.core.userdetails.UserDetails;
import org.springframework.security.core.userdetails.UserDetailsService;
import org.springframework.security.core.userdetails.UsernameNotFoundException;
import org.springframework.stereotype.Service;
import java.util.ArrayList;
import java.util.List;
/**
 * 获得对应的UserDetails，保存与认证相关的信息
 */
@Service
public class MyUserSecurityService implements UserDetailsService {
    @Autowired
    private MyUserRepository myUserRepository;
    /**
     * 通过重写loadUserByUsername 方法查询对应的用户
     * UserDetails 是Spring Security 的一个核心接口
     * UserDetails 定义了可以获取用户名、密码、权限等与认证相关信息的方法
     */
    @Override
    public UserDetails loadUserByUsername(String username) throws UsernameNotFoundException {
        //根据用户名（页面接收的用户名）查询当前用户
        MyUser myUser = myUserRepository.findByUsername(username);
        if(myUser == null) {
            throw new UsernameNotFoundException("用户名不存在");
        }
        //GrantedAuthority 代表赋予当前用户的权限（认证权限）
        List<GrantedAuthority> authorities = new ArrayList<GrantedAuthority>();
        //获得当前用户权限集合
        List<Authority> roles = myUser.getAuthorityList();
        //将当前用户的权限保存为用户的认证权限
        for (Authority authority: roles) {
            GrantedAuthority sg = new SimpleGrantedAuthority(authority.getName());
            authorities.add(sg);
        }
        //org.springframework.security.core.userdetails.User 是Spring Security
        //的内部实现，专门用于保存用户名、密码、权限等与认证相关的信息
        User su = new User(myUser.getUsername(), myUser.getPassword(), authorities);
        return su;
    }
}
```

第 8 章　Spring Boot 的安全控制

MyAuthenticationSuccessHandler 继承了 SimpleUrlAuthenticationSuccessHandler 类，并重写了 handle(HttpServletRequest request, HttpServletResponse response, Authentication authentication) 方法，根据当前认证用户的角色指定对应的 URL。

MyAuthenticationSuccessHandler 的代码如下：

```java
package com.ch.ch8.security;
import org.springframework.security.core.Authentication;
import org.springframework.security.core.GrantedAuthority;
import org.springframework.security.web.DefaultRedirectStrategy;
import org.springframework.security.web.RedirectStrategy;
import org.springframework.security.web.authentication.SimpleUrlAuthenticationSuccessHandler;
import org.springframework.stereotype.Component;
import jakarta.servlet.ServletException;
import jakarta.servlet.http.HttpServletRequest;
import jakarta.servlet.http.HttpServletResponse;
import java.io.IOException;
import java.util.ArrayList;
import java.util.Collection;
import java.util.List;
/**
 * 用户授权、认证成功处理类
 */
@Component
public class MyAuthenticationSuccessHandler extends SimpleUrlAuthenticationSuccessHandler {
    //Spring Security 的重定向策略
    private RedirectStrategy redirectStrategy = new DefaultRedirectStrategy();
    /**
     * 重写 handle 方法，通过 RedirectStrategy 重定向到指定的 URL
     */
    @Override
    protected void handle(HttpServletRequest request, HttpServletResponse response, Authentication authentication) throws IOException, ServletException {
        //根据当前认证用户的角色返回适当的 URL
        String tagetURL = getTargetURL(authentication);
        //重定向到指定的 URL
        redirectStrategy.sendRedirect(request, response, tagetURL);
    }
    /**
     * 从 Authentication 对象中提取当前登录用户的角色，并根据其角色返回适当的 URL
     */
    protected String getTargetURL(Authentication authentication) {
        String url = "";
        //获得当前登录用户的权限（角色）集合
        Collection<? extends GrantedAuthority> authorities = authentication.getAuthorities();
        List<String> roles = new ArrayList<String>();
        //将权限（角色）名称添加到 List 集合
        for (GrantedAuthority au: authorities) {
            roles.add(au.getAuthority());
        }
        //判断不同角色的用户跳转到不同的 URL
        //这里的 URL 是控制器的请求匹配路径
        if(roles.contains("ROLE_USER")) {
            url = "/user/loginSuccess";
        }else if(roles.contains("ROLE_ADMIN")) {
            url = "/admin/main";
```

```
        }else {
            url = "/deniedAccess";
        }
        return url;
    }
}
```

SpringSecurityConfig 类是认证和授权处理的配置类，需要使用@Configuration 和 @EnableWebSecurity 注解。在该类中配置了加密规则、认证实现方式、认证管理器以及授权操作。MySecurityConfigurerAdapter 的代码如下：

```
package com.ch.ch8.security;
import org.springframework.beans.factory.annotation.Autowired;
import org.springframework.context.annotation.Bean;
import org.springframework.context.annotation.Configuration;
import org.springframework.security.authentication.AuthenticationManager;
import org.springframework.security.authentication.AuthenticationProvider;
import org.springframework.security.authentication.dao.DaoAuthenticationProvider;
import org.springframework.security.config.annotation.authentication.configuration.AuthenticationConfiguration;
import org.springframework.security.config.annotation.web.builders.HttpSecurity;
import org.springframework.security.config.annotation.web.configuration.EnableWebSecurity;
import org.springframework.security.crypto.bcrypt.BCryptPasswordEncoder;
import org.springframework.security.crypto.password.PasswordEncoder;
import org.springframework.security.web.SecurityFilterChain;
/**
 * 认证和授权处理类
 */
@Configuration
@EnableWebSecurity
public class SpringSecurityConfig {
    //依赖注入通用的用户服务类
    @Autowired
    private MyUserSecurityService myUserSecurityService;
    @Autowired
    private MyAuthenticationSuccessHandler myAuthenticationSuccessHandler;
    /**
     * BCryptPasswordEncoder 是 PasswordEncoder 的接口实现，用于实现加密功能
     */
    @Bean
    public PasswordEncoder passwordEncoder() {
        return new BCryptPasswordEncoder();
    }
    /**
     * DaoAuthenticationProvider 是 AuthenticationProvider 的实现，是认证的实现方式
     */
    @Bean
    public AuthenticationProvider authenticationProvider() {
        DaoAuthenticationProvider provide = new DaoAuthenticationProvider();
        //不隐藏用户未找到异常
        provide.setHideUserNotFoundExceptions(false);
        //设置自定义认证方式，用户登录认证
        provide.setUserDetailsService(myUserSecurityService);
        //设置密码加密程序认证
        provide.setPasswordEncoder(passwordEncoder());
        return provide;
    }
    /**
```

```
    * 获取AuthenticationManager（认证管理器）
    */
    @Bean
    public AuthenticationManager authenticationManager(AuthenticationConfiguration authenticationConfiguration) throws Exception{
        return authenticationConfiguration.getAuthenticationManager();
    }
    /**
    * 请求授权，用户授权操作
    */
    @Bean
    public SecurityFilterChain filterChain(HttpSecurity http) throws Exception {
        return http
                //设置权限
                .authorizeHttpRequests(authorize -> authorize
                //首页、登录、注册页面、登录注册功能，以及静态资源过滤
                .requestMatchers("/toLogin","/toRegister","/","/login",
"/register","/css/**","/fonts/**","/js/**").permitAll()
                //这里默认追加ROLE_，/user/**是控制器的请求匹配路径
                .requestMatchers("/user/**").hasRole("USER")
                .requestMatchers("/admin/**").hasAnyRole("ADMIN", "DBA")
                //其他所有请求登录后才能访问
                .anyRequest().authenticated()
                )
                //将输入的用户名与密码和授权进行比较
                .formLogin()
                    .loginPage("/login").successHandler(myAuthenticationSuccessHandler)
                    .usernameParameter("username")
                    .passwordParameter("password")
                    //登录失败
                    .failureUrl("/login?error")
                .and()
                //注销行为可任意访问
                .logout().permitAll()
                .and()
                //指定异常处理页面
                .exceptionHandling().accessDeniedPage("/deniedAccess")
                .and().build();
    }
}
```

⑩ 创建用于测试的视图页面

在src/main/resources/templates目录下创建应用首页、注册、登录以及拒绝访问页面；在src/main/resources/templates/admin目录下创建管理员用户认证成功后访问的页面；在src/main/resources/templates/user目录下创建普通用户认证成功后访问的页面。相关视图页面的代码参见本书提供的源程序ch8。

⑪ 测试应用

运行Ch8Application的主方法启动项目。在Spring Boot应用启动后，用户可以通过"http://localhost:8080/ch8"访问首页面，如图8.2所示；然后单击"去注册"超链接打开注册页面，如图8.3所示。成功注册后，打开登录页面进行用户登录。

图8.2　首页面

图 8.3 注册页面

如果在图 8.3 中输入的用户名不是 admin，那么就注册了一个普通用户，其权限为"ROLE_USER"；如果在图 8.3 中输入的用户名是 admin，那么就注册了一个管理员用户，其权限为"ROLE_ADMIN"和"ROLE_DBA"。

在登录页面中任意输入用户名和密码，单击"登录"按钮，会提示用户名或密码错误，如图 8.4 所示。

图 8.4 用户名或密码错误

在登录页面中输入管理员用户名和密码，成功登录后将打开管理员页面，如图 8.5 所示。

图 8.5 管理员成功登录页面

单击图 8.5 中的"去访问用户登录成功页面"显示拒绝访问页面，如图 8.6 所示。

图 8.6　管理员被拒绝访问页面

在登录页面中输入普通用户名和密码，成功登录后打开用户成功登录页面，如图 8.7 所示。

图 8.7　用户成功登录页面

单击图 8.7 中的"去访问管理员页面"显示拒绝访问页面，如图 8.8 所示。

图 8.8　用户被拒绝访问页面

本章小结

本章首先介绍了 Spring Security 入门知识，然后详细介绍了实际开发中的 Spring Security 操作实例。通过本章的学习，读者应该了解 Spring Security 安全机制的基本原理，掌握如何在实际应用开发中使用 Spring Security 安全机制提供系统安全解决方案。

习题 8

1. 简述 Spring Security 的验证机制。
2. Spring Security 的用户认证和请求授权是如何实现的？请举例说明。

扫一扫

自测题

第 9 章 异步消息

学习目的与要求

本章主要讲解了企业级消息代理 JMS 和 AMQP。通过本章的学习，要求读者理解异步消息通信原理，掌握异步消息通信技术。

本章主要内容

- JMS
- AMQP

当跨越多个微服务进行通信时，异步消息显得至关重要。比如在电子商务系统中，订单服务在下单时需要和库存服务进行通信，完成库存的扣减操作，这时就需要基于异步消息和最终一致性的通信方式来进行这样的操作，并且能够在发生故障时正常工作。

9.1 消息模型

异步消息的主要目的是解决跨系统的通信。所谓异步消息，即消息发送者无须等待消息接收者的处理及返回，甚至无须关心消息是否发送与接收成功。在异步消息中有两个极其重要的概念，即消息代理和目的地。当消息发送者发送消息后，消息将由消息代理管理，消息代理保证消息传递到目的地。

异步消息的目的地主要有两种形式，即队列和主题。队列用于点对点式的消息通信，即端到端通信（单接收者）；主题用于发布/订阅式的消息通信，即广播通信（多接收者）。

▶ 9.1.1 点对点式

在点对点式的消息通信中，消息代理获得发送者发送的消息后，将消息存入一个队列中，当有消息接收者接收消息时，将从队列中取出消息并传递给接收者，这时队列中清除该消息。

在点对点式的消息通信中，确保的是每一条消息只有唯一的发送者和接收者，但并不能说明只有一个接收者可以从队列中接收消息。这是因为队列中有多个消息，点对点式的消息通信只保证每一条消息有唯一的发送者和接收者。

▶ 9.1.2 发布/订阅式

多接收者是消息通信中一种更加灵活的方式，而点对点式的消息通信只保证每一条消息有唯一的接收者，这时可以使用发布/订阅式的消息通信解决多接收者的问题。和点对点式不同，发布/订阅式是消息发送者将消息发送到主题，多个消息接收者监听这个主题。此时的消息发送者叫作发布者，接收者叫作订阅者。

9.2 企业级消息代理

常用的异步消息传递技术有 JMS 和 AMQP。JMS 是面向基于 Java 的企业应用的异步消息代理。AMQP 是面向所有应用的异步消息代理。

9.2.1 JMS

JMS（Java Messaging Service）即 Java 消息服务，是 Java 平台上有关面向消息中间件的技术规范，它便于消息系统中的 Java 应用程序进行消息交换，并且通过提供标准的产生、发送、接收消息的接口简化企业应用的开发。

❶ JMS 元素

JMS 由以下元素组成。

（1）JMS 消息代理实现：连接面向消息中间件的，JMS 消息代理接口的一个实现。JMS 的消息代理实现可以是 Java 平台的 JMS 实现，也可以是非 Java 平台的面向消息中间件的适配器。开源的 JMS 实现由 Apache ActiveMQ Artemis、JBoss 社区所研发的 HornetQ、The OpenJMS Group 的 OpenJMS 等实现。

（2）JMS 客户：生产或消费基于消息的 Java 应用程序或对象。

（3）JMS 生产者：创建并发送消息的 JMS 客户。

（4）JMS 消费者：接收消息的 JMS 客户。

（5）JMS 消息：包括可以在 JMS 客户之间传递的数据对象。JMS 定义了 5 种不同的消息正文格式，以及调用的消息类型，允许用户发送并接收一些不同形式的数据，提供现有消息格式的一些级别的兼容性。常见的消息格式有 StreamMessage（指 Java 原始值的数据流消息）、MapMessage（映射消息）、TextMessage（文本消息）、ObjectMessage（一个序列化的 Java 对象消息）、BytesMessage（字节消息）。

（6）JMS 队列：一个容纳被发送的等待阅读的消息的区域。与队列名字所暗示的意思不同，消息的接收顺序并不一定要与消息的发送顺序相同。一旦一个消息被阅读，该消息将被从队列中移走。

（7）JMS 主题：一种支持发送消息给多个订阅者的机制。

❷ JMS 的应用接口

JMS 的应用接口包括以下接口类型。

（1）ConnectionFactory 接口（连接工厂）：用来创建到 JMS 消息代理实现的连接的被管对象。JMS 客户通过可移植的接口访问连接，这样当下层的实现改变时，代码不需要进行修改。管理员在 JNDI 名字空间中配置连接工厂，这样 JMS 客户才能够查找到它们。根据目的地的不同，用户将使用队列连接工厂或者主题连接工厂。

（2）Connection 接口（连接）：连接代表了应用程序和消息服务器之间的通信链路。在获得连接工厂以后，就可以创建一个与 JMS 消息代理实现（提供者）的连接。根据不同的连接类型，连接允许用户创建会话，以发送和接收队列和主题到目的地。

（3）Destination 接口（目的地）：目的地是一个包装了消息目的地标识符的被管对象，消息目的地是指消息发布和接收的地点，或者是队列，或者是主题。JMS 管理员创建这些对象，然后用户通过 JNDI 发现它们。和连接工厂一样，管理员可以创建两种类型的目的地，即点对点模型的队列以及发布者/订阅者模型的主题。

（4）Session 接口（会话）：表示一个单线程的上下文，用于发送和接收消息。由于会话是单线程的，所以消息是连续的，也就是说消息是按照发送的顺序一个一个接收的。会话的好处是它支持事务。如果用户选择了事务支持，会话上下文将保存一组消息，直到事务被提交才发送这些消息。在提交事务之前，用户可以使用回滚操作取消这些消息。一个会话允许用户创建消息，生产者来发送消息，消费者来接收消息。

（5）MessageConsumer 接口（消息消费者）：由会话创建的对象，用于接收发送到目的地的消息。消费者可以同步地（阻塞模式）或异步地（非阻塞模式）接收队列和主题类型的消息。

（6）MessageProducer 接口（消息生产者）：由会话创建的对象，用于发送消息到目的地。用户可以创建某个目的地的发送者，也可以创建一个通用的发送者，在发送消息时指定目的地。

（7）Message 接口（消息）：在消费者和生产者之间传送的对象，也就是说从一个应用程序传送到另一个应用程序。一个消息有以下 3 个主要部分。

消息头（必需）：包含用于识别和为消息寻找路由的操作设置。

一组消息属性（可选）：包含额外的属性，支持其他消息代理实现和用户的兼容，可以创建定制的字段和过滤器（消息选择器）。

一个消息体（可选）：允许用户创建 5 种类型的消息（文本消息、映射消息、字节消息、流消息和对象消息）。

JMS 各接口角色间的关系如图 9.1 所示。

图 9.1 JMS 各接口角色间的关系

9.2.2 AMQP

AMQP（Advanced Message Queuing Protocol）即高级消息队列协议，它是一个提供统一消息服务的应用层标准高级消息队列协议，是应用层协议的一个开放标准，为面向消息的中间件设计。基于此协议的客户端与消息中间件可传递消息，并且不受客户端/中间件的不同产品、不同开发语言等条件的限制。

AMQP 的技术术语如下。

（1）AMQP 模型（AMQP Model）：一个由关键实体和语义表示的逻辑框架，遵从 AMQP 规范的服务器必须提供这些实体和语义。为了实现本规范中定义的语义，客户端可以发送命令来控制 AMQP 服务器。

（2）连接（Connection）：一个网络连接，比如 TCP/IP 套接字连接。

（3）会话（Session）：端点之间的命名对话。在一个会话上下文中保证"恰好传递一次"。

（4）信道（Channel）：多路复用连接中的一条独立的双向数据流通道，为会话提供物理传输介质。

（5）客户端（Client）：AMQP 连接或者会话的发起者。AMQP 是非对称的，客户端生产和消费消息，服务器存储和路由这些消息。

（6）服务器（Server）：接收客户端连接，实现 AMQP 消息队列和路由功能的进程，也称为"消息代理"。

（7）端点（Peer）：AMQP 对话的任意一方。一个 AMQP 连接包括两个端点（一个是客户端，另一个是服务器）。

（8）搭档（Partner）：当描述两个端点之间的交互过程时，使用术语"搭档"来表示"另一个"端点的简记法。比如定义端点 A 和端点 B，当它们进行通信时，端点 B 是端点 A 的搭档，端点 A 是端点 B 的搭档。

（9）片段集（Assembly）：段的有序集合，形成一个逻辑工作单元。

（10）段（Segment）：帧的有序集合，形成片段集中的一个完整子单元。

（11）帧（Frame）：AMQP 传输的一个原子单元。一个帧是一个段中的任意分片。

（12）控制（Control）：单向指令，AMQP 规范假设这些指令的传输是不可靠的。

（13）命令（Command）：需要确认的指令，AMQP 规范规定这些指令的传输是可靠的。

（14）异常（Exception）：在执行一个或者多个命令时可能发生的错误状态。

（15）类（Class）：一批用来描述某种特定功能的 AMQP 命令或者控制。

（16）消息头（Header）：描述消息数据属性的一种特殊段。

（17）消息体（Body）：包含应用程序数据的一种特殊段。消息体对于服务器来说完全透明，服务器不能查看或者修改消息体。

（18）消息内容（Content）：包含在消息体中的消息数据。

（19）交换器（Exchange）：服务器中的实体，用来接收生产者发送的消息并将这些消息路由给服务器中的队列。

（20）交换器类型（Exchange Type）：基于不同路由语义的交换器类。

（21）消息队列（Message Queue）：一个命名实体，用来保存消息直到发送给消费者。

（22）绑定器（Binding）：消息队列和交换器之间的关联。

（23）绑定器关键字（Binding Key）：绑定的名称。一些交换器类型可能使用这个名称作为定义绑定器路由行为的模式。

（24）路由关键字（Routing Key）：一个消息头，交换器可以用这个消息头决定如何路由某条消息。

（25）持久存储（Durable）：一种服务器资源，当服务器重启时，保存的消息数据不会丢失。

（26）临时存储（Transient）：一种服务器资源，当服务器重启时，保存的消息数据会丢失。

（27）持久化（Persistent）：服务器将消息保存在可靠的磁盘中，当服务器重启时，消息不会丢失。

（28）非持久化（Non-Persistent）：服务器将消息保存在内存中，当服务器重启时，消息可能丢失。

（29）消费者（Consumer）：一个从消息队列中请求消息的客户端应用程序。

（30）生产者（Producer）：一个向交换器发布消息的客户端应用程序。

（31）虚拟主机（Virtual Host）：一批交换器、消息队列和相关对象。虚拟主机是共享相同的身份认证和加密环境的独立服务器域。客户端应用程序在登录到服务器之后可以选择一个虚拟主机。

9.3 Spring Boot 的支持

9.3.1 JMS 的自动配置

Spring Boot 对 JMS 的自动配置位于 org.springframework.boot.autoconfigure.jms 包下，在 Spring Boot 3.0 中支持 JMS 的实现有 ActiveMQ Artemis（下一代 ActiveMQ）。Spring Boot 为用户定义了 ArtemisConnectionFactoryFactory 的 Bean 作为连接，并通过以 "spring.artemis" 为前缀的属性配置 ActiveMQ Artemis 的连接属性，主要包含以下内容：

```
spring.artemis.broker-url= tcp://localhost:61616    #消息代理路径
spring.artemis.user=
spring.artemis.password=
spring.artemis.mode=
```

另外，Spring Boot 在 JmsAutoConfiguration 自动配置类中为用户配置了 JmsTemplate，并且在 JmsAnnotationDrivenConfiguration 配置类中为用户开启了注解式消息监听的支持，即自动开启 @EnableJms。

9.3.2 AMQP 的自动配置

Spring Boot 对 AMQP 的自动配置位于 org.springframework.boot.autoconfigure.amqp 包下，RabbitMQ 是 AMQP 的主要实现。在 RabbitAutoConfiguration 自动配置类中为用户配置了连接的 RabbitConnectionFactoryBean 和 RabbitTemplate，并且在 RabbitAnnotationDrivenConfiguration 配置类中开启了 @EnableRabbit。从 RabbitProperties 类中可以看出 RabbitMQ 的配置能够通过以 "spring.rabbitmq" 为前缀的属性进行配置，主要包含以下内容：

```
spring.rabbitmq.host=localhost          #RabbitMQ 服务器地址，默认为 localhost
spring.rabbitmq.port=5672               #RabbitMQ 端口，默认为 5672
spring.rabbitmq.username=guest          #默认用户名
spring.rabbitmq.password =guest         #默认密码
```

9.4 异步消息通信实例

本节通过两个实例讲解异步消息通信的实现过程。

9.4.1 JMS 实例

下面使用 JMS 的一种实现 ActiveMQ Artemis 讲解 JMS 实例，因此需要事先安装 ActiveMQ Artemis（注意需要安装 Java 11+）。读者可以访问 "http://activemq.apache.org/" 下

载适合自己的 ActiveMQ Artemis。在编写本书时，作者下载了 apache-artemis-2.27.1-bin.zip。对于该版本的 ActiveMQ Artemis，解压缩即可完成安装。

在解压缩后需要创建一个代理服务器，这里以 Windows 10 为例，具体步骤如下：

（1）以管理员身份运行 cmd，进入 ActiveMQ Artemis 解压缩后的 bin 目录，例如执行"D:\>cd D:\soft\Java EE\apache-artemis-2.27.1\bin"。

（2）执行"artemis.cmd create 代理所在目录"命令，例如执行"D:\soft\Java EE\apache-artemis-2.27.1\bin>artemis.cmd create D:\soft\ActiveMQArtemis"。

（3）输入默认用户名和密码（admin），并输入 y 允许匿名访问。在代理服务器创建成功后，cmd 信息显示如图 9.2 所示。

图 9.2 创建 ActiveMQ Artemis 的代理服务器

在代理服务器创建成功后，执行 "D:\soft\ActiveMQArtemis\bin\artemis" run 命令启动 Apache ActiveMQ Artemis 服务，如图 9.3 所示。

图 9.3 启动 Apache ActiveMQ Artemis 服务

然后通过"http://localhost:8161"运行 ActiveMQ Artemis 的管理界面，管理员账号和密码默认为 admin、admin，如图 9.4 所示。

图 9.4　ActiveMQ Artemis 的管理界面

在启动 Apache ActiveMQ Artemis 服务后，下面通过实例讲解如何使用 JMS 的实现 ActiveMQ Artemis 进行两个应用系统间的点对点式通信。

【例 9-1】使用 JMS 的实现 ActiveMQ Artemis 进行两个应用系统间的点对点式通信。

其具体实现步骤如下：

❶ 创建基于 Spring for Apache ActiveMQ Artemis 的 Spring Boot 应用 ch9_1Sender（消息发送者）

创建基于 Spring for Apache ActiveMQ Artemis 的 Spring Boot 应用 ch9_1Sender，该应用作为消息发送者。

❷ 配置 ActiveMQ Artemis 的消息代理地址

在应用 ch9_1Sender 的配置文件 application.properties 中配置 ActiveMQ Artemis 的消息代理地址，具体如下：

```
spring.artemis.broker-url= tcp://localhost:61616
```

❸ 定义消息

在应用 ch9_1Sender 的 com.ch.ch9_1sender 包下创建消息定义类 MyMessage，该类需要实现 MessageCreator 接口，并重写接口方法 createMessage 进行消息定义，具体代码如下：

```java
package com.ch.ch9_1sender;
import java.util.ArrayList;
import jakarta.jms.JMSException;
import jakarta.jms.MapMessage;
import jakarta.jms.Message;
import jakarta.jms.Session;
import org.springframework.jms.core.MessageCreator;
public class MyMessage implements MessageCreator{
    @Override
    public Message createMessage(Session session) throws JMSException {
        MapMessage mapm = session.createMapMessage();
        ArrayList<String> arrayList = new ArrayList<String>();
        arrayList.add("陈恒 1");
        arrayList.add("陈恒 2");
        mapm.setString("mesg1", arrayList.toString());   //只能存储 Java 的基本对象
        mapm.setString("mesg2", "测试消息 2");
```

```
        return mapm;
    }
}
```

❹ 发送消息

在应用 ch9_1Sender 的主类 Ch91SenderApplication 中实现 Spring Boot 的 CommandLineRunner 接口,并重写 run 方法,用于程序启动后执行的代码。在该 run 方法中,使用 JmsTemplate 的 send 方法向目的地 mydestination 发送 MyMessage 的消息,也相当于在消息代理上定义了一个目的地叫 mydestination。其具体代码如下:

```
package com.ch.ch9_1sender;
import org.springframework.beans.factory.annotation.Autowired;
import org.springframework.boot.CommandLineRunner;
import org.springframework.boot.SpringApplication;
import org.springframework.boot.autoconfigure.SpringBootApplication;
import org.springframework.jms.core.JmsTemplate;
@SpringBootApplication
public class Ch91SenderApplication implements CommandLineRunner{
    @Autowired
    private JmsTemplate jmsTemplate;
    public static void main(String[] args) {
        SpringApplication.run(Ch91SenderApplication.class, args);
    }
    /**
    *这里为了方便使用 run 方法发送消息,
    *完全可以使用控制器通过 Web 访问
    */
    @Override
    public void run(String... args) throws Exception {
        //new MyMessage()回调接口方法 createMessage 产生消息
        jmsTemplate.send("mydestination", new MyMessage());
    }
}
```

❺ 创建消息接收者

按照步骤 1 创建 Spring Boot 应用 ch9_1Receive,该应用作为消息接收者,并按照步骤 2 配置 ch9_1Receive 的 ActiveMQ Artemis 的消息代理地址。

❻ 定义消息监听器接收消息

在应用 ch9_1Receive 的 com.ch.ch9_1receive 包中创建消息监听器类 ReceiverMsg。在该类中使用@JmsListener 注解不停地监听目的地 mydestination 是否有消息发送过来,如果有就获取消息。其具体代码如下:

```
package com.ch.ch9_1receive;
import java.util.ArrayList;
import jakarta.jms.JMSException;
import jakarta.jms.MapMessage;
import org.springframework.jms.annotation.JmsListener;
import org.springframework.stereotype.Component;
@Component
public class ReceiverMsg {
    @JmsListener(destination="mydestination")
    public void receiverMessage(MapMessage mapm) throws JMSException {
        System.out.println(mapm.getString("mesg1"));
        System.out.println(mapm.getString("mesg2"));
    }
}
```

❼ 测试运行

首先启动消息接收者 ch9_1Receive 应用，然后单击图 9.4 中的 JMX，可以看到如图 9.5 所示的界面。

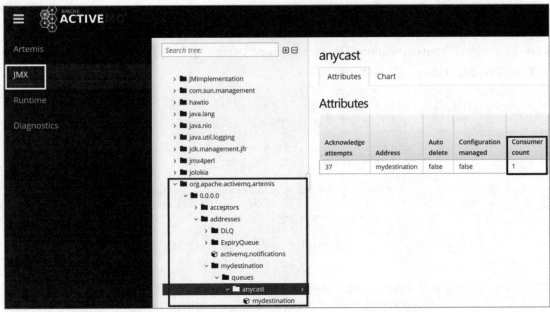

图 9.5　ActiveMQ Artemis 的 Queues

从图 9.5 可以看出目的地 mydestination 有一个消费者，正在等待接收消息。启动消息发送者 ch9_1Sender 应用后，可在接收者 ch9_1Receive 应用的控制台上看到有消息打印，如图 9.6 所示。此时刷新图 9.5，可以看到如图 9.7 所示的界面。

图 9.6　消息接收者接收的消息

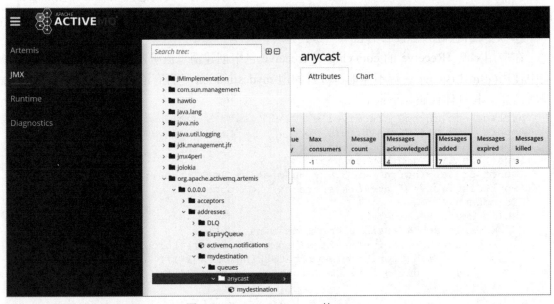

图 9.7　ActiveMQ Artemis 的 Queues

第 9 章 异步消息

从图 9.7 可以看出目的地 mydestination 的 Messages added 有数据增加（表示发送成功），同时 Messages acknowledged 有数据增加（表示接收成功）。

▶ 9.4.2 AMQP 实例

本小节使用 AMQP 的主要实现 RabbitMQ 讲解 AMQP 实例，因此需要事先安装 RabbitMQ。又因为 RabbitMQ 是基于 Erlang 语言开发的，所以在安装 RabbitMQ 之前先下载、安装 Erlang。Erlang 的下载地址为"https://www.erlang.org/downloads"，RabbitMQ 的下载地址为"https://www.rabbitmq.com/download.html"。在编写本书时，作者下载的 Erlang 版本是"otp_win64_25.2.2.exe"，下载的 RabbitMQ 版本是"rabbitmq-server-3.11.8.exe"。

运行 Erlang 语言安装包"otp_win64_25.2.2.exe"，按照提示完成安装。在安装 Erlang 后需要配置环境变量 ERLANG_HOME 以及在 path 中新增%ERLANG_HOME%\bin，如图 9.8 和图 9.9 所示。

图 9.8 ERLANG_HOME

图 9.9 在 path 中新增%ERLANG_HOME%\bin

运行 RabbitMQ 安装包"rabbitmq-server-3.11.8.exe"，按照提示完成安装。在安装 RabbitMQ 后需要配置环境变量 RABBITMQ_SERVER 为安装目录\RabbitMQ Server\rabbitmq_server-3.11.8 以及在 path 中新增%RABBITMQ_SERVER%\sbin，操作界面与图 9.8 和图 9.9 类似。

在 cmd 命令行窗口中进入 RabbitMQ 的 sbin 目录下，运行 rabbitmq-plugins.bat enable rabbitmq_management 命令，打开 RabbitMQ 的管理组件，如图 9.10 所示。

图 9.10 打开 RabbitMQ 的管理组件

以管理员方式打开 cmd 命令，运行 net start RabbitMQ 命令，提示 RabbitMQ 服务已经启动，如图 9.11 所示。

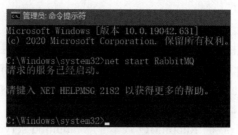

图 9.11　启动 RabbitMQ 服务

在浏览器的地址栏中输入 "http://localhost:15672"，账号和密码默认为 guest、guest，进入 RabbitMQ 的管理界面（如果进不去界面，请重启计算机服务中的 RabbitMQ），如图 9.12 所示。

图 9.12　RabbitMQ 的管理界面

至此完成了 RabbitMQ 服务器的搭建。

在例 9-1 中，不管是消息发送者（生产者）还是消息接收者（消费者）都必须知道一个指定的目的地（队列）才能发送、获取消息。如果是同一个消息，要求每个消费者都处理，则需要发布/订阅式的消息分发模式。

下面通过实例讲解如何使用 RabbitMQ 实现发布/订阅式异步消息通信。在本例中创建一个发布者应用、两个订阅者应用。该实例中的 3 个应用都是使用 Spring Boot 默认为用户配置的 RabbitMQ，主机为 localhost、端口号为 5672，所以无须在配置文件中配置 RabbitMQ 的连接信息。另外，3 个应用需要使用 Weather 实体类封装消息，并且使用 JSON 数据格式发布和订阅消息。

第9章 异步消息

【例9-2】使用 RabbitMQ 实现发布/订阅式异步消息通信（天气预报的发布与接收）。
其具体实现步骤如下：

❶ 创建发布者应用 ch9_2Sender

创建发布者应用 ch9_2Sender 有以下几个步骤。

（1）创建基于 Lombok、Spring for RabbitMQ 的 Spring Boot 应用 ch9_2Sender。

（2）在 ch9_2Sender 应用的 pom.xml 中添加 spring-boot-starter-json 依赖，代码如下：

```xml
<dependency>
    <groupId>org.springframework.boot</groupId>
    <artifactId>spring-boot-starter-json</artifactId>
</dependency>
```

（3）在 ch9_2Sender 应用中创建名为 com.ch.ch9_2sender.entity 的包，并在该包中创建 Weather 实体类，具体代码如下：

```java
package com.ch.ch9_2sender.entity;
import lombok.Data;
import java.io.Serializable;
@Data
public class Weather implements Serializable{
    private static final long serialVersionUID = -8221467966772683998L;
    private String id;
    private String city;
    private String weatherDetail;
    @Override
    public String toString() {
        return "Weather [id=" + id + ", city=" + city + ", weatherDetail=" + weatherDetail + "]";
    }
}
```

（4）在应用 ch9_2Sender 的主类 Ch92SenderApplication 中实现 Spring Boot 的 CommandLineRunner 接口，并重写 run 方法，用于程序启动后执行的代码。在该 run 方法中，使用 RabbitTemplate 的 convertAndSend 方法将 Weather 消息发达到指定交换机 "weather-exchange" 的指定路由 "weather.message"。在发布消息前需要使用 ObjectMapper 将 Weather 对象转换成 byte[]类型的 JSON 数据。其具体代码如下：

```java
package com.ch.ch9_2sender;
import org.springframework.amqp.core.Message;
import org.springframework.amqp.core.MessageBuilder;
import org.springframework.amqp.core.MessageDeliveryMode;
import org.springframework.amqp.rabbit.connection.CorrelationData;
import org.springframework.amqp.rabbit.core.RabbitTemplate;
import org.springframework.amqp.support.converter.Jackson2JsonMessageConverter;
import org.springframework.beans.factory.annotation.Autowired;
import org.springframework.boot.CommandLineRunner;
import org.springframework.boot.SpringApplication;
import org.springframework.boot.autoconfigure.SpringBootApplication;
import com.ch.ch9_2sender.entity.Weather;
import com.fasterxml.jackson.databind.ObjectMapper;
@SpringBootApplication
public class Ch92SenderApplication implements CommandLineRunner{
    @Autowired
    private ObjectMapper objectMapper;
    @Autowired
    RabbitTemplate rabbitTemplate;
    public static void main(String[] args) {
```

```
            SpringApplication.run(Ch92SenderApplication.class, args);
    }
    /**
     * 定义发布者
     */
    @Override
    public void run(String... args) throws Exception {
        //定义消息对象
        Weather weather = new Weather();
        weather.setId("010");
        weather.setCity("北京");
        weather.setWeatherDetail("今天晴到多云,南风5~6级,温度19~26℃");
        //指定 JSON 转换器,Jackson2JsonMessageConverter 默认将消息转换成 byte[]
类型的消息
        rabbitTemplate.setMessageConverter(new Jackson2JsonMessageConverter());
        //objectMapper 将 Weather 对象转换为 JSON 字节数组
        Message msg=MessageBuilder.withBody(objectMapper.writeValueAsBytes(weather))
                .setDeliveryMode(MessageDeliveryMode.NON_PERSISTENT)
                .build();
        //消息的唯一 ID
        CorrelationData correlationData = new CorrelationData(weather.getId());
        //发送消息到指定的交换机
        rabbitTemplate.convertAndSend(
                "weather-exchange",      //分发消息的交换机名称
                "weather.message",       //用来匹配消息的路由 Key
                msg,                     //消息体
                correlationData);
    }
}
```

❷ 创建订阅者应用 ch9_2Receiver-1

创建订阅者应用 ch9_2Receiver-1 有以下几个步骤。

(1) 创建基于 Lombok、Spring for RabbitMQ 的 Spring Boot 应用 ch9_2Receiver-1。

(2) 在 ch9_2Receiver-1 应用的 pom.xml 中添加 spring-boot-starter-json 依赖。

(3) 将 ch9_2Sender 中的 Weather 实体类复制到 com.ch.ch9_2Receiver1 包中。

(4) 在 com.ch.ch9_2receiver1 包中创建订阅者类 Receiver1,在该类中使用@RabbitListener 和@RabbitHandler 注解监听发布者并接收消息,具体代码如下:

```
package com.ch.ch9_2receiver1;
import com.ch.ch9_2receiver1.entity.Weather;
import org.springframework.amqp.rabbit.annotation.Exchange;
import org.springframework.amqp.rabbit.annotation.Queue;
import org.springframework.amqp.rabbit.annotation.QueueBinding;
import org.springframework.amqp.rabbit.annotation.RabbitHandler;
import org.springframework.amqp.rabbit.annotation.RabbitListener;
import org.springframework.beans.factory.annotation.Autowired;
import org.springframework.messaging.handler.annotation.Payload;
import org.springframework.stereotype.Component;
import com.fasterxml.jackson.databind.ObjectMapper;
/**
 * 定义订阅者 Receiver1
 */
@Component
public class Receiver1 {
    @Autowired
    private ObjectMapper objectMapper;
```

```
        @RabbitListener(
                bindings =
                @QueueBinding(
                        //队列名 weather-queue1 保证和其他订阅者不一样，可以随机起名
                        value = @Queue(value = "weather-queue1",durable = "true"),
                        //weather-exchange 与发布者的交换机名相同
                        exchange = @Exchange(value = "weather-exchange",durable = "true",type = "topic"),
                        //weather.message 与发布者的消息的路由 Key 相同
                        key = "weather.message"
                )
        )
        @RabbitHandler
        public void receiveWeather(@Payload byte[] weatherMessage)throws Exception{
            System.out.println("----------订阅者 Receiver1 接收到消息--------");
            //将 JSON 字节数组转换为 Weather 对象
            Weather w=objectMapper.readValue(weatherMessage, Weather.class);
            System.out.println("Receiver1 收到的消息内容："+w);
        }
}
```

❸ 创建订阅者应用 ch9_2Receiver-2

与创建订阅者应用 ch9_2Receiver-1 的步骤一样，这里不再赘述，需要注意的是两个订阅者的队列名不同。

❹ 测试运行

首先运行发布者应用 ch9_2Sender 的主类 Ch92SenderApplication。

然后运行订阅者应用 ch9_2Receiver-1 的主类 Ch92Receiver1Application，此时接收到的消息如图 9.13 所示。

图 9.13 订阅者 ch9_2Receiver-1 接收到的消息

最后运行订阅者应用 ch9_2Receiver-2 的主类 Ch92Receiver2Application，此时接收到的消息如图 9.14 所示。现在再看图 9.12 中的 Connections 为 3。

图 9.14 订阅者 ch9_2Receiver-2 接收到的消息

从例 9-2 可以看出，一个发布者发布的消息可以被多个订阅者订阅，这就是所谓的发布/订阅式异步消息通信。

本章小结

本章主要介绍了多个应用系统间的异步消息。通过本章的学习,读者应该了解 Spring Boot 对 JMS 和 AMQP 的支持,掌握如何在实际应用开发中使用 JMS 或 AMQP 提供异步通信解决方案。

扫一扫

自测题

习题 9

1. 在多个应用系统间的异步消息中有哪些消息模型?
2. JMS 和 AMQP 有什么区别?

第 10 章　Spring Boot 单元测试

学习目的与要求

本章重点讲解 Spring Boot 单元测试的相关内容，包括 JUnit 5 的注解、断言以及单元测试用例。通过本章的学习，要求读者掌握 JUnit 5 的注解与断言机制的用法，掌握单元测试用例的编写。

本章主要内容

- ❖ JUnit 5 注解
- ❖ JUnit 5 断言
- ❖ 单元测试用例
- ❖ 使用 Postman 测试 Controller 层

单元测试（Unit Testing）是指对软件中的最小可测试单元进行检查和验证，是对开发人员所编写的代码进行测试。本章将重点讲解 JUnit 单元测试框架的应用过程。

10.1　JUnit 5

10.1.1　JUnit 5 简介

JUnit 是一个基于 Java 语言的单元测试框架，是由 Erich Gamma 和 Kent Beck 编写的一个回归测试框架（Regression Testing Framework）。JUnit 测试是程序员测试，即所谓的白盒测试，因为程序员知道被测试的软件如何（How）完成功能和完成什么样（What）的功能。多数 Java 开发环境（例如 Eclipse、IntelliJ IDEA）都已经集成了 JUnit 作为单元测试工具。

JUnit 5 由 JUnit Platform、JUnit Jupiter 以及 JUnit Vintage 三部分组成，Java 运行环境的最低版本是 Java 8。

JUnit Platform：JUnit 提供的平台功能模块，通过 JUnit Platform，其他的测试引擎都可以接入 JUnit 实现接口和执行。

JUnit Jupiter：JUnit 5 的核心，是一个基于 JUnit Platform 的引擎实现，JUnit Jupiter 包含许多丰富的新特性，使得自动化测试更加方便和强大。

JUnit Vintage：兼容 JUnit 3、JUnit 4 版本的测试引擎，使得旧版本的自动化测试也可以在 JUnit 5 下正常运行。

JUnit 5 利用了 Java 8 或更高版本的特性，例如 lambda 函数，使测试更强大，更容易维护。JUnit 5 可以同时使用多个扩展，能够轻松地将 Spring 扩展与其他扩展（例如自定义扩展）结合起来，包容性强，可以接入其他的测试引擎。JUnit 的功能更强大，提供了新的断言机制、参数化测试、重复性测试等功能。

10.1.2　JUnit 5 注解

在单元测试中，JUnit 5 有以下应用在方法上的常用注解。

❶ @Test

@Test 注解表示方法是单元测试方法（返回值都是 void），但是与 JUnit 4 的@Test 不同，它的职责非常单一，不能声明任何属性，拓展的测试将会由 Jupiter 提供额外测试。示例代码如下：

```
@Test
void testSelectAllUser() {}
```

❷ @RepeatedTest

@RepeatedTest 注解表示单元测试方法可重复执行，示例代码如下：

```
@Test
@RepeatedTest(value = 5)
void firstTest() {      //该测试方法重复执行 5 次
    System.out.println(55555);
}
```

❸ @DisplayName

@DisplayName 注解为单元测试方法设置展示名称（默认为方法名），示例代码如下：

```
@Test
@DisplayName("测试用户名查询方法")
void findByUname() {}
```

❹ @BeforeEach

@BeforeEach 注解表示在每个单元测试方法之前执行，示例代码如下：

```
@BeforeEach
void setUp() {}
```

❺ @AfterEach

@AfterEach 注解表示在每个单元测试方法之后执行，示例代码如下：

```
@AfterEach
void tearDown() {}
```

❻ @BeforeAll

@BeforeAll 注解表示在所有单元测试方法之前执行。被@BeforeAll 注解的方法必须为静态方法，该静态方法将在当前测试类的所有@Test 方法之前执行一次。示例代码如下：

```
@BeforeAll
static void superBefore(){
    System.out.println("最前面执行");
}
```

❼ @AfterAll

@AfterAll 注解表示在所有单元测试方法之后执行。被@AfterAll 注解的方法必须为静态方法，该静态方法将在当前测试类的所有@Test 方法之后执行一次。示例代码如下：

```
@AfterAll
static void superAfter(){
    System.out.println("最后面执行");
}
```

❽ @Disabled

@Disabled 注解表示单元测试方法不执行，类似于 JUnit 4 中的@Ignore。

❾ @Timeout

@Timeout 注解表示单元测试方法运行时，如果超过了指定时间将会返回错误。示例代码如下：

```
@Test
@Timeout(value = 500, unit = TimeUnit.MILLISECONDS)
void testTimeout() throws InterruptedException {
    Thread.sleep(600);
}
```

▶ 10.1.3　JUnit 5 断言

断言，简单地理解就是用来判断的语句，判断待测试代码的结果和期望的结果是否一致。如果不一致，则说明单元测试失败。JUnit 5 的断言方法都是 org.junit.jupiter.api.Assertions 的静态方法（返回值为 void）。JUnit 5 有以下常用的断言方法。

❶ assertEquals 和 assertNotEquals

Assertions.assertEquals(Object expected, Object actual, String message)方法的第一个参数是期望值，第二个参数是待测试方法的实际返回值，第三个参数 message 是可选的，表示判断失败的提示信息。其功能是判断两者的值是否相等，换而言之不判断类型是否相等。示例代码如下：

```
@Test
void myTest(){
    int a = 1;
    long b = 1L;
    //虽然a和b的类型不同，但判断依旧是成功的，当a和b不相等时测试不通过
    assertEquals(a, b, "a 与 b 不相等");
    MyUser au = new MyUser();
    MyUser bu = new MyUser();
    //虽然au和bu指向不同的对象，但它们的值相同，判断依旧是成功的
    assertEquals(au, bu, "au 与 bu 的对象属性值不相等");
    bu.setUname("ch");
    assertEquals(au, bu, "au 与 bu 的对象属性值不相等");
}
```

❷ assertSame 和 assertNotSame

assertSame 和 assertEquals 的区别是，assertSame 不仅判断值是否相同，还判断类型是否相同。对于对象，判断两者的引用是否为同一个。示例代码如下：

```
@Test
void yourTest(){
    int a = 1;
    long b = 1L;
    long c = 1L;
    //b 和 c 比较，判断成功，因为它们的类型也相同
    assertSame(b, c, "测试失败");
    //a 和 b 比较，判断失败，因为它们的类型不相同
    assertSame(a, b, "测试失败");
    MyUser au = new MyUser();
    MyUser bu = new MyUser();
    MyUser cu = bu;
    //bu 和 cu 比较，判断成功，因为它们的引用是同一个
    assertSame(bu, cu, "测试失败");
    //au 和 bu 比较，判断失败，因为它们的引用不相同
    assertSame(au, bu, "测试失败");
}
```

❸ assertNull 和 assertNotNull

Assertions.assertNull(Object actual)的实际测试值是 null，则单元测试成功。

❹ assertTrue 和 assertFalse

Assertions.assertTrue(boolean condition)的实际测试值是 true，则单元测试成功。

❺ assertThrows

Assertions.assertThrows(Class<T> expectedType, Executable executable, String message) 判断 executable 方法在执行过程中是否抛出指定异常 expectedType。如果没有抛出异常，或者抛出的异常类型不对，则单元测试失败。示例代码如下：

```
@Test
void testAssertThrows() {
    assertThrows(ArithmeticException.class, () -> errorMethod());
}
private void errorMethod() {
    int a[] = {1,2,3,4,5};
    for(int i = 0; i <= 5; i++){
        System.out.println(a[i]);
    }
}
```

❻ assertDoesNotThrow

assertDoesNotThrow(Executable executable)判断测试方法是否抛出异常，如果没有抛出任何异常，则单元测试成功。示例代码如下：

```
@Test
void testAssertDoesNotThrow() {
    assertDoesNotThrow(() -> rightMethod());
}
private void rightMethod() {
    int a = 1/1;
}
```

❼ assertAll

assertAll(Executable... executables)判断一组断言是否都成功，如果都成功，则整个单元测试成功。示例代码如下：

```
@Test
void testAll(){
    assertAll(
        () -> assertEquals(1, 1),
        () -> assertNotEquals(1, 2),
        () -> assertNull(null)
    );
}
```

扫一扫

视频讲解

10.2 单元测试用例

本节将讲解如何使用 JUnit 5 的注解与断言机制编写单元测试用例。在编写单元测试用例前，首先进行测试环境的构建。

▶ 10.2.1 测试环境的构建

在 Spring Boot Web 应用中已经集成了 JUnit 5 和 JSON 的相关 JAR 包，所以可以直接进行单元测试。为了方便本章后续的单元测试，下面构建一个 Spring Boot Web 应用，具体步骤如下。

❶ 创建 Spring Boot Web 应用

创建基于 Lombok 依赖的 Spring Boot Web 应用 ch10。

❷ 修改 pom.xml 文件

在 pom.xml 文件中添加 MySQL 连接器与 MyBatis-Plus 依赖，由于和例 7-5 的相同，这里不再赘述。

❸ 设置 Web 应用 ch10 的上下文路径及数据源配置信息

在应用 ch10 的 application.properties 文件中配置上下文路径及数据源配置信息，配置内容与例 7-5 的基本相同，这里不再赘述。

❹ 创建实体类

创建名为 com.ch.ch10.entity 的包，并在该包中创建 MyUser 实体类，MyUser 的代码与例 7-5 的相同，这里不再赘述。

❺ 创建数据访问接口

创建名为 com.ch.ch10.mapper 的包，并在该包中创建 UserMapper 接口。UserMapper 接口通过继承 BaseMapper<MyUser>接口对实体类 MyUser 对应的数据表 user 进行 CRUD 操作。UserMapper 接口的代码与例 7-5 的相同，这里不再赘述。

❻ 创建 Service 接口及实现类

创建名为 com.ch.ch10.service 的包，并在该包中创建 UserService 接口及实现类 UserServiceImpl。UserService 接口及实现类 UserServiceImpl 的代码与例 7-5 的相同，这里不再赘述。

❼ 创建控制器类 MyUserController

创建名为 com.ch.ch10.controller 的包，并在该包中创建控制器类 MyUserController。MyUserController 的代码如下：

```java
package com.ch.ch10.controller;
import com.ch.ch10.entity.MyUser;
import com.ch.ch10.mapper.UserMapper;
import com.ch.ch10.service.UserService;
import org.springframework.beans.factory.annotation.Autowired;
import org.springframework.web.bind.annotation.*;
import java.util.List;
@RestController
public class MyUserController {
    @Autowired
    private UserMapper userMapper;
    @Autowired
    private UserService userService;
    @GetMapping("/selectAllUsers")
    public List<MyUser> selectAllUsers(){
        return userMapper.selectList(null);
    }
    @PostMapping("/addAUser")
    public MyUser addAUser(MyUser mu){
        //实体类主键属性使用@TableId注解后，主键自动回填
        int result = userMapper.insert(mu);
        return mu;
    }
```

```
    @PutMapping("/updateAUser")
    public boolean updateAUser(MyUser mu){
        return userService.updateById(mu);
    }
    @DeleteMapping("/deleteAUser")
    public boolean deleteAUser(MyUser mu){
        return userService.removeById(mu);
    }
    @GetMapping("/getOne")
    public MyUser getOne(int id){
        return userService.getById(id);
    }
}
```

❽ 在应用程序的主类中扫描 Mapper 接口

在应用程序的 Ch10Application 主类中使用@MapperScan 注解扫描 MyBatis 的 Mapper 接口，核心代码如下：

```
@SpringBootApplication
@MapperScan(basePackages={"com.ch.ch10.mapper"})
public class Ch10Application {
    public static void main(String[] args) {
        SpringApplication.run(Ch10Application.class, args);
    }
}
```

经过以上步骤完成测试环境的构建。下面使用 JUnit 5 测试框架对应用 ch10 中的 Mapper 接口、Service 层进行单元测试。

▶ 10.2.2 测试 Mapper 接口

在 IntelliJ IDEA 中选中类或接口的名字，按快捷键 Ctrl+Shift+T 创建测试类，此时生成的测试类在 test 文件夹中，测试方法都是 void 方法。Create Test 窗口如图 10.1 所示。

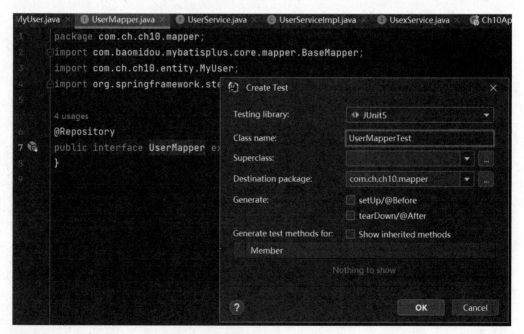

图 10.1 创建 Mapper 接口测试类

第 10 章 Spring Boot 单元测试

单击图 10.1 中的 OK 按钮，完成 Mapper 接口 UserMapper 的单元测试类 UserMapperTest 的创建。

@SpringBootTest 用于 Spring Boot 应用测试，它默认根据包名逐级往上找，一直找到 Spring Boot 主程序（包含@SpringBootApplication 注解的类），并在单元测试时启动该主程序来创建 Spring 上下文环境，所以需要在单元测试类上使用@SpringBootTest 注解标注后才能进行单元测试。

在测试类 UserMapperTest 中使用 JUnit 5 的注解与断言进行 Mapper 接口方法的测试。UserMapperTest 的代码具体如下：

```java
package com.ch.ch10.mapper;
import com.ch.ch10.entity.MyUser;
import org.junit.jupiter.api.Test;
import org.springframework.beans.factory.annotation.Autowired;
import org.springframework.boot.test.context.SpringBootTest;
import static org.junit.jupiter.api.Assertions.*;
@SpringBootTest
class UserMapperTest {
    @Autowired
    private UserMapper userMapper;
    @Test
    void getOne(){
        MyUser mu = userMapper.selectById(1);
        assertEquals(mu.getUid(), 1, "a与b不相等");
    }
}
```

运行上述测试类 UserMapperTest 中的测试方法即可进行 Mapper 接口方法的测试。例如右击 getOne 方法名，选择 Run 'getOne()'运行测试方法 getOne，如果测试成功，如图 10.2 所示。

图 10.2　测试方法 getOne 的运行效果

▶ 10.2.3　测试 Service 层

测试 Service 层与测试 Mapper 接口类似，需要特别考虑 Service 是否依赖其他还未开发完毕的 Service（第三方接口）。如果依赖其他还未开发完毕的 Service，需要使用 Mockito（Java Mock 测试框架，用于模拟任何 Spring 管理的 Bean）来模拟未完成的 Service。

假设应用 ch10 的 UserServiceImpl 类依赖一个还未开发完毕的第三方接口 UsexService，在接口 UsexService 中有一个获得用户性别的接口方法 getUsex，具体代码如下：

```java
package com.ch.ch10.service;
public interface UsexService {
    String getUsex(int id);
}
```

现在创建 UserServiceImpl 的测试类 UserServiceImplTest，在 UserServiceImplTest 类中使用 Mockito.mock 方法模拟第三方接口 UsexService 的对象，并进行测试。UserServiceImplTest 类的代码具体如下：

```java
package com.ch.ch10.service;
import org.junit.jupiter.api.Test;
import org.mockito.BDDMockito;
import org.mockito.Mockito;
import org.springframework.beans.factory.annotation.Autowired;
import org.springframework.boot.test.context.SpringBootTest;
import static org.junit.jupiter.api.Assertions.*;
import static org.mockito.ArgumentMatchers.anyInt;
@SpringBootTest
class UserServiceImplTest {
    @Autowired
    private UserService userService;
    //模拟第三方接口 UsexService 的对象
    private UsexService usexService = Mockito.mock(UsexService.class);
    @Test
    void testGetOne() {
        int uid = 1;
        String expectedUsex = "女";
        /*given 是 BDDMockito 的一个静态方法，用来模拟一个 Service 方法调用返回，
anyInt()表示可以传入任何参数，willReturn 方法说明这个调用将返回"女" */
        BDDMockito.given(usexService.getUsex(anyInt())).willReturn(expectedUsex);
        assertEquals(usexService.getUsex(1), userService.getById(uid).getUsex(),
"测试失败，与期望值不一致");
    }
}
```

运行上述 testGetOne 测试方法，运行效果如图 10.3 所示。

图 10.3 testGetOne 测试方法的运行效果

扫一扫

视频讲解

10.3 使用 Postman 测试 Controller 层

在 Spring 框架中可以使用 org.springframework.test.web.servlet.MockMvc 类进行 Controller 层的测试。MockMvc 类的核心方法如下：

```java
public ResultActions perform(RequestBuilder requestBuilder)
```

通过 perform 方法调用 MockMvcRequestBuilders 的 get、post、multipart 等方法来模拟 Controller 请求，进而测试 Controller 层，模拟请求示例如下。

模拟一个 get 请求：

```java
mvc.peform(get("/getCredit/{id}", uid));
```

模拟一个 post 请求：
```
mvc.peform(post("/getCredit/{id}", uid));
```
模拟文件上传：
```
mvc.peform(multipart("/upload").file("file", "文件内容".getBytes("UTF-8")));
```
模拟请求参数：
```
//模拟提交 errorMessage 参数
mvc.peform(get("/getCredit/{id}/{uname}", uid, uname).param("errorMessage",
"用户名或密码错误"));
//模拟提交 check
mvc.peform(get("/getCredit/{id}/{uname}", uid, uname).param("job", "收银员",
"IT"));
```

综上可知，在通过 MockMvc 类的 perform 方法测试 Controller 时需要编写复杂的单元测试程序。为了提高 Controller 层的测试效率，本节将讲解一个针对 Controller 层测试的接口测试工具 Postman。

Postman 是一个接口测试工具，在做接口测试时 Postman 相当于一个客户端，它可以模拟用户发起的各类 HTTP 请求，将请求数据发送至服务器端，获取对应的响应结果，从而验证响应中的结果数据是否和预期值相匹配。

Postman 主要用来模拟各种 HTTP 请求（例如 get、post、delete、put 等），Postman 与浏览器的区别在于有的浏览器不能输出 JSON 格式，而 Postman 能更直观地显示接口返回的结果。

用户可以从官网"https://www.postman.com/"下载对应的 Postman 安装程序，在安装成功后不需要创建账号即可使用。

下面讲解如何使用 Postman 测试 ch10 应用的 Controller 类 MyUserController。首先将 ch10 应用的主类启动，然后打开 Postman 客户端即可测试 Controller 类中的请求方法。

❶ 测试 selectAllUsers 方法

在 Postman 客户端中输入 Controller 请求方法 selectAllUsers 对应的 URL，并选择对应的请求方式 GET，然后单击 Send 按钮即可完成查询所有用户的测试，如图 10.4 所示。

图 10.4　selectAllUsers 方法的测试结果

❷ 测试 addAUser 方法

在 Postman 客户端中输入 Controller 请求方法 addAUser 对应的 URL 与表单数据，并选择对应的请求方式 POST，然后单击 Send 按钮即可完成添加用户的测试，如图 10.5 所示。

图 10.5　addAUser 方法的测试结果

❸ 测试 updateAUser 方法

在 Postman 客户端中输入 Controller 请求方法 updateAUser 对应的 URL 与表单数据，并选择对应的请求方式 PUT，然后单击 Send 按钮即可完成修改用户的测试，如图 10.6 所示。

图 10.6　updateAUser 方法的测试结果

❹ 测试 deleteAUser 方法

在 Postman 客户端中输入 Controller 请求方法 deleteAUser 对应的 URL 与表单数据，并选择对应的请求方式 DELETE，然后单击 Send 按钮即可完成删除用户的测试，如图 10.7 所示。

❺ 测试 getOne 方法

在 Postman 客户端中输入 Controller 请求方法 getOne 对应的 URL 与表单数据，并选择对应的请求方式 GET，然后单击 Send 按钮即可完成查询一个用户的测试，如图 10.8 所示。

图 10.7 deleteAUser 方法的测试结果

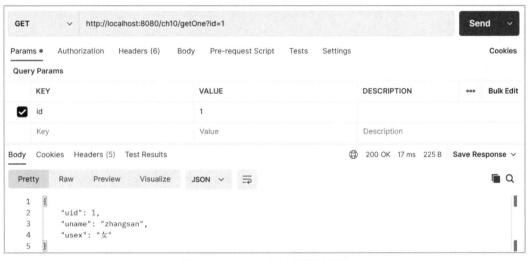

图 10.8 getOne 方法的测试结果

本章小结

本章首先重点讲解了 JUnit 5 的常用注解与断言；然后详细讲解了在 IntelliJ IDEA 中如何使用 JUnit 5 进行 Mapper 接口与 Service 层的单元测试；最后介绍了一个针对 Controller 层测试的接口测试工具 Postman，此测试工具在前后端分离开发中广泛应用于 RESTful 接口测试。

习题 10

扫一扫

自测题

1. 在 JUnit 5 测试框架中，测试方法的返回值是（　　）类型。
   A. boolean　　　　　B. int　　　　　C. void　　　　　D. 以上都不是
2. 在 JUnit 5 测试框架中，使用（　　）注解标注的方法是测试方法。
   A. @Test　　　　　B. @RepeatedTest　　　C. @BeforeEach　　　D. @AfterEach
3. 简述断言方法 assertSame 与 assertEquals 的区别。

第 11 章　Vue.js 基础

学习目的与要求

本章主要讲解 Vue.js 的基础知识，包括 Vue.js 的安装方法、集成开发环境、生命周期、插值与表达式、计算属性、监听器属性、内置指令、组件、自定义指令以及 setup 组件选项等内容。通过本章的学习，要求读者掌握 Vue.js 的基础知识，为学习 Vue.js 的进阶知识做准备。

本章主要内容

- Vue.js 是什么
- 如何安装 Vue.js
- 如何安装 Visual Studio Code 及其插件
- Vue.js 的生命周期
- 插值与表达式
- 计算属性和监听器属性
- 内置指令
- 组件
- 自定义指令
- 响应性
- setup 组件选项

目前，广泛应用的 Web 前端三大主流框架是 Angular.js、React.js 和 Vue.js。

Angular.js 由 Google 公司开发，诞生于 2009 年，之前多用 jQuery 开发，它最大的特点是把后端的一些开发模式移植到前端实现，例如 MVC、依赖注入等。

React.js 由 Facebook 公司开发，正式版推出于 2013 年，比 Angular.js 晚 4 年，采用函数式编程，门槛稍高，但更灵活，开发具有更多可能性。

Vue.js 作为后起之秀（2014 年），借鉴了前辈 Angular.js 和 React.js 的特点，并做了相关优化，使其更加方便，更容易上手，比较适合初学者。

不管学习哪种前端框架，建议大家事先了解 HTML、CSS 和 JavaScript 知识。在学习本章之前，假设读者已经了解有关 HTML、CSS 和 JavaScript 的知识。

11.1　网站交互方式

Web 网有单页应用程序（Single-page Application，SPA）和多页应用程序（Multi-page Application，MPA）两种交互方式。

▶ 11.1.1　多页应用程序

多页应用程序，顾名思义是由多个页面组成的站点。在多页应用程序中，每个网页在每次收到相应的请求时都会重新加载。多页应用程序很大，由于包含了不同页面的数量和层数，有时甚至认为很麻烦，大家可以在大多数电子商务网站上找到 MPA 的示例。

多页应用程序以服务器端为主导，前后端混合开发，例如.php、.aspx、.jsp。技术堆栈包括 HTML、CSS、JavaScript、jQuery，有时还包括 AJAX。

❶ 多页应用程序的优点

多页应用程序的优点如下。

（1）搜索引擎优化效果好：搜索引擎在做网页排名时需要根据网页内容给网页添加权重，进行网页排名。搜索引擎可以识别 HTML 内容，而多页应用程序的每个页面的所有内容都放在 HTML 中，所以排名效果较好。

（2）更容易扩展：添加到现有应用程序的页面数几乎没有限制。如果需要显示很多信息，建议使用 MPA，因为它可以确保将来能够更轻松地扩展。

（3）深入的数据分析：有许多数据分析工具可以为 MPA 提供有关客户行为、系统功能和其他重要内容的深刻解析，可以分析每个功能的性能，每个网页的受欢迎程度，每个功能所花费的时间，每日和每月的用户数量，以及按年龄、城市、国家/地区划分的受众群体等。与 MPA 相比，单页应用程序通常只提供有关访问者数量及其会话持续时间的信息。

❷ 多页应用程序的缺点

多页应用程序的缺点如下。

（1）开发及维护更加困难且昂贵：与单页解决方案相比，多页应用程序具有更多功能，因此创建它们需要更多的精力和资源，开发时间与要构建的页面数和实现的功能成比例地增加。

Web 应用程序开发工具更新快，同时市场上还引入了其他库、框架、编程语言或者至少发布了新版本，而多页应用程序通常需要在开发过程中使用多种技术，因此当维护 Web 系统时变得更加困难且昂贵。

（2）页面切换慢：多页应用程序每次跳转时都会发出一个 HTTP 请求。如果网速较慢，在页面之间来回跳转时将发生明显的卡顿现象。

（3）较低的绩效指标：多页应用程序中的内容会不断重新加载，增加了服务器的负载，对网页速度和整体系统性能产生了负面影响。

▶ 11.1.2 单页应用程序

单页应用程序就是只有一张 Web 页面的应用。单页应用程序是加载单个 HTML 页面并在用户与应用程序交互时动态更新该页面的 Web 应用程序。浏览器一开始会加载必需的 HTML、CSS 和 JavaScript，所有的操作都在这张页面上完成，都由 JavaScript 来控制，因此对单页应用来说模块化的开发和设计显得相当重要。单页应用开发技术复杂，所以诞生了许多前端开发框架，例如 Angular.js、React.js、Vue.js 等。

在选择单页应用程序开发时，软件工程师通常采用 HTML 5、Angular.js、React.js、Vue.js、Ember.js、AJAX 等技术堆栈。

❶ 单页应用程序的优点

单页应用程序的优点如下。

（1）用户体验好：单页应用程序就像一个原生的客户端软件一样使用，在切换过程中不会频繁有被"打断"的感觉。

（2）前后端分离：开发效率高，可维护性强。服务器端不关心页面，只关心数据；客户端不关心数据及数据操作,只关心通过接口得到数据和服务器端交互、处理页面。

(3)局部刷新:只需要加载渲染局部视图即可,不需要整页刷新。

(4)完全的前端组件化:前端开发不再以页面为主,更多采用组件化的思想,代码结构和组织方式更加规范化,便于修改和调整。

(5)API 共享:如果服务是多端的(浏览器端、Andriod、iOS、微信等),单页应用的模式便于在多端共用 API,可以显著地减少服务器端的工作量。

(6)组件共享:在某些对性能体验要求不高的场景下或者产品处于快速试错阶段时,借助于 Hybrid、React Native 等技术在多端共享组件,便于产品的快速迭代,能够节约资源。

❷ 单页应用程序的缺点

单页应用程序的缺点如下。

(1)首次加载大量资源:需要在一个页面上为用户提供产品的所有功能,在加载该页面时首先加载大量的静态资源,加载时间相对较长。

(2)对搜索引擎不友好:单页应用的界面数据绝大部分都是异步加载的,很难被搜索引擎搜索到。

(3)安全问题:单页应用更容易受到所谓的跨站点脚本(XSS)攻击,这意味着黑客可以将各种恶意脚本注入应用程序中,原因是缺乏经验的 Web 开发人员将某些功能和逻辑移至客户端。

11.2 MVVM 模式

MVVM 是 Model-View-ViewModel 的缩写,它是一种基于前端开发的架构模式,其核心是提供对 View 和 ViewModel 的双向数据绑定,这使得 ViewModel 的状态改变可以自动传递给 View,即所谓的数据双向绑定。

MVVM 由 Model、View、ViewModel 三部分构成,Model 层代表数据模型,也可以在 Model 中定义数据修改和操作的业务逻辑;View 代表 UI 组件,它负责将数据模型转化成 UI 展现出来;ViewModel 是一个同步 View 和 Model 的对象。

在 MVVM 架构下,View 和 Model 之间并没有直接的联系,而是通过 ViewModel 进行交互,Model 和 ViewModel 之间的交互是双向的,因此 View 数据的变化会同步到 Model 中,而 Model 数据的变化也会立即反映到 View 上。Model、View、ViewModel 三者之间的关系如图 11.1 所示。

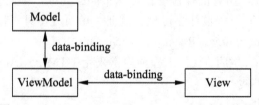

图 11.1　Model、View、ViewModel 三者之间的关系

ViewModel 通过双向数据绑定把 View 层和 Model 层连接了起来,而 View 和 Model 之间的同步工作完全是自动的,无须人为干涉,因此开发者只需要关注业务逻辑,不需要手动操作 DOM,不需要关注数据状态的同步问题,复杂的数据状态维护完全由 MVVM 来统一管理。

11.3 Vue.js 是什么

Vue（读音/vju:/，类似于 View）是一套构建用户界面的渐进式框架。与其他重量级框架不同，Vue.js 是采用自底向上增量开发的设计。Vue.js 的核心库仅关注视图层，它不仅易于上手，还便于和第三方库或既有项目整合。另一方面，Vue.js 采用单文件组件和 Vue.js 生态系统支持的库开发复杂的单页应用。

Vue.js 本身只是一个 JS 库，它的目标是通过尽可能简单的 API 实现响应的数据绑定和组合的视图组件。Vue.js 可以轻松地构建 SPA（Single Page Application）应用程序，通过指令扩展 HTML，通过表达式将数据绑定到 HTML，最大程度地解放 DOM 操作。

Vue.js 具有简单、易用、灵活、高效等特点，读者在具备 HTML、CSS、JavaScript 的基础上可以快速上手。

11.4 安装 Vue.js

将 Vue.js 添加到项目中有 4 种主要方法，即本地独立版本方法、CDN 方法、NPM 方法以及命令行工具（Vue CLI）方法。

❶ 本地独立版本方法

用户可通过网址"https://unpkg.com/vue@next"将最新版本的 Vue.js 库（vue.global.js）下载到本地，然后在界面文件中引入 Vue.js 库，示例代码如下：

```
<script src="js/vue.global.js"></script>
```

❷ CDN 方法

读者在学习或开发时可以在界面文件中通过 CDN（Content Delivery Network，内容分发网络）引入最新版本的 Vue.js 库，示例代码如下：

```
<script src="https://unpkg.com/vue@next"></script>
```

对于生产环境，建议使用固定版本，以免因版本不同带来兼容性问题，示例代码如下：

```
<script src="https://unpkg.com/vue@3.2.36/dist/vue.global.js"></script>
```

❸ NPM 方法

在使用 Vue.js 构建大型应用时推荐使用 NPM 安装最新稳定版的 Vue.js，因为 NPM 能很好地和 webpack 模块打包器配合使用，示例代码如下：

```
npm install vue@next
```

❹ 命令行工具（Vue CLI）方法

Vue.js 提供了一个官方命令行工具（Vue CLI），为单页应用快速搭建繁杂的"脚手架"。对于初学者来说不建议使用 NPM 和 Vue CLI 方法安装 Vue.js。NPM 和 Vue CLI 方法的安装过程将在本书第 12 章 Vue.js 进阶中介绍。

11.5 第一个 Vue.js 程序

前端开发工具极少人使用记事本，大多数程序员使用 JetBrains WebStorm 和 Visual Studio Code（VSCode）。JetBrains WebStorm 是收费的，本书推荐使用 VSCode。

扫一扫

视频讲解

11.5.1 安装 Visual Studio Code 及其插件

用户可通过网址"https://code.visualstudio.com"下载 VSCode，本书使用的安装文件是 VSCodeUserSetup-x64-1.52.1.exe（双击即可安装）。在 VSCode 中许多插件需要用户安装，例如安装 Vue.js 的插件 Vetur。打开 VSCode，单击左侧最下面的图标，按照图 11.2 所示的步骤安装即可。

图 11.2　VSCode 插件的安装

其他插件的安装方式与图 11.2 类似，这里不再赘述。在 VSCode 中，有关 Vue.js 的部分插件描述如下。

（1）Vetur：此插件能够在.vue 文件中实现语法错误检查、语法高亮显示以及代码自动补全。

（2）ESLint：此插件能够检测代码的语法问题与格式问题，对统一项目代码风格至关重要。

（3）EditorConfig for Visual Studio Code：EditorConfig 是一种被各种编辑器广泛支持的配置，使用此配置有助于项目在整个团队中保持一致的代码风格。

（4）Path Intellisense：此插件能够在编辑器中输入路径时实现自动补全。

（5）View In Browser：在 VSCode 中使用浏览器预览运行静态文件。

（6）Live Server：此插件很有用，在安装后可以打开一个简单的服务器，而且还会自动更新。在安装后，在文件上右击，选择 Open with Live Server 命令，自动打开浏览器，默认端口号是 5500。

（7）GitLens：此插件可查看.git 文件提交的历史。

（8）Document This：此插件生成注释文档。

（9）HTML CSS Support：在编写样式表的时候自动补全，缩短编写时间。

（10）JavaScript Snippet Pack：针对 JS 的插件，包含 JS 的常用语法关键字。

（11）HTML Snippets：此插件包含 HTML 标签。

（12）One Monokai Theme：此插件能够让编者选择自己喜欢的颜色主题编写代码。

（13）vscode-icons：此插件能够让编者选择自己喜欢的图标主题。

11.5.2 创建第一个 Vue.js 应用

每个 Vue.js 应用都是从使用 createApp 函数创建一个新实例开始，具体语法如下：

```
const app = Vue.createApp({ /*选项*/ })
```

传递给 createApp 的选项用于配置根组件（渲染的起点）。在 Vue.js 应用创建后，调用 mount 方法将 Vue.js 应用挂载到一个 DOM 元素（HTML 元素或 CSS 选择器）中。例如，如果把一个 Vue.js 应用挂载到<div id="app"></div>上，应传递#app。示例代码如下：

```
const HelloVueApp = {}                          //配置根组件
const vueApp = Vue.createApp(HelloVueApp)       //创建 Vue 实例
const vm = vueApp.mount('#app')                 //将 Vue 实例挂载到#app
```

下面使用 VSCode 开发第一个 Vue.js 程序。

【例 11-1】使用 VSCode 新建一个名为 hellovue.html 的页面，在此页面中使用"<script src="js/vue.global.js"></script>"语句引入 Vue.js。

hellovue.html 的具体代码如下：

```
<div id="hello-vue" class="demo">
    {{message}}
</div>
<script src="js/vue.global.js"></script>
<script>
    const HelloVueApp = {
        data() {            //Vue 实例的数据对象，ES 语法，等价于 data: function() {}
            return {
                message: 'Hello Vue!!'
            }
        }
    }
    //每个 Vue.js 应用都是从使用 createApp 函数创建一个新的应用实例开始
    //mount 函数把一个 Vue.js 应用实例挂载到<div id="hello-vue"></div>上
    Vue.createApp(HelloVueApp).mount('#hello-vue')
</script>
<style>
.demo {
    font-family: sans-serif;
}
</style>
```

从上述代码中可以看出，hellovue.html 文件内容由 HTML、JavaScript 和 CSS 三部分组成，所以读者在学习 Vue.js 之前应掌握 HTML、JavaScript 和 CSS 等内容。

从上述代码中还可以看出，创建一个 Vue.js 应用程序只需要 3 个步骤，即引入 Vue.js 库文件、创建一个 Vue 实例、渲染 Vue 实例。

在 VSCode 中，按照 11.5.1 小节的步骤安装 Live Server 插件后，在 hellovue.html 代码上右击，选择 Open with Live Server 命令，自动打开浏览器，默认端口号是 5500，如图 11.3 所示。

图 11.3　hellovue.html 的运行效果

在 VSCode 中可以设置 Live Server 插件默认打开的浏览器，具体操作步骤如下：首先选择 File→Preferences→Settings 命令，打开 Search settings 对话框；然后单击 User→Extensions→Live Server Config，通过界面右侧的 Custom Browser 修改默认打开的浏览器，如图 11.4 所示。

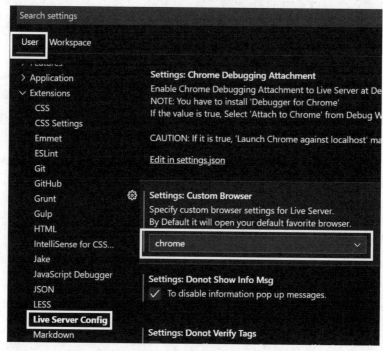

图 11.4　设置 Live Server 插件默认打开的浏览器

▶ 11.5.3　声明式渲染

Vue.js 的核心是采用简洁的模板将数据渲染到 DOM 中，例如在 11.5.2 小节的 hellovue.html 文件中，通过模板 `<div id="hello-vue" class="demo">{{message}}</div>` 声明将属性变量 message 的值 "Hello Vue!!" 渲染到页面显示。

Vue.js 框架在实现声明式渲染时做的主要工作就是将数据和 DOM 建立关联，一切皆响应。例如例 11-2 中的 counter 属性每秒递增。

【例 11-2】使用 VSCode 新建一个名为 ch11_2.html 的页面，在该页面中使用时钟函数 setInterval 来演示响应式程序。

ch11_2.html 的具体代码如下：

```html
<div id="counter" class="demo">
    <!--通过模板获取变量 counter 的值-->
    {{counter}}
</div>
<script src="js/vue.global.js"></script>
<script>
    const CounterApp = {
        data() {
            return {
                counter: 0
            }
        },
```

```
            /*mounted 是一个钩子函数,在挂载到实例上后(初始化页面后)调用该函数,一般是第
一个业务逻辑从这里开始*/
            mounted() {
                //定时器
                setInterval(() => {
                    this.counter++
                }, 1000)
            },
            /*在 Vue.js 实例销毁前调用(离开页面前调用),这是一个常用的生命周期钩子函数,一
般在此时做一些重置操作,比如清除定时器和监听的 DOM 事件*/
            beforeUnmount(){
                //清除定时器
                clearTimeout(this.counter)
            }
        }
        Vue.createApp(CounterApp).mount('#counter')
    </script>
    <style>
        .demo {
            font-family: sans-serif;
        }
    </style>
```

▶ 11.5.4 Vue.js 的生命周期

每个 Vue.js 实例在被创建时都要经过一系列的初始化过程,例如数据监听、编译模板、将实例挂载到 DOM 并在数据变化时更新 DOM 等。同时在这个过程中也会调用一些叫生命周期钩子的函数,在适当的时机执行程序的业务逻辑。

例如,created 钩子函数可以用来在一个 Vue.js 实例被创建后执行代码(Vue.js 实例创建后被立即调用,即 HTML 加载完成前):

```
Vue.createApp({
    data() {
        return {
            message: '测试钩子函数'
        }
    },
    created() {
        //this 指向调用它的 Vue.js 实例
        console.log('message 是: ' + this.message)   // "message 是: 测试钩子函数"
    }
})
```

Vue.js 的生命周期共分 8 个阶段(如图 11.5 所示),对应 8 个与 created 类似的钩子函数。

beforeCreate(创建前):在 Vue.js 实例初始化后,数据观测和事件配置前调用,此时 el 和 data 并未初始化,因此无法访问 methods、data、computed 等上的方法和数据。

created(创建后):Vue.js 实例创建后被立即调用,即 HTML 加载完成前,此时 Vue.js 实例已完成数据观测、属性和方法的运算、watch/event 事件回调、data 数据的初始化,然而挂载阶段还没有开始,el 属性目前不可见。这是一个常用的生命周期钩子函数,可以调用 methods 中的方法、改变 data 中的数据、获取 computed 中的计算属性等,通常在此钩子函数中对实例进行预处理。

beforeMount(载入前):挂载开始前被调用,Vue.js 实例已完成编译模板、把 data 里面的数据和模板生成 HTML、el 和 data 初始化,注意此时还没有挂载 HTML 到页面上。

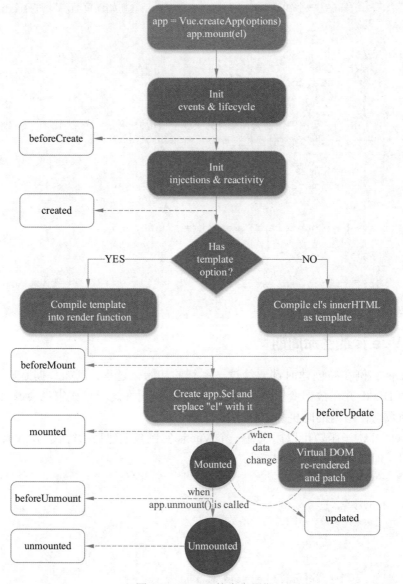

图 11.5　Vue.js 的生命周期

mounted（载入后）：页面加载后调用该函数，这是一个常用的生命周期钩子函数，一般第一个业务逻辑从此钩子开始，mounted 只会执行一次。

beforeUpdate（更新前）：在数据更新前被调用，发生在虚拟 DOM 重新渲染和打补丁之前，可以在该钩子中进一步地更改状态，不会触发附加的重渲染过程。

updated（更新后）：在由数据更改导致虚拟 DOM 重新渲染和打补丁时调用，在调用时DOM 已经更新，所以可以执行依赖于 DOM 的操作，应该避免在此期间更改状态，这可能会导致更新无限循环。

beforeUnmount（销毁前）：Vue.js 实例销毁前调用（离开页面前调用），这是一个常用的生命周期钩子函数，一般在此时做一些重置操作，比如清除定时器和监听的 DOM 事件。

unmounted（销毁后）：在实例销毁后调用，在调用后事件监听器被移出，所有子实例也被销毁。

图 11.5 展示了 Vue.js 实例的生命周期，现在大家不需要弄明白所有阶段的钩子函数，随着不断地学习和使用，大家会慢慢了解它们。

11.6 插值与表达式

Vue 的插值表达式"{{ }}"的作用是读取 Vue.js 中的 data 数据，显示在视图中，数据更新，视图也随之更新。"{{ }}"中只能放表达式（有返回值），不可以放语句，例如{{ var a = 1 }}和{{ if (ok) { return message } }}都是无效的。

▶ 11.6.1 文本插值

数据绑定最常见的形式就是使用"Mustache（小胡子）"语法（双花括号）的文本插值，它将绑定的数据实时显示出来。例如，例 11-2 中的{{ counter }}，无论何时，绑定的 Vue.js 实例的 counter 属性值发生改变，插值处的内容都将更新。

用户可以通过使用 v-once 指令执行一次性插值，即当数据改变时插值处的内容不会更新。示例代码如下：

```
<span v-once>{{ counter }}</span>
```

▶ 11.6.2 原始 HTML 插值

"{{ }}"将数据解释为普通文本，而非 HTML 代码。当需要输出真正的 HTML 代码时，可以使用 v-html 指令。动态渲染任意的 HTML 是非常危险的，因为很容易导致 XSS 攻击，最好只对可信内容使用 HTML 插值，绝不可将用户提供的 HTML 作为插值。v-html 指令的示例如下：

假如 Vue.js 实例的 data 为：

```
data() {
    return {
        rawHtml: '<hr>'
    }
}
```

则"\<p\>无法显示 HTML 元素内容: {{ rawHtml }}\</p\>"显示的结果是\<hr\>；而"\<p\>可正常显示 HTML 元素内容: \<span v-html="rawHtml"\>\</span\>\</p\>"显示的结果是一条水平线。

▶ 11.6.3 JavaScript 表达式

在前面的学习中，仅用表达式绑定简单的属性值。实际上，对于所有的数据绑定，Vue.js 都提供了完全的 JavaScript 表达式支持。示例如下：

```
{{ number + 1 }}
{{ isLogin? 'True' : 'False' }}
{{ message.split('').reverse().join('') }}
```

11.7 计算属性和监听器属性

Vue 模板的插值表达式"{{ }}"用起来非常便利，设计它们的初衷是用于简单运算，但在表达式中放入太多逻辑将会难以维护。计算属性就是用来解决该类问题的。

11.7.1 计算属性 computed

在 Vue 实例的 computed 选项中定义一些属性（可以使用 this 引用），这些属性称作"计算属性"。所有的计算属性都是以方法（函数）的形式定义，但仅当作属性来使用。在一个计算属性中可以完成各种复杂的逻辑，包括运算、方法调用等，但最终必须返回一个结果。计算属性的结果还可以依赖于多个数据，只要其中任一数据发生变化，计算属性都将重新执行，视图也会更新。下面以计算购物车中商品的总价为例讲解计算属性的用法。

【例 11-3】使用计算属性计算购物车中商品的总价。

其具体代码如下：

```html
<div id="myCart">
    <!-- v-model 指令在表单元素上实现双向数据绑定，在后续章节讲解-->
    商品1数量：<input type="text" v-model="num1">
    商品1价格：<input type="text" v-model="price1"><br>
    商品2数量：<input type="text" v-model="num2">
    商品2价格：<input type="text" v-model="price2"><br>
    商品总价：{{ total }}
</div>
<script src="js/vue.global.js"></script>
<script>
    const CounterApp = {
        data() {
            return {
                num1: 1,
                price1: 10,
                num2: 2,
                price2: 20
            }
        },
        computed: {
            total() {
                return this.num1*this.price1 + this.num2*this.price2
            }
        }
    }
    Vue.createApp(CounterApp).mount('#myCart')
</script>
```

商品数量或商品价格只要有一个发生变化，购物车中商品的总价都将自动更新。例 11-3 的运行结果如图 11.6 所示。

图 11.6 计算购物车中商品的总价

Vue 的每一个计算属性都包含一个 getter 方法和一个 setter 方法，在例 11-3 中只是使用了计算属性的默认用法，即仅使用了 getter 方法来读取计算属性。当然，用户也可以使用计算属性的 setter 方法来修改其值。下面讲解计算属性的 setter 方法。

第 11 章　Vue.js 基础

【例 11-4】通过单击按钮重新为计算属性赋值。

其具体代码如下：

```html
<div id="app">
    <!--调用计算属性的getter方法获取计算属性的值-->
    姓名：{{fullName}}<br>
    <!-- v-on 指令给 HTML 元素添加一个事件监听器，在后续章节讲解-->
    <button v-on:click="changeName">修改计算属性</button>
</div>
<script src="js/vue.global.js"></script>
<script>
    const CounterApp = {
        data() {
            return {
                firstName: '陈',
                lastName: '恒'
            }
        },
        methods:{       //方法的定义
            changeName(){
                //计算属性可以使用 this 引用，当计算属性 fullName 的值发生变化时调用 setter 方法
                this.fullName = '张三'
            }
        },
        computed: {
            fullName: {//计算属性
                get(){
                    return this.firstName + this.lastName
                },
                set(newValue){
                    var names = newValue.split(' ')
                    this.firstName = names[0]
                    this.lastName = names[1]
                }
            }
        }
    }
    Vue.createApp(CounterApp).mount('#app')
</script>
```

从例 11-4 的代码可以看出，当单击"修改计算属性"按钮时执行 changeName()方法；在 changeName()方法中执行 "this.fullName = '张三'" 语句时调用计算属性 fullName 的 setter 方法；在 setter 方法中数据 firstName 和 lastName 相继更新，视图同样也会更新。

在大多数情况下使用默认的 getter 方法来读取计算属性的值即可，不必声明 setter，例如例 11-3。

用户可以在表达式中调用方法达到与计算属性同样的效果。下面通过修改例 11-3 的程序演示的 methods 达到与计算属性同样的效果。

【例 11-5】用 methods 改写例 11-3。

修改后的代码如下：

```html
<div id="myCart">
    商品1数量：<input type="text" v-model="num1">
    商品1价格：<input type="text" v-model="price1"><br>
    商品2数量：<input type="text" v-model="num2">
```

```
        商品 2 价格：<input type="text" v-model="price2"><br>
        商品总价：{{ total() }}
    </div>
    <script src="js/vue.global.js"></script>
    <script>
        const CounterApp = {
            data() {
                return {
                    num1: 1,
                    price1: 10,
                    num2: 2,
                    price2: 20
                }
            },
            methods: {
                total() {
                    return this.num1*this.price1 + this.num2*this.price2
                }
            }
        }
        Vue.createApp(CounterApp).mount('#myCart')
    </script>
```

在例 11-5 中没有使用计算属性，在 methods 选项中定义了一个方法实现同样的效果，方法还可以接收参数，使用起来更方便。那么为什么还使用计算属性呢？这是因为计算属性是基于它的依赖缓存的。也就是说，一个计算属性所依赖的数据发生变化时，它才会重新取值，所以只要商品数量和商品价格不发生改变，计算属性 total 也就不更新。下面分别使用计算属性和 methods 显示时间，演示它们的区别。

【例 11-6】分别使用计算属性和 methods 显示时间，每次显示时间后都会弹出暂停警告框。其具体代码如下：

```
    <div id="app">
        时间 1：{{ mytime() }}{{ mystop() }}<br>
        时间 2：{{ mytime() }}{{ mystop() }}<br>
        时间 3：{{ yourtime }}{{ mystop() }}<br>
        时间 4：{{ yourtime }}
    </div>
    <script src="js/vue.global.js"></script>
    <script>
        const CounterApp = {
            methods: {
                mytime() {
                    return Date.now()
                },
                mystop() {
                    alert('暂停一下')
                }
            },
            computed: {
                yourtime(){
                    return Date.now()
                }
            }
        }
        Vue.createApp(CounterApp).mount('#app')
    </script>
```

运行例 11-6 的程序，连续 3 次弹出暂停警告框后显示结果如图 11.7 所示。

从图 11.7 可以看出，使用 methods 显示时间 1 和时间 2 时执行了两次 Date.now()，说明调用了两次 methods；而使用计算属性显示时间 3 和时间 4 时，即使中间弹出了暂停警告框，显示的时间也相同，说明从缓存中获取计算属性（Date.now()不是响应式依赖）。

图 11.7　分别使用计算属性和 methods 显示时间

使用计算属性还是 methods 取决于是否需要缓存，当遍历大数组或做大量计算时应该使用计算属性，从缓存中获取计算结果，提高执行效率。

▶ 11.7.2　监听器属性 watch

虽然计算属性在大多数情况下更合适，但有时也需要一个监听器来响应数据的变化。Vue 通过 watch 选项提供监听数据属性的方法（方法名与属性名相同）来响应数据的变化。当被监视的数据发生变化时触发 watch 中对应的处理方法。下面通过实例讲解 watch 属性的用法。

【例 11-7】使用 watch 属性监视 data 中 question 的变化（watch 中需要提供与 question 同名的方法）。

其具体代码如下：

```html
<div id="watch-example">
    <p>
        请问一个问题，包含英文字符?：
        <input v-model="question"/>
    </p>
    <p>{{ answer }}</p>
</div>
<script src="js/vue.global.js"></script>
<script src="https://cdn.jsdelivr.net/npm/axios@0.12.0/dist/axios.min.js"></script>
<script>
    const watchExampleVM = Vue.createApp({
        data() {
            return {
                question: '',
                answer: '这是一个好问题。'
            }
        },
        watch: {      //watch 选项提供监听数据属性的方法
            //question 方法名与数据属性名 question 一致
            question(newQuestion, oldQuestion) {
                //newQuestion 是改变的值，oldQuestion 是没改变的值
                if (newQuestion.indexOf('?') > -1) {
                    //包含英文字符?时执行 getAnswer()方法
                    this.getAnswer()
                }
            }
        },
        methods: {
            getAnswer() {
                this.answer = '让我想一想'      //设置中间状态，即答案返回前
                axios
                    .get('https://yesno.wtf/api')      //使用axios实现AJAX异步请求
```

```
                        .then(response => {
                            this.answer = response.data.answer
                        })
                        .catch(error => {
                            this.answer = '错误,不能访问 API. ' + error
                        })
                }
            }
        }).mount('#watch-example')
    </script>
```

在例 11-7 运行结果的界面中,当输入没有 "?" 的问题时,watch 选项监视到 question 数据发生变化,但没有 "?",不执行 this.getAnswer()操作,运行结果如图 11.8 所示;当输入带有 "?" 的问题时,运行结果如图 11.9 所示,在显示图 11.9 之前瞬间显示了 answer 的中间状态——"让我想一想"。

图 11.8　输入没有 "?" 的问题

图 11.9　输入带有 "?" 的问题

在例 11-7 的代码中,使用 watch 选项执行异步操作 getAnswer(),限制执行该操作的频率,并在得到最终结果前设置中间状态(this.answer = '让我想一想')。这些都是计算属性无法做到的。

从前面的学习可知:

(1) computed 属性所依赖的数据发生变化时将自动重新计算,并把计算结果缓存起来。另外,computed 属性最后一定有一个返回值,而且不带参数。

(2) watch 属性用来监听某些特定数据的变化,从而进行具体的业务逻辑操作。另外,watch 选项中的方法可以传入被监听属性的新、旧值,通过这两个值可以做一些特定的操作;computed 属性通常是做简单的计算。

那么,应该如何选择 computed 属性和 watch 属性呢?如果一个值依赖多个属性,建议使用 computed 属性;如果一个值发生变化后引起一系列业务逻辑操作或者引起一系列值的变化,建议使用 watch 属性。

11.8　内置指令

指令是 Vue.js 模板中最常用的一项功能,它带有特殊前缀 "v-"。指令的主要职责是当其表达式的值改变时相应地将某些行为应用到 DOM 上。Vue.js 内置了许多指令,可以快速地完成常见的 DOM 操作,例如循环渲染、条件渲染等,本节将学习这些内置指令。

▶ 11.8.1 v-bind 指令

扫一扫

视频讲解

在 HTML 元素的属性中不能使用表达式动态更新属性值。幸运的是，Vue.js 提供了 v-bind 指令绑定 HTML 元素的属性，并可动态更新属性值，例如 id、class 等。下面通过实例讲解 v-bind 指令的用法。

【例 11-8】使用 v-bind 指令绑定超链接的 href 属性、图片的 src 属性以及 class 属性。

其具体代码如下：

```html
<div id="app">
    <a v-bind:href="myurl.baiduUrl">去百度</a>
    <img v-bind:src="myurl.imgUrl"/><br>
    <!-- v-bind:可缩写为 ":"，这种缩写称为语法糖-->
    <a :href="myurl.baiduUrl">去百度</a>
    <img :src="myurl.imgUrl"/><br>
    <div :class="mycolor">对象语法</div>
    <div class="static" :class="{'active': isActive, 'text-danger': hasError}">在对象中传入更多字段</div>
    <div :class="[activeClass, errorClass]">数组语法</div>
    <div :class="[isActive ? activeClass : ' ', errorClass]">数组中使用三元表达式</div>
    <div :class="[{ 'active': isActive }, errorClass]">数组中嵌套对象</div>
</div>
<script src="js/vue.global.js"></script>
<script>
    Vue.createApp({
        data() {
            return {
                myurl: {
                    baiduUrl: 'https://www.baidu.com/',
                    imgUrl:'images/ok.gif'
                },
                mycolor: 'my',
                isActive: true,
                hasError: false,
                activeClass:'your',
                errorClass:'his'
            }
        }
    }).mount('#app')
</script>
<style>
    .my {
        background-color: red
    }
    .your {
        font-size: 20px
    }
    .his {
        background-color: blue
    }
    .static {
        background-color: yellow
    }
    .active {
        font-size: 40px
    }
</style>
```

11.8.2 条件渲染指令 v-if 和 v-show

与 JavaScript 的条件语句 if、else、else if 类似，Vue.js 的条件指令 v-if 也可以根据表达式的值渲染或销毁元素/组件。下面通过具体实例讲解 v-if 指令的用法。

【例 11-9】使用条件渲染指令判断成绩的等级。

其具体代码如下：

```
<div id="if-handling">
    <div v-if="score >= 90">优秀</div>
    <div v-else-if="score >= 80">良好</div>
    <div v-else-if="score >= 70">中等</div>
    <div v-else-if="score >= 60">及格</div>
    <div v-else>不及格</div>
</div>
<script src="js/vue.global.js"></script>
<script>
    Vue.createApp({
        data() {
            return {
                score: 87
            }
        }
    }).mount('#if-handling')
</script>
```

从上述示例代码可以看出，v-else 元素必须紧跟在 v-if 或者 v-else-if 元素的后面；v-else-if 必须紧跟在 v-if 或者 v-else-if 元素的后面。

v-if 条件渲染指令必须添加到一个元素上。但是如果想包含多个元素呢？此时可以使用 <template> 元素（模板占位符）帮助用户包裹元素，并在上面使用 v-if。最终的渲染结果不包含 <template> 元素。示例代码如下：

```
<template v-if="ok">
    <h1>Title</h1>
    <p>Paragraph 1</p>
    <p>Paragraph 2</p>
</template>
```

v-show 指令的用法基本上和 v-if 一样，也是根据条件展示元素，例如 <h1 v-show="yes"> 一级标题 </h1>。不同的是，v-if 每次都会重新删除或创建元素，而带有 v-show 的元素始终会被渲染并保留在 DOM 中，只是切换元素的 display:none 样式。所以，v-if 有更大的切换消耗，而 v-show 有更大的初始渲染消耗。因此，如果需要频繁切换，用 v-show 较好；如果在运行时条件不太可能改变，用 v-if 较好。另外，v-show 不支持 <template> 元素，也不支持 v-else。

【例 11-10】演示 v-if 与 v-show 的区别。

其具体代码如下：

```
<div id="event-handling">
    <div v-if="flag">一直显示</div>
    <div v-show="flag">反复无常</div>
    <!-- v-on:click 调用事件方法，v-on:可缩写为@-->
    <button @click="flag=!flag">隐藏/显示</button>
</div>
<script src="js/vue.global.js"></script>
<script>
```

第 11 章　Vue.js 基础

```
        Vue.createApp({
            data() {
                return {
                    flag: true
                }
            }
        }).mount('#event-handling')
    </script>
```

使用谷歌浏览器第一次运行程序时（按 F12 键），页面的初始效果如图 11.10 所示。
单击"隐藏/显示"按钮后，页面如图 11.11 所示。

图 11.10　页面的初始效果

图 11.11　单击"隐藏/显示"按钮后的效果

从图 11.11 可以看出，通过 v-if 控制的元素如果隐藏，从 DOM 中移除，而通过 v-show 控制的元素并没有真正移除，只是给其添加了 CSS 样式 display:none。

▶ 11.8.3　列表渲染指令 v-for

用户可以使用 v-for 指令遍历一个数组或对象，它的表达式需要结合 in 来使用，形式为 item in items，其中 items 是源数据，而 item 是被迭代集合中元素的别名。v-for 还支持一个可选的参数作为当前项的索引。v-for 指令的常用方式如下：

❶ 遍历普通数组

```
<ul>
    <li v-for="(item,index) in items">
        {{index}} - {{ item }}
    </li>
</ul>
```

❷ 遍历对象数组

```
<ul>
<li v-for="user in users">
    {{ user.uname }}
</li>
</ul>
```

❸ 遍历对象属性

```
<li v-for="(value, key, index) in myObject">
    {{ ++index }}. {{ key }}: {{ value }}
</li>
```

❹ 迭代数字

```
<li v-for="i in 100">
```

```
        {{ i }}
    </li>
```

【例 11-11】演示 v-for 指令的常用方式。

其具体代码如下：

```
<div id="myfor">
    <ul>
        <li v-for="(book, index) in books">
            {{++index}}. {{book}}
        </li>
    </ul>
    <ul>
        <li v-for="(author, index) in authors">
            {{++index}}. {{author.name}} - {{author.sex}}
        </li>
    </ul>
    <p v-for="(value, key, index) in userinfo">键是：{{key}}，值是：{{value}}，索引是：{{index}}</p>
    <p v-for="i in 5">这是第{{i}}段。</p>
</div>
<script src="js/vue.global.js"></script>
<script>
    Vue.createApp({
        data() {
            return {
                //普通数组
                books: ['Java Web 开发从入门到实战',
                'Java EE 框架整合开发从入门到实战——Spring+Spring MVC+MyBatis',
                'Spring Boot 从入门到实战'],
                //对象数组
                authors:[{name: '陈恒', sex: '男'}, {name: '陈恒 11', sex: '女'},
                        {name: '陈恒 22', sex: '男'}],
                //对象
                userinfo:{
                    uname: '陈恒 3',
                    age: 88
                }
            }
        }
    }).mount('#myfor')
</script>
```

▶ 11.8.4 事件处理

扫一扫

视频讲解

所有的事件处理都离不开事件监听器，在 Vue.js 中可以使用 v-on 指令给 HTML 元素添加一个事件监听器，通过该指令调用在 Vue 实例中定义的方法。下面使用 v-on 指令监听按钮事件，实现字符串的反转。

【例 11-12】使用 v-on 指令监听按钮事件，实现字符串的反转。

其具体代码如下：

```
<div id="event-handling">
    <p>{{ message }}</p>
    <button v-on:click="reverseMessage">反转 Message</button>
    <!-- v-on:可缩写为 "@"，是一个语法糖-->
    <button @click="reverseMessage">反转 Message</button>
</div>
```

```
<script src="js/vue.global.js"></script>
<script>
    const EventHandling = {
        data() {
            return {
                message: 'Hello Vue.js!'
            }
        },
        methods: {    //方法的定义
            reverseMessage() {
                this.message = this.message.split('').reverse().join('')
            }
        }
    }
    Vue.createApp(EventHandling).mount('#event-handling')
</script>
```

在上述代码中，@click 等同于 v-on:click，是一个语法糖；在 methods 属性中定义了触发事件时调用的方法。@click 调用的方法如果没有参数，在方法名后可以不写括号"()"。

在 Vue.js 中有时需要访问原生的 DOM 事件。Vue.js 提供了一个特殊变量$event，用于访问原生的 DOM 事件，例如下面的实例阻止打开链接。

【例 11-13】阻止打开链接。

其具体代码如下：

```
<div id="event-handling">
    <a href="https://www.baidu.com/" @click="warn('考试期间禁止百度！', $event)">
去百度</a>
</div>
<script src="js/vue.global.js"></script>
<script>
    Vue.createApp({
        methods: {
            warn(message, event) {
                //event 访问原生的 DOM 事件
                event.preventDefault()
                alert(message)
            }
        }
    }).mount('#event-handling')
</script>
```

例 11-13 的运行效果如图 11.12 所示。

图 11.12　例 11-13 的运行效果

在事件处理中调用 event.preventDefault()或 event.stopPropagation()是非常常见的需求。尽管可以在方法中轻松实现这类需求，但方法最好只有纯粹的数据逻辑，而不是去处理 DOM 事件细节。

为了解决该问题，Vue.js 为 v-on 提供了事件修饰符。修饰符由点开头的指令后缀表示。Vue.js 支持的修饰符有.stop、.prevent、.capture、.self、.once 以及.passive。其用法是在@绑定的事件后加小圆点"."，再跟修饰符，具体如下：

```
<!--阻止单击事件-->
<a @click.stop="doThis"></a>
<!--提交事件不再重载页面-->
<form @submit.prevent="onSubmit"></form>
<!--修饰符可以串联-->
<a @click.stop.prevent="doThat"></a>
<!--只有修饰符-->
<form @submit.prevent></form>
<!--添加事件监听器时使用事件捕获模式，即内部元素触发的事件先在此处理，然后再交给内部元素
进行处理-->
<div @click.capture="doThis">...</div>
<!--当事件在该元素自身触发时触发回调，即事件不是从内部元素触发的-->
<div @click.self="doThat">...</div>
<!--只触发一次-->
<a @click.once="doThis"></a>
<!--滚动事件的默认行为（即滚动行为）将会立即触发，而不会等待"onScroll"完成-->
<div @scroll.passive="onScroll">...</div>
```

扫一扫

视频讲解

▶ 11.8.5　表单与 v-model

表单控件是网页数据交互的必备手段，用于向服务器传输数据，较为常见的表单控件有单选下拉列表、多选下拉列表、输入框等，用表单控件可以完成数据的输入、校验、提交等。Vue.js 用 v-model 指令在表单<input>、<textarea>及<select>元素上创建双向数据绑定（Model 到 View 以及 View 到 Model）。使用 v-model 指令的表单元素将忽略该元素的 value、checked、selected 等属性的初始值，而是将当前活动的 Vue 实例的数据作为数据来源，所以在使用 v-model 指令时应该通过 JavaScript 在 Vue 实例的 data 选项中声明初始值。

从 Model 到 View 的数据绑定，即 ViewModel 驱动将数据渲染到视图；从 View 到 Model 的数据绑定，即 View 中元素上的事件被触发后导致数据变更将通过 ViewModel 驱动修改数据层。下面通过一个实例演示 v-model 指令在表单元素上实现双向数据绑定。

【例 11-14】v-model 指令在表单元素上实现双向数据绑定。

其具体代码如下：

```
<div id="vmodel-databinding">
    用户名：<input v-model="uname" />
    <p>输入的用户名是：{{ uname }}</p>
    <textarea v-model="introduction"></textarea>
    <p>输入的个人简介是：{{ introduction }}</p>
    <p>
        备选歌手：
        <input type="checkbox" id="zhangsan" value="张三" v-model="singers" />
        <label for="zhangsan">张三</label>
        <input type="checkbox" id="lisi" value="李四" v-model="singers" />
        <label for="lisi">李四</label>
        <input type="checkbox" id="wangwu" value="王五" v-model="singers" />
        <label for="wangwu">王五</label>
        <input type="checkbox" id="chenheng" value="陈恒" v-model="singers" />
```

```html
            <label for="chenheng">陈恒</label>
            <br />
            <span>你喜欢的歌手：{{ singers }}</span>
    </p>
    <p>
        性别：
        <input type="radio" id="male" value="男" v-model="sex" />
        <label for="male">男</label>
        <input type="radio" id="female" value="女" v-model="sex" />
        <label for="female">女</label>
        <br />
        <span>你的性别：{{ sex }}</span>
    </p>
    <p>
        <!--单选下拉列表-->
        备选国籍：
        <select v-model="single">
            <option v-for="option in options" :value="option.value">
                {{ option.text }}
            </option>
        </select><br>
        <span>你的国籍：{{ single }}</span>
    </p>
    <p>
        <!--多选下拉列表-->
        备选国家：
        <select v-model="moreselect" multiple>
            <option v-for="option in moreoptions" :value="option.value">
                {{ option.text }}
            </option>
        </select><br>
        <span>你去过的国家：{{ moreselect }}</span>
    </p>
</div>
<script src="js/vue.global.js"></script>
<script>
    Vue.createApp({
        data() {
            return {
                uname: '陈恒',
                introduction: '我是一个好少年',
                //多个复选框，绑定到同一个数组，默认选择歌手'陈恒'
                singers: ['陈恒'],
                sex: '女',            //默认性别为女
                //单选下拉列表绑定变量，默认国籍为中国
                single: '中国',
                options: [
                    { text: '中国', value: '中国' },
                    { text: '日本', value: '日本' },
                    { text: '美国', value: '美国' }
                ],
                moreselect: ['中国'],      //多选下拉列表绑定一个数组，默认去过中国
                moreoptions: [
                    { text: '中国', value: '中国' },
                    { text: '英国', value: '英国' },
                    { text: '日本', value: '日本' },
```

```
                    { text: '美国', value: '美国' }
                ],
            }
        }
    }).mount('#vmodel-databinding')
</script>
```

例 11-14 的运行效果如图 11.13 所示。

在默认情况下，v-model 在每次 input 事件触发后都将输入框的值与数据进行同步。如果不想在每次 input 事件触发后同步，可以添加 lazy 修饰符，从而转为在 change 事件触发后进行同步。示例代码如下：

```
<!--在"change"时更新-->
<input v-model.lazy="msg"/>
```

如果需要将用户的输入值自动转为数值类型，可以给 v-model 添加 number 修饰符，示例代码如下：

```
<input v-model.number="age" type="number"/>
```

如果需要将用户输入的首尾空格自动去除，可以给 v-model 添加 trim 修饰符，示例代码如下：

```
<input v-model.trim="msg"/>
```

图 11.13　例 11-14 的运行效果

扫一扫

视频讲解

11.8.6　实战：购物车实例

本小节以 Vue.js 的计算属性、内置指令、方法等技术为基础，完成一个在电商平台中具有代表性的小功能——购物车。购物车的具体需求如下：

（1）展示已加入购物车的商品列表，包括商品名称、商品单价、购买数量以及实时购买的总价。

（2）购买数量可以增加或减少，每类商品还可以从购物车中删除。

最终实现的效果如图 11.14 所示。

图 11.14　购物车效果图

因购物车的代码实现稍微复杂，这里将 JavaScript 从 HTML 中分离出来，以便于阅读和维护。

具体代码文件如下：

（1）cart.html（引入资源、模板及 CSS）。

（2）cart.js（Vue 实例及业务代码）。

在 cart.html 中引入 Vue.js 和 cart.js，创建一个根元素（<div id="cart">）来挂载 Vue 实例。cart.html 的具体代码如下：

```html
<div id="cart">
    <template v-if="cartList.length">
        <table>
            <thead>
                <tr>
                    <th></th>
                    <th>商品名称</th>
                    <th>商品单价</th>
                    <th>购买数量</th>
                    <th>操作</th>
                </tr>
            <tbody>
                <tr v-for="(item, index) in cartList">
                    <td>{{ index + 1 }}</td>
                    <td>{{ item.gname }}</td>
                    <td>{{ item.gprice }}</td>
                    <td>
                        <button @click="reduce(index)" :disabled="item.count === 1">-</button>
                        {{ item.count }}
                        <button @click="add(index)">+</button>
                    </td>
                    <td>
                        <button @click="remove(index)">删除</button>
                    </td>
                </tr>
            </tbody>
            </thead>
        </table>
        <div>总价：¥ {{ totalPrice }} <button @click="removeAll">清空购物车</button></div>
    </template>
    <div v-else>购物车为空</div>
</div>
<script src="js/vue.global.js"></script>
<script src="cart.js"></script>
<style>
    table{
    border: 1px solid #E9E9E9;
    border-collapse:collapse;
    }
    th, td{
        padding: 8px 16px;
        border: 1px solid #E9E9E9;
    }
    th{
        background: #F7F7F7;
    }
</style>
```

在 cart.js 中首先初始化 Vue 实例，然后在 data 选项初始化商品列表 cartList，再次使用计算属性 totalPrice 计算购物车中商品的总价，最后在 methods 选项定义事件处理方法。cart.js 的具体代码如下：

```javascript
Vue.createApp({
    data() {
```

```
            return {
                cartList:[
                    {
                        id: 1,
                        gname: 'Spring Boot 从入门到实战',
                        gprice: 79.8,
                        count: 5
                    },
                    {
                        id: 2,
                        gname: 'Java Web 开发从入门到实战',
                        gprice: 69.8,
                        count: 10
                    },
                    {
                        id: 3,
                        gname: 'Java EE 框架整合开发入门到实战',
                        gprice: 69.8,
                        count: 100
                    }
                ]
            }
        },
        computed: {
            totalPrice() {
                var total = 0
                for (var i = 0; i < this.cartList.length; i++) {
                    var item = this.cartList[i]
                    total = total + item.gprice * item.count
                }
                return total
            }
        },
        methods: {
            reduce(index) {
                if(this.cartList[index].count === 1)
                    return
                this.cartList[index].count--
            },
            add(index) {
                this.cartList[index].count++
            },
            remove(index) {
                this.cartList.splice(index, 1)
            },
            removeAll() {
                this.cartList.splice(0, this.cartList.length)
            }
        }
}).mount('#cart')
```

11.9 组件

组件（Component）是 Vue.js 的最核心的功能，是可扩展的 HTML 元素（可以看作自定义的 HTML 元素），是封装可重用的代码，同时也是 Vue 实例，可以接受与 Vue 相同的选项对象并提供相同的生命周期钩子。

第 11 章 Vue.js 基础

组件系统是 Vue.js 中的一个重要概念，它提供了一种抽象，让用户可以使用独立可复用的小组件来构建大型应用，任意类型的应用界面都可以抽象为一个组件树。这种前端组件化方便 UI 组件的重用。

本节将和读者一起由浅入深地学习组件的相关内容。

扫一扫

视频讲解

▶ 11.9.1 组件的注册

为了能在 UI 模板中使用组件，必须先注册组件以便 Vue 识别。组件的注册类型通常有两种，即全局注册和局部注册。

❶ 全局注册

组件可通过 component 方法实现全局注册，全局注册的示例代码如下：

```
const app = Vue.createApp({})
app.component('component-a', {
    //选项
})
app.component('component-b', {
    //选项
})
app.component('component-c', {
    //选项
})
app.mount('#app')
```

app.component 的第一个参数 component-a 是组件的名称（自定义标签），组件的名称推荐全部小写，包含连字符（即有多个单词），避免与 HTML 元素相冲突。

在注册后任何 Vue 实例都可以使用这些组件，示例代码如下：

```
<div id="app">
    <component-a></component-a>
    <component-b></component-b>
    <component-c></component-c>
</div>
```

❷ 局部注册

全局注册往往是不够理想的，比如使用打包工具构建系统时，全局注册的组件即使不再使用，仍然被包含在最终的构建结果中，造成用户无意义地下载 JavaScript。在这些情况下可以通过一个普通的 JavaScript 对象来定义组件：

```
const ComponentA = {
    /* ... */
}
const ComponentB = {
    /* ... */
}
const ComponentC = {
    /* ... */
}
```

然后使用 Vue 实例的 components 选项局部注册组件：

```
const app = Vue.createApp({
    components: {
        'component-a': ComponentA,  //component-a 为局部组件的名称
        'component-b': ComponentB
    }
})
```

局部注册的组件只在该组件的作用域下有效。例如希望 ComponentA 在 ComponentB 中可用，需要在 ComponentB 中使用 components 选项局部注册 ComponentA：

```
const ComponentA = {
    /* ... */
}
const ComponentB = {
    components: {
        'component-a': ComponentA
    }
    //...
}
```

【例 11-15】定义一个名为 button-counter 的全局组件，以及两个局部组件 ComponentA 和 ComponentB，运行效果如图 11.15 所示。

其具体代码如下：

```
<template id="button-counter">
    <button @click="count++">You clicked me {{ count }} times.</button>
</template>
<div id="components-demo">
    <!--在模板中任意使用组件-->
    <!--每用组件各自独立维护它的 count，因为每用一次组件就会有一个它的新实例被创建-->
    <button-counter></button-counter><br><br>
    <button-counter></button-counter><br><br>
    <button-counter></button-counter><br><br>
    <component-b></component-b>
    <!--component-a 失效-->
    <component-a></component-a>
</div>
<script src="js/vue.global.js"></script>
<script>
    const ComponentA = {
        template: '<span>这是私有组件A！</span>'
    }
    const ComponentB = {
        //使用 components 选项局部注册 ComponentA
        components: {
            'component-a': ComponentA
        },
        template: '<span>这是私有组件B！</span>'
    }
    //创建一个 Vue 应用
    const app = Vue.createApp({
        //使用 Vue 实例的 components 选项局部注册组件
        components: {
            'component-b': ComponentB
        }
    })
    //定义一个名为 button-counter 的全局组件（注册）
    app.component('button-counter', {
        data() {
            return {
                count: 0
            }
        },
        //组件显示的内容
        template: '#button-counter'
    })
```

```
        app.mount('#components-demo')
</script>
```

从图 11.15 可以看出，component-a 组件没有显示，这是因为它不是 Vue 实例 app 的组件，而是 ComponentB 的私有组件。

在例 11-15 组件定义的代码中，template 定义组件显示的内容，必须被一个 HTML 元素包含（例如<button>），否则无法渲染。除了 template 选项以外，组件可以像 Vue 实例那样使用 data、computed、methods 选项，例如上述 button-counter 组件。

图 11.15　例 11-15 的运行效果

▶ 11.9.2　使用 props 传递数据

组件除了把模板内容复用以外，更重要的是向组件传递数据。在组件中，使用 props 选项来声明从父级组件接收的数据，props 的值可以是字符串数组，也可以是对象。

使用 props 实现数据传递都是单向的，即父组件的数据变化时，子组件中所有的 prop 将刷新为最新的值，但是反过来不行。这样设计的原因是尽可能将父子组件解耦，避免子组件无意中修改父组件的状态。如果在业务中需要改变 prop，一种是父组件传递初始值进来，子组件将它作为初始值保存起来，在子组件自己的作用域下随意修改；另一种是使用计算属性修改。

【例 11-16】构造两个数组 props，一个数组接收来自父组件的数据 message（实现静态传递），另一个数组接收来自父组件的数据 id 和 title（实现动态传递），将它们在组件模板中渲染。然后在子组件中声明数据 count 保存来自父组件的 mycount，count 的改变不影响 mycount。该例的运行效果如图 11.16 所示。

其具体代码如下：

```
<!--父组件的显示-->
<template id="parent">
    <h4>{{ message }}</h4>
    <!--使用 v-bind 将父组件 parent 的 data(posts)动态传递给 props, children 组件只能在 parent 中-->
    <children v-for="post in posts" :id="post.id" :title="post.title"></children>
    <!--将一个对象的所有属性都作为 prop 传入，与上一句等价-->
    <children v-for="post in posts" v-bind="post" ></children>
</template>
<!--子组件的显示-->
<template id="children">
    <h4>{{id}} : {{ title }}</h4>
</template>
<template id="child-app">
    <button @click="count++">You clicked me {{ count }} times.</button>
</template>
<div id="message-post-demo">
    <!--静态传递字符串，父组件就是 Vue 当前的实例-->
    <parent message="来自父组件的消息"></parent>
    父组件的计数器：{{mycount}}<br>
    <child-counter :parent-count="mycount"></child-counter>
</div>
```

```
<script src="js/vue.global.js"></script>
<script>
    const messageApp = Vue.createApp({
        data() {
            return {
                mycount: 10
            }
        }
    })
    messageApp.component('parent', {
        data() {
            return {
                //posts 是对象数组
                posts: [
                    { id: 1, title: 'My journey with Vue' },
                    { id: 2, title: 'Blogging with Vue' },
                    { id: 3, title: 'Why Vue is so fun' }
                ]
            }
        },
        props: ['message'],         //接收父组件 messageApp 传递的数据
        components: {               //创建子组件 children
            'children':{
                props: ['id','title'],   //接收父组件 parent 传递的数据
                template: '#children'
            }
        },
        template: '#parent'
    })
    messageApp.component('child-counter', {
        props: ['parentCount'],
        data() {
            return {
                count: this.parentCount
            }
        },
        //组件显示的内容
        template: '#child-app'
    })
    messageApp.mount('#message-post-demo')
</script>
```

在例 11-16 中，因为 HTML 不区分大小写，所以当使用 DOM 模板时，以驼峰命名的 props 名称 parentCount 要转换为短横线命名 parent-count。在图 11.16 中，单击按钮改变子组件计数器 count 的值，而父组件计数器 mycount（通过 props 传递给子组件）不受影响。

▶ 11.9.3 组件的通信

通过学习 11.9.2 小节，大家知道 props 可以实现父组件向子组件传递数据，即通信。Vue 组件通信的场景有多种，例如父子组件通信、兄弟组件通信、组件链通信。本小节将介绍各种组件通信的方法。

图 11.16 例 11-16 的运行效果

❶ 使用自定义事件通信

用户可以通过 props 从父组件向子组件传递数据,并且这种传递是单向的。当需要从子组件向父组件传递数据时,可以首先给子组件自定义事件并使用$emit(事件名,要传递的数据)方法触发事件,然后父组件使用 v-on 或@监听子组件的事件。下面通过一个实例讲解使用自定义事件实现通信的方法。

【例 11-17】子组件触发两个事件,分别实现字体变大、变小。

其具体代码如下:

```
<template id="blog">
    <!--0.1 是传递给父组件blogApp的数据,可以不填。当在父组件监听这个事件时,可以通过
$event 访问该数据。如果事件处理函数是一个方法,那么该数据将会作为第一个参数传入该方法(例如
onEnlargeText) -->
    <h4>{{id}} : {{ title }}</h4>
    <button @click="$emit('enlarge-text', 0.1)">变大</button>
    <button @click="$emit('ensmall-text', 0.1)">变小</button>
</template>
<div id="blog-post-demo">
    <div v-bind:style="{ fontSize: postFontSize + 'em' }">
    <!--将一个对象的所有属性作为 prop 传递给子组件,@父组件监听事件并更新
postFontSize 的值-->
        <!--$event 接收子组件传递过来的数据 0.1-->
        <blog-post v-for="post in posts" v-bind="post" @ensmall-text="postFontSize -= $event" @enlarge-text="onEnlargeText"></blog-post>
    </div>
</div>
<script src="js/vue.global.js"></script>
<script>
    const blogApp = Vue.createApp({
        data() {
            return {
                //posts是对象数组
                posts: [
                    { id: 1, title: 'My journey with Vue' },
                    { id: 2, title: 'Blogging with Vue' },
                    { id: 3, title: 'Why Vue is so fun' }
                ],
                postFontSize: 1
            }
        },
        methods: {
            onEnlargeText(enlargeAmount) {
                this.postFontSize += enlargeAmount
            }
        }
    })
    blogApp.component('blog-post', {          //定义子组件
        props: ['id', 'title'],               //接收父组件blogApp的两个参数id和title
        template: '#blog'
    })
    blogApp.mount('#blog-post-demo')
</script>
```

注意:在上述代码中,事件名推荐使用短横线命名(例如 enlarge-text),这是因为 HTML 是不区分大小写的。如果事件名为 enlargeText,@enlargeText 将变成@enlargetext,事件 enlargeText 不可能被父组件监听到。

❷ 使用 v-model 通信

除了可以自定义事件实现子组件向父组件传值以外，还可以在子组件上使用 v-model 向父组件传值，实现双向绑定。

下面通过一个实例讲解如何在子组件上使用 v-model 实现向父组件传值。

【例 11-18】使用 v-model 实现子组件向父组件传值，并实现双向绑定，运行效果如图 11.17 所示。

其具体代码如下：

```
<template id="custom">
<!--为了让子组件正常工作，子组件内的 <input>必须将其value 属性绑定到一个名为modelValue
的props 上，在其 input 事件被触发时将新的值通过自定义的update:modelValue 事件传递-->
    <input :value="modelValue" @input="$emit('update:modelValue',
$event.target.value)" >
</template>
<div id="vmodel-post-demo">
    {{searchText}}<br>
    <custom-input v-model="searchText"></custom-input><br>
    <!--这两个子组件等价-->
    <custom-input :model-value="searchText" @update:model-value="searchText =
$event"></custom-input>
</div>
<script src="js/vue.global.js"></script>
<script>
    const blogApp = Vue.createApp({
        data() {
            return {
                searchText: '陈恒'
            }
        }
    })
    blogApp.component('custom-input', {
        props: ['modelValue'],
        template: '#custom'
    })
    blogApp.mount('#vmodel-post-demo')
</script>
```

在例 11-18 中实现了一个具有双向绑定的 v-model 组件，需要满足以下两个要求：

（1）接收一个 value 属性。

（2）在有新的 value 时触发 input 事件。

❸ 使用 mitt 实现非父子组件通信

在 Vue.js 中，推荐使用一个空的 Vue 实例作为媒介（中央事件总线）实现父子组件、兄弟组件及组件链通信。比如人们生活中的例子——买房卖房中介帮忙，买卖双方通过房产中介（中央事件总线）实现需求对接。

图 11.17　例 11-18 的运行效果

在 Vue 2.x 中，Vue 实例可以通过事件触发 API（$on、$off 和 $once）实现中央事件总线，但是在 Vue 3.x 中移除了 $on、$off 和 $once 实例方法，推荐使用外部库 mitt 来代替 $on、$emit 和 $off 实例方法。使用 TypeScript 编写的 mitt 源代码具体如下：

```
/**
 * @param 入参为 EventHandlerMap 对象
 * @returns 返回一个对象，对象包含 all 属性以及 on、off、emit 方法
```

```js
    */
function mitt(all) {
    all = all || new Map();
    return {
        //事件键值对映射对象
        all,
        /**
         * 注册一个命名的事件处理
         * @param type 事件名
         * @param handler 事件处理函数
         */
        on(type, handler) {
            //根据type 去查找事件
            const handlers = all.get(type);
            //如果找到有相同的事件,则继续添加,Array.prototype.push 返回值为添加后
            //的新长度
            const added = handlers && handlers.push(handler);
            //如果已添加了type 事件,则不再执行set 操作
            if (!added) {
                all.set(type, [handler]);   //注意此处的值是数组类型,可以添加多
                                            //个相同的事件
            }
        },
        /**
         * 移除指定的事件处理
         * @param type 事件名,和第二个参数一起用来移除指定的事件
         * @param handler 事件处理函数
         */
        off(type, handler) {
            //根据type 去查找事件
            const handlers = all.get(type);
            //如果找到则进行删除操作
            if (handlers) {
                handlers.splice(handlers.indexOf(handler) >>> 0, 1);
            }
        },
        /**
         * 触发所有type 事件,如果有type 为*的事件,则最后执行
         * @param type 事件名
         * @param evt 传递给处理函数的参数
         */
        emit(type, evt) {
            //找到type 的事件循环执行
            (all.get(type) || []).slice().map((handler) => { handler(evt); });
            //找到所有为*的事件,循环执行
            (all.get('*') || []).slice().map((handler) => { handler(type, evt); });
        }
    };
}
```

下面通过一个实例讲解如何使用 mitt 实现非父子组件通信。

【例 11-19】使用 mitt 新建一个中央事件总线 bus,然后分别创建两个 Vue 实例 buyer(买方)和 seller(卖方),买、卖双方互相通信。

其具体代码如下:

```html
<div id="buyer">
    <h1>{{ message1 }}</h1>
```

```
        <button @click="transferb">我是买方，向卖方传递信息</button>
</div>
<div id="seller">
    <h1>{{ message2 }}</h1>
    <button @click="transfers">我是卖方，向买方传递信息</button>
</div>
<script src="js/vue.global.js"></script>
<script src="js/mitt.js"></script>
<script>
    const bus = mitt()
    //买方
    const buyer = Vue.createApp({
        data() {
            return {
                message1: ''
            }
        },
        methods: {
            transferb() {
                //用emit触发事件传值
                bus.emit('on-message1', '来自买方的信息')
            }
        },
        mounted(){
            //监听
            bus.on('on-message2', (msg) => {     //(msg)相当于function(msg)
                this.message1 = msg
            })
        }
    })
    buyer.mount('#buyer')
    //卖方
    const seller = Vue.createApp({
        data() {
            return {
                message2: ''
            }
        },
        methods: {
            transfers() {
                //用emit触发事件传值
                bus.emit('on-message2', '来自卖方的信息')
            }
        },
        mounted(){
            //监听
            bus.on('on-message1', (msg) => {     //(msg)相当于function(msg)
                this.message2 = msg
            })
        }
    })
    seller.mount('#seller')
</script>
```

❹ 提供/注入（组件链传值）

当需要将数据从父组件传递到子组件时，可以使用 props 实现，但有时一些子组件是深嵌套的，如果将 props 传递到整个组件链中将很麻烦，更不可取。对于这种情况，可以使用 provide 和 inject 实现组件链传值。父组件可以作为其所有子组件的依赖项提供程序，而不管组件的层次结构有多深，父组件有一个 provide 选项来提供数据，子组件有一个 inject 选项来使用这个数据。下面通过一个实例演示组件链传值的用法。

【例 11-20】创建 Vue 实例为祖先组件，并使用 provide 提供一个数据供其子孙组件 inject。其具体代码如下：

```
<template id="son">
    <div>{{ todos.length }}</div>
    <!--todo-son 是 todo-list 的私有组件-->
    <todo-son></todo-son>
</template>
<template id="grandson">
    <div>
        <!--使用注入的数据-->
        {{ todoLength }}
    </div>
</template>
<div id="vmodel-post-demo">
    <!--父组件 Vue 实例传递数据 todos 给子组件 todo-list-->
    <todo-list:todos="todos"></todo-list>
</div>
<script src="js/vue.global.js"></script>
<script>
    const app = Vue.createApp({
        data() {
            return {
                todos: ['Feed a cat', 'Buy tickets']
            }
        },
        provide() {    //祖先组件 app 提供一个数据 todoLength
            return {
                todoLength: this.todos.length
            }
        }
    })
    app.component('todo-list', {
        props: ['todos'],
        components:{    //在父组件 todo-list 中定义子组件 todo-son
            'todo-son': {
                inject: ['todoLength'], //孙子组件注入数据 todoLength 供自己使用
                template: '#grandson'
            }
        },
        template: '#son'
    })
    app.mount('#vmodel-post-demo')
</script>
```

▶ 11.9.4 动态组件与异步组件

组件间切换或异步加载是常见的应用场景，本小节将介绍动态与异步组件的实现方法。

❶ 动态组件

在不同组件之间进行动态切换是常见的场景，比如在一个多标签的页面中进行内容的收纳和展现。Vue 可通过<component>元素动态加载不同的组件，进行组件切换。示例代码如下：

```
<!-- is 属性选择加载的组件，currentView 是已注册组件的名称或一个组件的选项对象-->
<component:is="currentView"></component>
```

扫一扫

视频讲解

下面通过一个实例讲解动态组件的用法。

【例 11-21】通过<component>元素动态切换组件，在该实例中有 3 个按钮，单击不同按钮展示不同组件的信息。

其具体代码如下：

```
<div id="app">
    <button @click="changeCom('add')">添加信息</button> 
    <button @click="changeCom('update')">修改信息</button> 
    <button @click="changeCom('delete')">删除信息</button>
    <component :is="currentCom"></component>
</div>
<script src="js/vue.global.js"></script>
<script>
    const blogApp = Vue.createApp({
        data() {
            return {
                currentCom: 'add'
            }
        },
        components:{      //在组件选项中定义 3 个局部组件供切换
            'add': {
                template: '<div>添加信息展示界面</div>'
            },
            'update': {
                template: '<div>修改信息展示界面</div>'
            },
            'delete': {
                template: '<div>删除信息展示界面</div>'
            }
        },
        methods:{
            changeCom(com) {
                this.currentCom = com
            }
        }
    })
    blogApp.mount('#app')
</script>
```

❷ 异步组件

在大型应用中可能需要将应用分割成许多小的代码块，并且只在需要时才从服务器异步加载一个模块，这样可以避免一开始就把所有组件加载，浪费不必要的开销。Vue 有一个 defineAsyncComponent 方法将组件定义为一个工厂函数，动态地解析组件。Vue 只在组件需要渲染时触发工厂函数，并把结果缓存起来，以备再次渲染。下面通过一个实例讲解异步组件的用法。

【例 11-22】实现 5 秒钟后加载组件信息。

其具体代码如下：

```
<div id="app">
    <!--5 秒钟后才下载组件并展示-->
    <async-example></async-example>
</div>
<script src="js/vue.global.js"></script>
<script>
    const blogApp = Vue.createApp({})
```

```
                //定义异步组件
                const AsyncComp = Vue.defineAsyncComponent(() =>   //定义 defineAsyncComponent
                                                                  //方法的返回值
                    new Promise((resolve, reject) => {   //返回 Promise 的工厂函数
                        window.setTimeout(() => {   //window.setTimeout 只是演示异步
                            resolve({   /*从服务器收到加载组件定义之后,调用 Promise 的 resolve
方法异步加载组件,也可以调用 reject(reason)指示加载失败*/
                                template: '<div>5 秒钟后才展示我!</div>'
                            })
                        }, 5000)
                    })
                )
                blogApp.component('async-example', AsyncComp)
                blogApp.mount('#app')
</script>
```

上述工厂函数返回 Promise 对象,当从服务器收到加载组件定义后,调用 Promise 的 resolve 方法异步加载组件,也可以调用 reject(reason)指示加载失败。这里 window.setTimeout 只是演示异步,具体的异步加载逻辑可由开发者自己决定。

▶ 11.9.5 实战:正整数数字输入框组件

本小节将普通输入框扩展成正整数数字输入框,用来快捷地输入一个标准的数字,如图 11.18 所示。

图 11.18 正整数数字输入框

其具体实现过程如下:

❶ 定义正整数数字输入框组件

在 ch11 的 js 目录中创建 input-num.js 文件。在 input-num.js 文件中定义正整数数字输入框组件 inputNumber,在该组件中定义方法 handleDown()、handleUp()和 handleChange()分别实现减一、加一和数值判断的功能。input-num.js 的具体代码如下:

```
function isValueNumber (value){
    return (/^[1-9]\d*$/).test(value+'')
}
//定义正整数数字输入框组件
const inputNumber = {
    //组件显示的内容
    template: '\
    <div>\
        <input type="text" :value="currentValue" @change="handleChange">\
        <button @click="handleDown" :disabled="currentValue <= 1">-</button>\
        <button @click="handleUp">+</button>\
    </div>',
    data() {
        return {
            currentValue: 1
        }
    },
    methods: {
        handleDown() {
            if (this.currentValue <= 1)
                return
```

```
            this.currentValue -= 1
        },
        handleUp() {
            this.currentValue += 1
        },
        handleChange(event) {
            var val = event.target.value.trim()
            if(isValueNumber(val)) {
                this.currentValue = Number(val)
            } else {
                event.target.value = this.currentValue
            }
        }
    }
}
```

❷ 引用正整数数字输入框组件

在 ch11 目录中创建 inputNumber.html 文件。在 inputNumber.html 文件中创建 Vue 实例 app，并引入正整数数字输入框组件 inputNumber。inputNumber.html 的具体代码如下：

```
<head>
    <meta charset="UTF-8">
    <title>数字输入框</title>
</head>
<body>
    <div id="app">
        <input-number></input-number>
    </div>
    <script src="js/vue.global.js"></script>
    <script src="js/input-num.js"></script>
    <script>
        const app = Vue.createApp({
            //引入数字输入框组件
            components: {
                'input-number': inputNumber
            }
        })
        app.mount('#app')
    </script>
</body>
```

扫一扫

视频讲解

11.10 自定义指令

Vue 为用户提供了功能丰富的内置指令，例如 v-model、v-show 等。这些内置指令可满足大部分业务需求，但有时用户需要一些特殊功能，比如对普通 DOM 元素进行底层操作。幸运的是，Vue 允许用户自定义指令，实现特殊功能。

▶ 11.10.1 自定义指令的注册

与组件类似，自定义指令的注册也分为全局注册和局部注册，比如注册一个名为 v-focus 的指令，用于输入元素（<input>和<textarea>）初始化时自动获得焦点，两种注册的示例代码如下：

```
const app = Vue.createApp({})
//注册一个全局自定义指令 v-focus
app.directive('focus', {
```

```
        //指令选项
    })
//注册一个局部自定义指令 v-focus
const app = Vue.createApp({
    directives: {
        focus: {
            //指令选项
        }
    }
})
```

一个自定义指令的选项是由几个钩子函数（均为可选）组成的。

（1）beforeMount：只调用一次，当指令第一次绑定到元素时调用，并进行初始化设置。

（2）mounted：在加载绑定元素的父组件时调用。

（3）beforeUpdate：在元素本身更新前调用。

（4）updated：在元素本身更新后调用。

（5）beforeUnmount：在卸载绑定元素的父组件前调用。

（6）unmounted：只调用一次，当指令与元素解除绑定时调用。

用户可以根据业务需求在不同的钩子函数中完成业务逻辑代码，例如下面的实例。

【例 11-23】自定义名为 v-focus 的指令，用于输入元素<input>（加载绑定父组件调用 mounted 函数）初始化时自动获得焦点。

其具体代码如下：

```
<div id="app">
    <input v-focus type="text"/>
</div>
<script src="js/vue.global.js"></script>
<script>
    const app = Vue.createApp({})
    //自定义指令 focus，在模板中使用 v-focus
    app.directive('focus', {
        mounted(el) {
            el.focus()
        }
    })
    app.mount('#app')
</script>
```

例 11-23 的代码中，el 为钩子函数的参数，除了 el 参数以外，还有 binding、vnode 和 prevNode 参数，它们的含义具体如下。

（1）el：指令所绑定的元素，可以用来直接操作 DOM。

（2）binding：一个对象，包含以下常用属性。

- value：指令的绑定值。例如在 v-my-directive="1 + 1"中绑定值为 2。
- oldValue：指令绑定的前一个值，仅在 updated 钩子函数中可用，无论值是否改变都可用。
- arg：传给指令的参数，可选。例如在 v-my-directive:foo 中参数为 "foo"。
- modifiers：一个包含修饰符的对象。例如在 v-my-directive.foo.bar 中修饰符对象为 { foo: true, bar: true }。

（3）vnode：Vue 编译生成的虚拟节点。

（4）prevNode：上一个虚拟节点，仅在 updated 钩子函数中可用。

11.10.2 实战：实时时间转换指令

例如大家发布的朋友圈会有一个相对本机时间转换后的时间，如图 11.19 中方框的时间。

图 11.19 某朋友圈发文

为了显示实时性，在一些社交类软件中经常将 UNIX 时间戳转换为可读的时间格式，例如几分钟前、几小时前、几天前等。本实战将实现这样一个自定义指令 v-time，该指令将传入的时间戳实时转换为需要的时间格式。

其具体实现步骤如下：

（1）为了统一使用时间戳进行逻辑判断，在编写指令 v-time 前先编写一系列与时间相关的函数，将这些函数封装在 Time 对象中，time.js 的代码如下：

```
var Time = {
//获得当前时间戳
getUnix: function() {
    var date = new Date();
    return date.getTime();
},
//获得今天0点0分0秒的时间戳
getTodayUnix: function() {
    var date = new Date();
    date.setHours(0);
    date.setMinutes(0);
    date.setSeconds(0);
    date.setMilliseconds(0);
    return date.getTime();
},
//获取标准年月日
getNormalDate: function(time) {
    var date = new Date(time);
    var month = date.getMonth() + 1;
    var monthFormate = month < 10 ? ('0' + month) : month;
    var day = date.getDate() < 10 ? ('0' + date.getDate()) : date.getDate();
    return date.getFullYear() + '-' + monthFormate + '-' + day;
},
//自定义指令需要调用的函数，参数为毫秒级时间戳
getFormatTime: function (timeStamp) {
    //获得当前时间戳
    var now = this.getUnix();
    //获得今天0点0分0秒的时间戳
    var today = this.getTodayUnix();
    //转换秒级时间
    var timer = (now - timeStamp) / 1000;
    var timeFormat = '';
```

```
            if (timer <= 0) {
                timeFormat = '刚刚';
            } else if (Math.floor(timer / 60) <= 0) {    //1 分钟以前显示刚刚
                timeFormat = '刚刚';
            } else if (timer < 3600) {    //1 分钟~1 小时显示 XX 分钟前
                timeFormat = Math.floor(timer / 60) + '分钟前';
            } else if (timer >= 3600 && (timeStamp - today) >= 0) {    //1 小时~
1 天显示 XX 小时前
                timeFormat = Math.floor(timer / 3600) + '小时前';
            } else if (timer / 86400 <= 31) {    //1 天~1 个月显示 XX 天前
                timeFormat = Math.ceil(timer / 86400) + '天前';
            } else {    //超过 1 个月显示 XX 年 XX 月 XX 日
                timeFormat = this.getNormalDate(timeStamp);
            }
            return timeFormat;
        }
    }
```

（2）在 HTML 文件 v-time.html 中注册一个全局指令 v-time。在 v-time 的钩子函数 beforeMount 中，将指令表达式的值 binding.value 作为参数传入 Time.getFormatTime()方法得到格式化时间，再通过 el.innerHTML 写入指令元素，并且每分钟触发一次定时器 el.timeout，更新时间。同时，在 v-time 的钩子函数 unmounted 中清除定时器。v-time.html 的具体代码如下：

```
<div id="app">
    <div v-time="nowTime"></div>
    <div v-time="beforeTime"></div>
</div>
<script src="js/vue.global.js"></script>
<script src="js/time.js"></script>
<script>
    const app = Vue.createApp({
        data() {
            return {
                //nowTime 是目前的时间
                nowTime: (new Date()).getTime(),
                //beforeTime 是 2021-08-08（固定时间）
                beforeTime: 1628407242588
            }
        }
    })
    app.directive('time', {
        //绑定一次性事件等初始化操作
        beforeMount(el, binding) {
            el.innerHTML = Time.getFormatTime(binding.value)
            //定时器 1 分钟更新 1 次
            el.timeout = setInterval(function() {
                el.innerHTML = Time.getFormatTime(binding.value)
            }, 60000)
        },
        //解除相关绑定
        unmounted(el) {
            clearInterval(el.timeout)
            delete el.timeout
        }
    })
    app.mount('#app')
</script>
```

11.11 响应性

非侵入性的响应性系统是 Vue.js 最独特的特性之一。在本节将学习 Vue.js 响应性系统的实现细节。

11.11.1 什么是响应性

响应性是一种允许用户以声明式的方式去适应变化的一种编程范例。例如在某个 Excel 电子表格中，将数字 x 放在第一个单元格中，将数字 y 放在第二个单元格中，并要求自动计算 x + y 的值放在第三个单元格中。同时，如果更新数字 x 或 y，第三个单元格中的值也会自动更新。

Vue.js 如何追踪数据的变化呢？在生成 Vue.js 实例时使用带有 getter 和 setter 的处理程序遍历传入的 data，将其所有 property 转换为 Proxy 对象，见例 11-24。Proxy 代理对象，顾名思义是在访问对象前增加一个中间层，通过中间层做一个中转，通过操作代理对象实现目标对象的修改。Proxy 对象对于用户来说是不可见的，但在内部，它使 Vue.js 能够在 property 值被访问或修改的情况下进行依赖跟踪和变更通知。

【例 11-24】property 转换为 Proxy 对象。

其具体代码如下：

```
<script>
    const data = {
        uname: 'chenheng',
        age: 90
    }
    const handler = {
        get(target, name, receiver) {
            alert('执行 get 方法')
            //Reflect.get 方法查找并返回 target 对象的 name 属性，如果没有该属性，则
            //返回 undefined
            return Reflect.get(...arguments)
        },
        set(target, name, value, receiver) {
            alert('执行 set 方法')
            //Reflect.set 方法设置 target 对象的 name 属性等于 value
            return Reflect.set(...arguments)
        }
    }
    const proxy = new Proxy(data, handler)
    alert(proxy.uname)            //执行 get 方法
    proxy.uname = 'hhhhh'         //执行 set 方法
    alert(proxy.uname)            //执行 get 方法
</script>
```

11.11.2 响应性的原理

reactive()方法和 watchEffect()方法是 Vue.js 3 中响应式的两个核心方法，reactive()方法负责将数据变成响应式代理对象，watchEffect()方法的作用是监听数据变化从而更新视图或调用函数。reactive()方法和 watchEffect()方法的应用示例见例 11-25。

【例 11-25】reactive()方法和 watchEffect()方法的应用。

其具体代码如下：
```
<script src="js/vue.global.js"></script>
<script>
    //reactive()方法接收一个普通对象，然后返回该对象的响应式代理
    let book = Vue.reactive({
        title: 'SSM + Spring Boot + Vue.js 3全栈开发从入门到实战（微课视频版）',
        author: '陈恒'
    })
    //watchEffect()方法监听数据变化从而更新视图或调用函数
    Vue.watchEffect(() => {
        alert(book.title)
    })
    book.title = 'Vue.js 3从入门到实战（微课视频版）'
</script>
```

11.12 setup 组件选项

扫一扫

视频讲解

Vue 组件提供了 setup 选项，供开发者使用组合 API。setup 选项在创建组件前执行，一旦 props 被解析，便充当组合式 API 的入口点。由于在执行 setup 时尚未创建组件实例，所以在 setup 选项中没有 this。这意味着除了 props 以外，无法访问组件中声明的任何属性，包括本地状态、计算属性或方法。

setup 选项是一个接收 props 和 context 参数的函数。此外，从 setup 返回的所有内容都将暴露给组件的其余部分（计算属性、方法、生命周期钩子、模板等）。

▶ 11.12.1 setup 函数的参数

❶ setup 函数中的第一个参数（props）

setup 函数中的 props 是响应式的，当传入新的属性时它将被更新，见例 11-26。

【例 11-26】在 setup 函数中参数 props 是响应式的。

其具体代码如下：
```
<template id="stesting">
    {{mybook}}
</template>
<div id="app">
    <setup-testing :abook="book"></setup-testing>
</div>
<script src="js/vue.global.js"></script>
<script>
    const app = Vue.createApp({
        data() {
            return {
                book: {
                    id: 1,
                    title: 'My journey with Vue'
                }
            }
        }
    })
    app.component('setup-testing', {
        props: ['abook'],
        setup(props) {   // props 是响应式的
```

```
            console.log(props.abook.id)
            mybook = props.abook
            //暴露给 template
            return {
                mybook
            }
        },
        template: '#stesting'
    })
    app.mount('#app')
</script>
```

但是，因为 props 是响应式的，不能使用 ES6 解构，将会消除 props 的响应性。如果需要解构 props，可以在 setup 函数中使用 toRefs 函数来完成，见例 11-27。

【例 11-27】在 setup 函数中使用 toRefs 函数创建 props 属性的响应式引用。

其具体代码如下：

```
<template id="stesting">
    {{mybook}}
</template>
<div id="app">
    <setup-testing :mytitle="book.title"></setup-testing>
</div>
<script src="js/vue.global.js"></script>
<script>
    const app = Vue.createApp({
        data() {
            return {
                book: {
                    id: 1,
                    title: 'My journey with Vue'
                }
            }
        }
    })
    app.component('setup-testing', {
        props: ['mytitle'],
        setup(props) {
            //使用 toRefs 函数创建 props 属性的响应式引用
            title = Vue.toRefs(props)
            //使用 ES6 解构
            console.log(title.mytitle.value)
            mybook = title
            //暴露给 template
            return {
                mybook
            }
        },
        template: '#stesting'
    })
    app.mount('#app')
</script>
```

❷ setup 函数中的第二个参数（context）

context 上下文是一个普通的 JavaScript 对象，它暴露组件的 4 个属性，即 attrs、slots、emit 以及 expose。示例代码如下：

```
setup(props, context) {
    //Attribute（非响应式对象，等同于$attrs）
```

```
        console.log(context.attrs)
        //插槽（非响应式对象，等同于$slots）
        console.log(context.slots)
        //触发事件（方法，等同于$emit）
        console.log(context.emit)
        //暴露公共property（函数）
        console.log(context.expose)
}
```

context 是一个普通的 JavaScript 对象，也就是说它不是响应式的，这意味着可以安全地对 context 使用 ES6 解构。示例代码如下：

```
setup(props, { attrs, slots, emit, expose }) {
    ...
}
```

attrs 和 slots 是有状态的对象，它们随组件本身的更新而更新。这意味着应该避免对它们进行解构，并始终以 attrs.x 或 slots.x 的方式引用属性。

▶ 11.12.2　setup 函数的返回值

❶ 对象

如果 setup 返回一个对象，则可以在组件的模板中访问该对象的属性。下面通过一个实例讲解 setup 函数的使用方法。

【例 11-28】使用 setup 函数返回一个对象。

其具体代码如下：

```
<template id="stesting">
    <!--在模板中使用readersNumber对象会被自动开箱，所以不需要.value-->
    <div>{{ readersNumber }} {{ book.title }}</div>
</template>
<div id="app">
    <setup-testing></setup-testing>
</div>
<script src="js/vue.global.js"></script>
<script>
    const app = Vue.createApp({})
    app.component('setup-testing', {
        setup() {
            //使用ref函数，对值创建一个响应式引用，并返回一个具有value属性的对象
            const readersNumber = Vue.ref(1000)
            //reactive()接收一个普通对象，然后返回该对象的响应式代理
            const book = Vue.reactive({ title: '好书' })
            //暴露给template
            return {
                readersNumber,
                book
            }
        },
        template: '#stesting'
    })
    app.mount('#app')
</script>
```

❷ 渲染函数

setup 还可以返回一个渲染函数，该函数可以直接使用在同一作用域中声明的响应式状态。下面通过一个实例讲解 setup 返回渲染函数。

【例 11-29】实现例 11-28 的功能，要求 setup 返回渲染函数。

其具体代码如下：

```
<div id="app">
    <setup-testing></setup-testing>
</div>
<script src="js/vue.global.js"></script>
<script>
    const app = Vue.createApp({})
    app.component('setup-testing', {
        setup() {
            //使用 ref 函数，对值创建一个响应式引用，并返回一个具有 value 属性的对象
            const readersNumber = Vue.ref(1000)
            //reactive()接收一个普通对象，然后返回该对象的响应式代理
            const book = Vue.reactive({ title: '好书' })
            //返回渲染函数
            return () => Vue.h('div', [readersNumber.value, book.title])
        }
    })
    app.mount('#app')
</script>
```

▶ 11.12.3 使用 ref 创建响应式引用

❶ 声明响应式状态

如果要为 JavaScript 对象创建响应式状态，可以使用 reactive()方法。reactive()方法接收一个普通对象，然后返回该对象的响应式代理。示例代码如下：

```
const book = Vue.reactive({ title: '好书' })
```

reactive()方法的响应式转换是"深层的"，即影响对象内部的所有嵌套属性。基于 ES 的 Proxy 实现，返回的代理对象不等于原始对象。建议用户使用代理对象，以避免依赖原始对象。例如在例 11-28 中使用代理对象 book。

❷ 使用 ref 创建独立的响应式值对象

ref 接收一个参数值并返回一个响应式且可改变的 ref 对象。ref 对象拥有一个指向内部值的单一属性.value。示例代码如下：

```
const readersNumber = Vue.ref(1000)
console.log(readersNumber.value)        //1000
readersNumber.value++
console.log(readersNumber.value)        // 1001
```

当 ref 作为渲染上下文的属性返回（即在 setup()返回的对象中）并在模板中使用时，它会自动开箱，无须在模板内额外书写.value。例如，在例 11-28 中以{{ readersNumber }}的方式使用。

当嵌套在响应式对象中时 ref 才会自动开箱，从 Array 或者 Map 等原生集合类中访问 ref 时不会自动开箱。示例代码如下：

```
const map = reactive(new Map([['foo', ref(0)]]))
//这里需要.value
console.log(map.get('foo').value)
```

在例 11-28 和例 11-29 中，使用 ref 将值封装在一个对象中，看似没有必要，但为了保持 JavaScript 中不同数据类型的行为统一，这是必须的。也就是说，任何数据类型的值都有一个封装对象，这样就可以在整个应用中安全地传递它，而不必担心在某个地方失去它的响应性。

▶ 11.12.4　在 setup 内部调用生命周期钩子函数

在 setup 内部，可以通过在生命周期钩子函数的前面加上 "on" 来访问组件的生命周期钩子函数。因为 setup 是围绕 beforeCreate 和 created 生命周期钩子函数运行的，所以不需要显式地定义它们。换句话说，在这些钩子函数中编写的任何代码都应该直接在 setup 函数中编写。这些 on 函数接受一个回调函数，当钩子函数被组件调用时将会被执行。示例代码如下：

```
setup() {
    //mounted 时执行
    onMounted(() => {
        console.log('Component is mounted!')
    })
}
```

本章小结

本章主要介绍了 Vue.js 的基础知识，在学习本章后，读者对 Vue.js 的插值与表达式、生命周期、计算属性与监听器属性、指令、组件系统等基础知识应该有一个初步的了解。组件是 Vue 最核心的功能，是非常抽象的概念，也是本章的重点。在以后的学习中，大家需要慢慢理解组件的内在机制。

习 题 11

扫一扫

自测题

1. 下列选项中能够定义 Vue.js 根实例对象的元素是（　　）。
    A．template　　　　B．script　　　　C．style　　　　D．title
2. 定义 Vue.js 根实例需要调用的方法是（　　）。
    A．createApp()　　B．mount()　　　C．createVue()　　D．create()
3. MVVM 模式中的 VM 是指（　　）。
    A．Model　　　　B．View　　　　C．ViewModel　　D．VueModel
4. 下列选项中插值不正确的是（　　）。
    A．{{myValue}}　B．{{one.join(two)}}　C．{{const a = 1}}　D．{{x + y}}
5. 下列选项中能够动态渲染数据属性的是（　　）。
    A．methods　　　B．watch　　　　C．computed　　　D．data
6. 下列有关 computed 属性的描述错误的是（　　）。
    A．computed 属性默认只有 getter 方法，但可以为其提供一个 setter 方法
    B．computed 属性所依赖的数据发生变化时将自动重新计算，并把计算结果缓存起来
    C．computed 属性一定有返回值，而且不带参数
    D．computed 属性只有 getter 方法
7. 下列选项中能够实现绑定属性的指令是（　　）。
    A．v-once　　　　B．v-bind　　　　C．v-model　　　D．v-on
8. 下列选项中能够实现表单数据的双向绑定的指令是（　　）。
    A．v-once　　　　B．v-if　　　　　C．v-model　　　D．v-on
9. 下列指令中能够捕获事件的是（　　）。
    A．v-for　　　　　B．v-bind　　　　C．v-if　　　　　D．v-on

10．下列事件修饰符中能够阻止事件冒泡的是（　　）。
　　A．.stop　　　　B．.prevent　　　C．.capture　　　D．.self
11．下列事件修饰符中能够实现事件只被触发一次的是（　　）。
　　A．.prevent　　B．.once　　　　C．.capture　　　D．.self
12．下列 v-model 修饰符中能够将用户的输入值自动转换为数值类型的是（　　）。
　　A．.trim　　　　B．.lazy　　　　C．.number　　　D．所有选项都不是
13．下列指令中不属于条件渲染指令的是（　　）。
　　A．v-if　　　　B．v-show　　　C．v-else-if　　　D．v-for
14．在父组件中绑定 myData="[1,2,3,5]"传递给子组件，在子组件中显示 myData.length 的值为（　　）。
　　A．5　　　　　B．3　　　　　　C．4　　　　　　D．9
15．子组件可以通过（　　）属性声明使用父组件的变量。
　　A．data　　　　B．message　　　C．parent　　　　D．props
16．父组件可以作为其所有子组件的依赖项提供程序，而不管组件的层次结构有多深，父组件有一个（　　）选项来提供数据，子组件有一个（　　）选项来使用这个数据。
　　A．provide inject　B．set get　　C．push get　　　D．provide props
17．什么是计算属性？为什么要使用计算属性？
18．简述计算属性和监听器属性的区别。
19．参考例 11-3，编写一个 HTML 页面 practice11_1.html，在该页面中输入 3 个数字，使用计算属性判断这 3 个数字是否构成三角形，运行结果如图 11.20 所示。

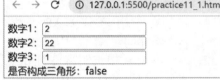

20．简述 Vue.js 的生命周期。
21．什么是 MVVM 模式？简述 Model、View、ViewModel 三者之间的关系。
22．如何创建一个 Vue.js 实例？又如何将 Vue.js 实例加载到一个 DOM 元素上？

图 11.20　是否构成三角形

23．参考 11.8.6 小节，将以下二维数组的数据显示在购物车界面 practice11_2.html（如图 11.21 所示）。

```
cartList:[
    {
        type:'图书',
        items:[
            {
                id: 1,
                gname: 'Spring Boot 从入门到实战',
                gprice: 79.8,
                count: 5
            },
            {
                id: 2,
                gname: 'Java Web 开发从入门到实战',
                gprice: 69.8,
                count: 10
            },
            {
                id: 3,
                gname: 'Java EE 框架整合开发入门到实战',
```

```
                    gprice: 69.8,
                    count: 100
                }
            ]
        },
        {
            type:'家电',
            items:[
                {
                    id: 1,
                    gname: '电视机X',
                    gprice: 888,
                    count: 5
                },
                {
                    id: 2,
                    gname: '冰箱Y',
                    gprice: 999,
                    count: 10
                }
            ]
        }
    ]
```

图 11.21　practice11_2.html 的运行效果

24．如何注册组件？如何区分父子组件？列出自己了解的组件传值，分析它们是如何实现的。

25．在 11.9.5 小节实战内容的基础上自定义一个奇偶数判定输入框组件，运行效果如图 11.22 所示。

图 11.22　奇偶数判定输入框

26．开发一个自定义指令 v-birthdayformat，接收一个出生日期（YYYY-MM-DD），将它转换为具体年龄，比如 8 岁 8 个月 8 天，运行效果如图 11.23 所示。

图 11.23　年龄计算自定义指令

第 12 章 Vue.js 进阶

学习目的与要求

本章主要讲解 Vue.js 的进阶知识，包括 Vue CLI、Vue Router、setup 语法糖以及 Element Plus UI 组件库。通过本章的学习，要求读者能够将 Vue.js 的进阶知识灵活地应用于综合实例中。

本章主要内容

- Vue Router 的基本用法与高级应用
- setup 语法糖
- Element Plus UI 组件库

高效的开发离不开高效的项目构建工具以及实用的插件，本章将学习 Vue CLI、Vue Router 插件以及 Element Plus UI 组件库。

12.1 单文件组件与 webpack

Vue.js 是一个渐进式的 JavaScript 框架，在用 webpack 构建 Vue 应用时可以使用一种新的构建模式——.vue。

.vue 是 Vue.js 自定义的一种文件格式，一个.vue 文件就是一个单独的组件，在文件内封装了组件的相关代码，例如 HTML、CSS、JS。

.vue 文件由三部分组成，即<template>、<style>、<script>，示例如下：

```
<template>
    html
</template>
<style>
    css
</style>
<script>
    js
</script>
```

但是浏览器本身并不识别.vue 文件，因此必须对.vue 文件进行加载解析，此时需要使用 webpack 的 vue-loader 加载器。

webpack 是一个用于 JavaScript 应用程序的静态模块打包工具。当 webpack 处理应用程序时，它会在内部从一个或多个入口点（即入口文件）构建一个依赖图（dependency graph），然后将项目中所需的每一个模块组合成一个或多个 bundles，它们均为静态资源，用于展示项目内容。

webpack 根据模块的依赖关系进行静态分析，然后将这些模块按照指定的规则生成对应的静态资源。图 12.1 是来自 webpack 官方网站（https://webpack.js.org/）的模块化示意图。

第 12 章　Vue.js 进阶

图 12.1　webpack 模块化示意图

图 12.1 的左边是业务中编写的各种类型文件，例如 TypeScript、JPG、Less、CSS，以及后续将要学习的 .vue 格式的文件。这些类型的文件通过特定的加载器（Loader）编译后最终统一生成 .js、.css、.jpg、.png 等静态资源文件。在 webpack 中，一张图片、一个 CSS 文件等都被称为模块，它们彼此存在依赖关系。使用 webpack 的目的就是处理模块间的依赖关系，并将它们进行打包。

webpack 的主要适用场景是单页面应用（SPA），SPA 通常是由一个 HTML 文件和一堆按需加载的 JS 文件组成的。

webpack 在搭建单页面应用程序时会安装许多插件并编写复杂的项目配置，大大降低了开发效率。为了提高单页面应用程序的开发效率，本章将使用 Vue CLI（Vue 脚手架）搭建 Vue.js 项目。

12.2　安装 Node.js 和 NPM

本章将使用 NPM 安装 Vue CLI，而 NPM 是 Node.js 的包管理器，集成在 Node.js 中，所以需要首先安装 Node.js。

▶ 12.2.1　安装 Node.js

通过访问官网"https://nodejs.org/en/"即可下载对应版本的 Node.js，本书下载的是"16.15.1 LTS"。

在下载完成后运行安装包 node-v16.15.1-x64.msi，一直单击"下一步"按钮即可完成安装。然后在命令行窗口中输入命令 node-v 检查是否安装成功，如图 12.2 所示。

图 12.2　查看 Node.js 的版本

如图 12.2 所示，出现了 Node.js 的版本号，说明 Node.js 已经安装成功。同时，NPM 包管理器也已经安装成功，可以输入 npm -v 查看版本号。最后输入 npm -g install npm 命令，可以将 NPM 更新至最新版本。

12.2.2 NPM 常用的命令

NPM（Node Package Manager）是和 Node.js 一起安装的包管理工具，方便下载、安装、上传以及管理包，解决了 Node 代码部署上的很多问题，经常用于以下 3 种场景。

（1）允许用户从 NPM 服务器下载别人编写的第三方包到本地使用。
（2）允许用户从 NPM 服务器下载并安装别人编写的命令行程序到本地使用。
（3）允许用户将自己编写的包或命令行程序上传到 NPM 服务器供别人使用。

NPM 有一个开源的面向文档的数据库管理系统支撑，详细记录了每个包的信息，包括作者、版本、依赖、授权等。它的作用是将开发者从繁琐的包管理工作中解放出来，使其更加专注于功能业务的开发。

下面讲解 NPM 的常用命令。

❶ 检测是否安装并显示相应版本

```
npm -v          #显示版本号说明已经安装了相应的版本
```

❷ 生成 package.json 文件

package.json 用来描述项目中用到的模块和其他信息，可以使用如下命令生成该文件：

```
npm init
```

❸ 安装模块

安装 package.json 定义好的模块，简写为 npm i。

```
npm install
```

安装包指定模块，具体命令如下：

```
npm i <ModuleName>
```

全局安装，具体命令如下：

```
npm i <ModuleName> -g
```

在安装包的同时将信息写入 package.json 的 dependencies 配置中，具体命令如下：

```
npm i <ModuleName> --save
```

在安装包的同时将信息写入 package.json 的 devDependencies 配置中，具体命令如下：

```
npm i <ModuleName> --save-dev
```

安装多模块，具体命令如下：

```
npm i <ModuleName1> <ModuleName2>
```

安装方式参数，具体如下：

```
-save           #简写-S，加入生产依赖中
-save-dev       #简写-D，加入开发依赖中
-g              #全局安装，将安装包放在/usr/local 下或者 Node 的安装目录中
```

❹ 查看

查看所有全局安装的包，具体命令如下：

```
npm ls -g
```

查看本地项目中安装的包，具体命令如下：

```
npm ls
```

查看包的 package.json 文件，具体命令如下：

```
npm view <ModuleName>
```

查看包的依赖关系，具体命令如下：

```
npm view <ModuleName> dependencies
```

查看包的源文件地址，具体命令如下：

```
npm view <ModuleName> repository.url
```

查看包所依赖的 Node 版本，具体命令如下：

```
npm view <ModuleName> engines
```

查看帮助，具体命令如下：

```
npm help
```

❺ 更新模块

更新本地模块，具体命令如下：

```
npm update <ModuleName>
```

更新全局模块，具体命令如下：

```
npm update -g <ModuleName>        #更新全局软件包
npm update -g                     #更新所有的全局软件包
npm outdated -g --depth=0         #找出需要更新的包
```

❻ 卸载模块

卸载本地模块，具体命令如下：

```
npm uninstall <ModuleName>
```

卸载全局模块，具体命令如下：

```
npm uninstall -g <ModuleName>     #卸载全局软件包
```

❼ 清空缓存

清空 NPM 缓存，具体命令如下：

```
npm cache clear
```

❽ 使用淘宝镜像

使用淘宝镜像，具体命令如下：

```
npm install -g cnpm --registry=https://registry.npm.taobao.org
```

❾ 其他

更改包内容后进行重建，具体命令如下：

```
npm rebuild <ModuleName>
```

检查包是否已经过时，此命令会列出所有已经过时的包，可以及时进行包的更新，具体命令如下：

```
npm outdated
```

访问 NPM 的 JSON 文件，此命令将会打开一个网页，具体命令如下：

```
npm help json
```

在发布一个包的时候需要检验某个包名是否存在，具体命令如下：

```
npm search <ModuleName>
```

撤销自己发布过的某个版本代码，具体命令如下：

```
npm unpublish <package> <version>
```

❿ 使用技巧

多次安装不成功尝试先清除缓存，具体命令如下：

```
npm cache clean -f
```

查看已安装的依赖包的版本号，具体命令如下：

```
npm ls <ModuleName>
```

注意：用此方法才能准确地知道项目使用的版本号，在查看 package.json 时有 "^" 符号表示新于此版本。

12.3 Vue Router

Vue Router 是 Vue.js 官方的路由管理器，它和 Vue.js 的核心深度集成，使构建单页面应用变得更加容易。Vue Router 包含的功能如下：

（1）嵌套路由映射。
（2）动态路由选择。
（3）模块化、基于组件的路由配置。
（4）路由参数、查询、通配符。
（5）展示由 Vue.js 的过渡系统提供的过渡效果。
（6）细粒度的导航控制。
（7）自动激活 CSS 类的链接。
（8）HTML5 History 模式或 Hash 模式。
（9）可定制的滚动行为。
（10）URL 的正确编码。

视频讲解

12.3.1 Vue Router 的安装

将 Vue Router 添加到项目中有 4 种主要方法，即本地独立版本方法、CDN 方法、NPM 方法以及命令行工具（Vue CLI）方法。

❶ 本地独立版本方法

用户可通过网址"https://unpkg.com/vue-router@next"将最新版本的 Vue Router 库（vue-router.global.js）下载到本地（在页面上右击，在弹出的快捷菜单中选择"另存为"命令），在编写本书时 Vue Router 库的最新版本是 4.0.13，然后在界面文件中引入 vue-router.global.js 库，示例代码如下：

```
<script src="js/vue-router.global.js"></script>
```

❷ CDN 方法

读者在学习或开发时可以在界面文件中通过 CDN（Content Delivery Network，内容分发网络）引入最新版本的 Vue Router 库，示例代码如下：

```
<script src="https://unpkg.com/vue-router@next"></script>
```

对于生产环境，建议使用固定版本，以免因版本不同带来兼容性问题，示例代码如下：

```
<script src="https://unpkg.com/vue-router@4.0.13/dist/vue-router.global.js"></script>
```

❸ NPM 方法

在使用 Vue.js 构建大型应用时推荐使用 NPM 安装最新稳定版的 Vue Router，因为 NPM 能很好地和 webpack 模块打包器配合使用，示例代码如下：

```
npm install vue-router@next
```

❹ 命令行工具（Vue CLI）方法

为了提高单页面应用程序的开发效率，这里使用 Vue CLI（Vue 脚手架）搭建 Vue.js 项目。Vue CLI 是一个基于 Vue.js 进行快速开发的完整系统，提供了如下功能：

（1）通过@vue/cli 实现交互式项目脚手架。
（2）通过@vue/cli + @vue/cli-service-global 实现零配置原型开发。

(3)一个运行时依赖@vue/cli-service,该依赖可升级,基于 webpack 构建,并带有合理的默认配置;可通过项目的配置文件进行配置;可通过插件进行扩展。

(4)一个丰富的官方插件集合,集成了前端生态工具。

(5)提供一套创建和管理 Vue.js 项目的用户界面。

Vue CLI 致力于将 Vue.js 生态工具基础标准化,确保各种构建工具平稳衔接,让开发者专注于撰写应用上,而不必纠结配置的问题。下面讲解如何安装 Vue CLI 以及如何使用 Vue CLI 创建 Vue.js 项目,具体步骤如下:

1)全局安装 Vue CLI

打开 cmd 命令行窗口,输入命令 npm install -g @vue/cli 全局安装 Vue 脚手架,输入命令 vue --version 查看版本(测试是否安装成功)。如果需要升级全局的 Vue CLI,在 cmd 命令行窗口中运行 npm update -g @vue/cli 命令即可。

2)打开图形化界面

安装成功后,在命令行窗口中输入命令 vue ui 打开一个浏览器窗口,并以图形化界面引导至项目创建的流程,如图 12.3 所示。

图 12.3　Vue CLI 图形化界面

3)创建项目

在图 12.3 中单击"创建"按钮进入创建项目界面,如图 12.4 所示。

图 12.4　创建项目界面

在图 12.4 中输入并选择项目位置信息，然后单击"+在此创建新项目"按钮进入项目详情界面，如图 12.5 所示。

图 12.5　项目详情界面

在图 12.5 中输入并选择项目相关信息，然后单击"下一步"按钮进入项目预设界面，选择手动，单击"下一步"按钮进入项目功能界面，在项目功能界面中激活 Router 按钮，安装 vue-router 插件，为本节后续内容做准备，如图 12.6 所示。

图 12.6　项目功能界面

在图 12.6 中单击"下一步"按钮进入项目配置界面，配置后单击"创建项目"按钮即可完成项目 router-demo 的创建（可能需要一定的创建时间），如图 12.7 所示。

图 12.7　项目配置界面

4）使用 VSCode 打开项目

使用 VSCode 打开（选择"文件"→"打开文件夹"命令，选择项目目录）第 3 步创建的项目 router-demo，打开后在 Terminal 终端输入 npm run serve 命令启动服务，如图 12.8 所示。

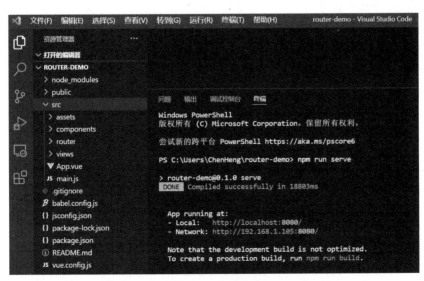

图 12.8　启动服务

5）运行项目

在浏览器的地址栏中访问"http://localhost:8080/"即可运行项目 router-demo，如图 12.9 所示。

图 12.9　运行项目

当通过 "http://localhost:8080/" 访问时,打开的页面是 public 目录下的 index.html。index.html 是一个普通的 HTML 文件,让它与众不同的是 "<div id="app"></div>" 这个语句,下面有一行注释,构建的文件将会被自动注入,也就是说编写的其他内容都将在这个 div 中展示。另外,整个项目只有这一个 HTML 文件,所以这是一个单页面应用,当浏览者打开这个应用,表面上可以看到很多页面,实际上它们都在这一个 div 中显示。

在 main.js 中创建了一个 Vue 对象。该 Vue 对象的加载目标是 "#app" (与 index.html 中的 id="app"对应); router 代表该对象包含 Vue Router,并使用项目中定义的路由 (在 src/router 目录下的 index.js 文件中定义)。

综上所述,main.js 和 index.html 是项目启动的首加载页面资源和 JS 资源,App.vue 则是 Vue 页面资源的首加载项,称为根组件。Vue 项目的具体执行过程如下:首先启动项目,找到 index.html 与 main.js,执行 main.js (入口程序),根据 import 加载 App.vue 根组件;然后将组件内容渲染到 index.html 中 id="app"的 DOM 元素上。

扫一扫

视频讲解

▶ 12.3.2　Vue Router 的基本用法

在使用 Vue Router 动态加载不同组件时,需要将组件 (Components) 映射到路由 (Routers),然后告诉 Vue Router 在哪里显示它们。

❶ 跳转

Vue Router 有两种跳转,第一种是使用内置的<router-link>组件,默认渲染一个<a>标签,示例代码如下:

```
<div id="nav">
    <router-link to="/">第一个页面</router-link> |
    <router-link to="/MView2">第二个页面</router-link>
</div>
```

<router-link>组件和一般组件一样,to 是一个 prop,指定跳转的路径。使用<router-link>组件,在 HTML5 的 History 模式下将拦截单击,避免浏览器重新加载页面。<router-link>组件还有以下常用属性:

（1）tag 属性：指定渲染的标签，例如<router-link to="/" tag="li">渲染的结果是<li>而不是<a>。

（2）replace 属性：使用 replace 不会留下 History 记录，所以导航后不能用后退键返回上一个页面，例如<router-link to="/" replace>。

Vue Router 的第二种跳转方式需要在 JavaScript 中进行，类似于 window.location.href。这种方式需要使用 router 实例方法 push 或 replace。例如在 MView1.vue 中通过单击事件跳转，示例代码如下：

```
<template>
    <div>第一个页面</div>
    <button @click="goto">去第二个页面</button>
</template>
<script>
export default {
    methods: {
        goto() {
            //也可以使用 replace 方法，和 replace 属性一样不会向 History 添加新记录
            this.$router.push('/MView2')
        }
    }
}
</script>
```

❷ 传参

路由传参一般有两种方式，即 query 和 params，不管哪种方式都是通过修改 URL 来实现。

1）query 传参

query 传递参数的示例代码如下：

```
<router-link to="/?id=888&pwd=999">
```

通过$route.query 获取路由中的参数，示例代码如下：

```
<h4>id: {{$route.query.id}}</h4>
<h4>pwd: {{$route.query.pwd}}</h4>
```

2）params 传参

在路由规则中定义参数，修改路由规则的 path 属性（动态匹配），示例代码如下：

```
{
path: '/:id/:pwd',
name: 'MView1',
component: MView1
}
<router-link to="/888/999">
```

通过$route.params 获取路由中的参数，示例代码如下：

```
<h4>id: {{$route.params.id}}</h4>
<h4>pwd: {{$route.params.pwd}}</h4>
```

❸ 配置路由

路由配置通常在前端工程项目的 src/router/index.js 文件中进行。首先需要在前端工程项目的 src/main.js 和 src/router/index.js 文件中分别导入 vue 和 vue-router 模块，并在 main.js 中执行 use 方法注册路由。

main.js 的示例代码如下：

```
import { createApp } from 'vue'
import App from './App.vue'
import router from './router'
```

```
//导入router目录中的index.js, 路由的创建与配置在该文件中
//将router注册到根实例App
createApp(App).use(router).mount('#app')
```

index.js 的示例代码如下：

```javascript
import { createRouter, createWebHistory } from 'vue-router'
//导入组件
import HomeView from '../views/HomeView.vue'
//定义路由
const routes = [
    {
        path: '/',
        name: 'home',
        component: HomeView
    },
    {
        path: '/about',
        name: 'about',
        //导入组件
        component: () => import('../views/AboutView.vue')
    }
]
//创建路由实例router(管理路由), 传入routes配置
const router = createRouter({
    //用createWebHistory()创建HTML5模式, 推荐使用这个模式
    history: createWebHistory(process.env.BASE_URL),
    routes
})
export default router
```

下面通过实例讲解路由的跳转、传参以及配置过程。

【例 12-1】 Vue Router 实战——3 个组件间的跳转与传参。

本实例在 12.3.1 节 Vue Router 项目 router-demo 的基础上进行，其他步骤具体如下：

1）创建共通组件 CommonView.vue

在 router-demo/src/components 目录下创建信息显示共通组件 CommonView.vue，具体代码如下：

```vue
<template>
    <div>
        <img alt="Vue logo" src="../assets/logo.png">
        <h1>{{ msg }}</h1>
    </div>
</template>
<script>
export default {
    name: 'CommonView',
    props: {
        msg: String
    }
}
</script>
```

2）创建视图组件 FirstView.vue

在 router-demo/src/views 目录下创建视图组件 FirstView.vue。在该视图组件中使用 CommonView.vue 组件显示信息，并通过$route.query 获取路由中的参数，具体代码如下：

```vue
<template>
    <div>
        <CommonView msg="欢迎来访第一个 View"/>
```

```
            <br>
            <h4>uname: {{$route.query.uname}}</h4>
            <h4>pwd: {{$route.query.pwd}}</h4>
        </div>
</template>
<script>
import CommonView from '@/components/CommonView.vue'
export default {
    name: 'FirstView',
    components: {
        CommonView
    }
}
</script>
```

3）创建视图组件 SecondView.vue

在 router-demo/src/views 目录下创建视图组件 SecondView.vue。在该视图组件中使用 CommonView.vue 组件显示信息，并通过$route.params 获取路由中的参数，具体代码如下：

```
<template>
    <div>
        <CommonView msg="欢迎来访第二个 View"/>
        <br>
        <h4>uname: {{$route.params.uname}}</h4>
        <h4>pwd: {{$route.params.pwd}}</h4>
    </div>
</template>
<script>
import CommonView from '@/components/CommonView.vue'
export default {
    name: 'SecondView',
    components: {
        CommonView
    }
}
</script>
```

4）修改根组件 App.vue

在根组件 App.vue 中通过路由跳转组件<router-link>跳转到视图组件 FirstView.vue 和 SecondView.vue，并传递参数，具体代码如下：

```
<template>
    <nav>
        <router-link to="/first?uname=chenheng&pwd=123456">第一个页面</router-link> |
        <router-link to="/second/chenheng1/654321">第二个页面</router-link>
    </nav>
    <!--router-view 表示路由出口,将匹配到的组件（相当于链接的页面）渲染在这里-->
    <router-view/>
</template>
```

5）配置路由

在 router-demo/src/router/index.js 文件中配置根组件 App.vue 中的路由，具体代码如下：

```
import { createRouter, createWebHistory } from 'vue-router'
//导入组件
import SecondView from '../views/SecondView.vue'
//定义路由
```

```
const routes = [
    {
        path: '/first',
        name: 'first',
        //导入组件
        component: () => import('../views/FirstView.vue')
    },
    {
        path: '/second/:uname/:pwd',
        name: 'second',
        //导入组件
        component: SecondView
    }
]
//创建路由实例 router（管理路由），传入 routes 配置
const router = createRouter({
    history: createWebHistory(process.env.BASE_URL),
    routes
})
export default router
```

6）测试运行

首先在项目 router-demo 的 Terminal 终端输入 npm run serve 命令启动服务，然后在浏览器的地址栏中输入"http://localhost:8080/"，运行效果如图 12.10 所示。

图 12.10　例 12-1 的首页面

单击图 12.10 中的"第一个页面"超链接，打开第一个视图组件，如图 12.11 所示。

图 12.11　例 12-1 的第一个视图组件

单击图 12.10 中的"第二个页面"超链接，打开第二个视图组件，如图 12.12 所示。

图 12.12　例 12-1 的第二个视图组件

注意：在本书提供的源代码中，读者可以使用 VS Code 打开本章对应的代码目录，并在 Terminal 终端执行 npm install 命令自动安装所有依赖，然后执行 npm run serve 命令启动服务即可运行本章实例。

12.3.3　Vue Router 的高级应用

扫一扫

视频讲解

❶ 动态路由匹配

当需要将符合某种匹配模式的所有路由映射到同一个组件时，可以在路由路径中使用动态路径参数（例如 path: '/user/:uname/:pwd'）实现。示例代码如下：

```
//定义路由
const routes = [
    {
        path: '/user/:uname/:pwd',
        name: 'user',
        component: UserView
    }
]
```

在上述示例代码中定义路由后，/user/zhangsan/123456、/user/lisi/654321 等用户都将映射到相同的路由。

每一个动态路径参数使用冒号（:）标记，冒号的后面是参数名。当匹配到一个路由时，参数值将被设置到$route.params 中，并可以在路由对应的组件内使用"$route.params.参数名"获得参数值。

如果有多个参数，即有多个冒号，在$route.params 中保存为对象。例如，路由路径 path 为/user/:uname/:pwd，则对应的访问路径为/user/zhangsan/123456，$route.params 中的对象为{ uname: 'zhangsan', pwd: '123456' }。另外，用户也可以使用 post 进行多个动态参数的传递，例如路由路径 path 为/user/:uname/post/:pwd/post/:age，则对应的访问路径为/user/:lisi/post/:654321/post/:18，$route.params 中的对象为{ uname: 'lisi', pwd: '123456', age: '18' }。

$route 路由信息对象表示当前激活的路由状态信息，每次成功导航后都将产生一个新的对象。除了$route.params 以外，$route 对象还提供了其他许多有用的信息，如表 12.1 所示。

表 12.1 $route 路由信息对象的属性

序号	属性名称	说明
1	$route.path	对应当前路由的路径，例如/third/:张三/post/:654321/post/:18
2	$route.params	一个 key:value 对象，包含了所有动态参数，如果没有参数，则是一个空对象，例如{ "uname": ":张三", "pwd": ":654321", "age": ":18" }
3	$route.query	一个 key:value 对象，表示 URL 查询参数。例如/first?uname=chenheng&pwd=123456，则有$route.query.uname 为 chenheng。如果没有查询参数，则是空对象
4	$route.hash	当前路由的哈希值（不带#），如果没有哈希值，则为空字符串
5	$route.fullPath	完成解析后的 URL，包含查询参数和哈希的完整路径
6	$route.matched	返回数组，包含当前匹配的路径中包含的所有片段所对应的配置
7	$route.name	当前路径名称
8	$route.meta	路由元信息

【例 12-2】Vue Router 实战——$route 对象的属性。

本实例在例 12-1 的基础上进行，其他步骤具体如下：

1）修改根组件 App.vue

在根组件 App.vue 中使用<router-link>添加路由链接，并在该路由链接中使用 post 进行多个动态参数的传递。修改后的 App.vue 的具体代码如下：

```
<template>
    <nav>
        <router-link to="/first?uname=chenheng&pwd=123456">第 一 个 页 面</router-link> |
        <router-link to="/second/chenheng1/654321">第二个页面</router-link>|
        <router-link to="/third/:张 三/post/:654321/post/:18">第 三 个 页 面</router-link>
    </nav>
    <router-view/>
</template>
```

2）创建视图组件 ThirdView.vue

在 router-demo/src/views 目录下创建视图组件 ThirdView.vue。在该视图组件中使用 CommonView.vue 组件显示信息，并显示$route 对象的各种属性值，具体代码如下：

```
<template>
    <div>
        <CommonView msg="欢迎来访第三个 View"/>
        <br>
        <h4>uname: {{$route.params.uname}}</h4>
        <h4>pwd: {{$route.params.pwd}}</h4>
        <h4>age: {{$route.params.age}}</h4>
        <h4>route 的 path: {{$route.path}}</h4>
        <h4>route 的 params: {{$route.params}}</h4>
        <h4>route 的 query: {{$route.query}}</h4>
        <h4>route 的 hash: {{$route.hash}}</h4>
        <h4>route 的 fullPath: {{$route.fullPath}}</h4>
        <h4>route 的 matched: {{$route.matched}}</h4>
        <h4>route 的 name: {{$route.name}}</h4>
```

```
            <h4>route 的 meta: {{$route.meta}}</h4>
        </div>
</template>
<script>
import CommonView from '@/components/CommonView.vue'
export default {
    name: 'ThirdView',
    components: {
        CommonView
    }
}
</script>
```

3）添加路由配置

在 router-demo/src/router/index.js 文件中添加路径/third/:张三/post/:654321/post/:18 对应的路由配置，具体代码如下：

```
import { createRouter, createWebHistory } from 'vue-router'
//导入组件
import SecondView from '../views/SecondView.vue'
import ThirdView from '../views/ThirdView.vue'
//定义路由
const routes = [
    {
        path: '/first',
        name: 'first',
        //导入组件
        component: () => import('../views/FirstView.vue')
    },
    {
        path: '/second/:uname/:pwd',
        name: 'second',
        //导入组件
        component: SecondView
    },
    {
        path: '/third/:uname/post/:pwd/post/:age',
        name: 'third',
        //导入组件
        component: ThirdView
    }
]
//创建路由实例 router（管理路由），传入 routes 配置
const router = createRouter({
    history: createWebHistory(process.env.BASE_URL),
    routes
})
export default router
```

4）测试运行

首先在项目 router-demo 的 Terminal 终端输入 npm run serve 命令启动服务，然后在浏览器的地址栏中输入 "http://localhost:8080/"，运行效果如图 12.13 所示。

图 12.13　例 12-2 的首页面

单击图 12.13 中的"第三个页面"超链接，打开第三个视图组件，如图 12.14 所示。

图 12.14　例 12-2 的第三个视图组件

❷ 嵌套路由

嵌套路由即路由的多层嵌套，也称为子路由。在实际应用中，嵌套路由相当于多级菜单，在一级菜单下有二级菜单，在二级菜单下有三级菜单，等等。

创建嵌套路由的一般步骤如下。

首先在根组件 App.vue 中定义基础路由（相当于一级菜单）导航，示例代码如下：

```
<nav>
    <router-link to="/">首页</router-link> |
    <router-link to="/about">关于我们</router-link> |
    <router-link to="/product">产品介绍</router-link>
</nav>
```

然后定义基础路由 product 对应的组件（ProductView.vue），示例代码如下：

```
<template>
    <div>
        <p>
            <!--为"产品介绍"定义了嵌套路由-->
            <router-link to="/product/alldev">全栈开发</router-link> |
            <router-link to="/product/JavaEE">Java EE 整合开发</router-link> |
            <router-link to="/product/SpringBoot">Spring Boot 开发</router-link>
        </p>
        <router-view/>
    </div>
</template>
```

最后完成所有嵌套路由组件的定义，并在 router/index.js 文件中定义嵌套路由。在基础路由 product 的定义中使用 children 属性定义嵌套的子路由，示例代码如下：

```
{
    path: '/product',
    name: 'product',
    component: ProductView,
```

```
        children:[    //子路由
            {
                path: '',    //空子路由为基础路由的默认显示
                component: () => import('../views/AlldevView.vue')
            },
            {
                path: 'alldev',    //注意这里没有'/'
                component: () => import('../views/AlldevView.vue')
            },
            {
                path: 'JavaEE',
                component: () => import('../views/JavaEEView.vue')
            },
            {
                path: 'SpringBoot',
                component: () => import('../views/SpringBoot.vue')
            }
        ]
}
```

下面通过一个实例讲解嵌套路由的实现过程。

【例 12-3】Vue Router 实战——嵌套路由的实现过程。

本实例在 Vue Router 项目 nested-routes 的基础上进行，其他步骤具体如下：

1）修改根组件 App.vue

在根组件 App.vue 中使用<router-link>添加基础路由链接，修改后的 App.vue 的具体代码如下：

```
<template>
    <h1>嵌套路由</h1>
    <nav>
        <router-link to="/">首页</router-link> |
        <router-link to="/about">关于我们</router-link> |
        <router-link to="/product">产品介绍</router-link>
    </nav>
    <router-view class="my-view"> </router-view>
</template>
```

2）创建视图组件 ProductView.vue

在 nested-routes/src/views 目录下创建视图组件 ProductView.vue。在该视图组件中定义嵌套路由，具体代码如下：

```
<template>
    <div>
        <p>
            <!--定义嵌套路由-->
            <router-link to="/product/alldev">全栈开发</router-link> |
            <router-link to="/product/JavaEE">Java EE 整合开发</router-link> |
            <router-link to="/product/SpringBoot">Spring Boot 开发</router-link>
        </p>
        <router-view/>
    </div>
</template>
<style scoped>
    p a {
        text-decoration: none;
    }
</style>
```

3)创建视图组件 AlldevView.vue、JavaEEView.vue 和 SpringBoot.vue

在 nested-routes/src/views 目录下分别创建子路由 alldev、JavaEE 和 SpringBoot 对应的视图组件 AlldevView.vue、JavaEEView.vue 和 SpringBoot.vue。

AlldevView.vue 的代码如下:

```
<template>
    <div>
        <img alt="alldev" src="../images/091883-all.jpg" width="200" height="300">
    </div>
</template>
```

JavaEEView.vue 的代码如下:

```
<template>
    <div>
        <img alt="javaee" src="../images/079720-javaee.jpg" width="200" height="300">
    </div>
</template>
```

SpringBoot.vue 的代码如下:

```
<template>
    <div>
        <img alt="springboot" src="../images/083960-springboot.jpg" width="200" height="300">
    </div>
</template>
```

4)添加路由配置

在 nested-routes/src/router/index.js 文件中定义基础路由 product。在基础路由 product 的定义中使用 children 属性定义嵌套的子路由,具体代码如下:

```
import { createRouter, createWebHistory } from 'vue-router'
import HomeView from '../views/HomeView.vue'
import ProductView from '../views/ProductView.vue'
const routes = [
    {
        path: '/',
        name: 'home',
        component: HomeView
    },
    {
        path: '/about',
        name: 'about',
        component: () => import('../views/AboutView.vue')
    },
    {
        path: '/product',
        name: 'product',
        component: ProductView,
        children:[           //子路由
            {
                path: '',    //空子路由为基础路由的默认显示
                component: () => import('../views/AlldevView.vue')
            },
            {
                path: 'alldev',   //注意这里没有'/'
                component: () => import('../views/AlldevView.vue')
            },
            {
                path: 'JavaEE',
                component: () => import('../views/JavaEEView.vue')
            },
            {
```

```
                path: 'SpringBoot',
                component: () => import('../views/SpringBoot.vue')
            }
        ]
    }
]
const router = createRouter({
    history: createWebHistory(process.env.BASE_URL),
    routes
})
export default router
```

5）测试运行

首先在项目 nested-routes 的 Terminal 终端输入 npm run serve 命令启动服务，然后在浏览器的地址栏中输入"http://localhost:8080/"，运行效果如图 12.15 所示。

单击图 12.15 中的"产品介绍"超链接，打开空子路由对应的 AlldevView.vue 视图组件，如图 12.16 所示。

图 12.15 例 12-3 的首页面

图 12.16 默认显示空子路由对应的视图组件

单击图 12.16 中的"Java EE 整合开发"超链接在子路由上切换导航，导航到"Java EE 整合开发"视图组件，如图 12.17 所示。

单击图 12.16 中的"Spring Boot 开发"超链接在子路由上切换导航，导航到"Spring Boot 开发"视图组件，如图 12.18 所示。

图 12.17 "Java EE 整合开发"视图组件

图 12.18 "Spring Boot 开发"视图组件

❸ 编程式导航

除了可以使用内置的<router-link>组件渲染一个<a>标签定义导航链接以外，还可以通过编程调用路由（router 或 this.$router）的实例方法实现导航链接。导航常用的路由实例方法如表 12.2 所示。

表 12.2　导航常用的路由实例方法

序　号	方 法 名 称	功　能　说　明
1	push()	跳转到由参数指定的新路由地址，在历史记录中添加一条新记录
2	replace()	跳转到由参数指定的新路由地址，替换当前的历史记录
3	go(n)	n 为整数，在历史记录中向前或后退 n 步
4	forward()	在历史记录中向前一步，相当于 this.$router.go(1)
5	back()	在历史记录中后退一步，相当于 this.$router.go(-1)

push()方法和 replace()方法的用法相似，唯一不同的是 push()方法在历史记录中添加一条新记录，replace()方法不会添加新记录，而是替换当前记录。下面介绍 push()方法和 replace()方法的参数。

push()方法和 replace()方法的参数可以是字符串、对象、命名路由以及带查询参数等多种形式，示例如下：

```
//字符串路由 path
this.$router.push('/')
//对象
this.$router.push({path: '/product'})
//命名路由及 params 传参，params 更像 post，是隐性传参
this.$router.push({name: 'home', params:{uname:'123', pwd:'abc'} })
//带查询参数，/product?uname=123&pwd=abc，更像 get 传参，是显性传参
this.$router.push({path: '/product', query:{uname:'123', pwd:'abc'} })
```

注意：Vue Router 更新后弃用了 params 传参，可以使用 History API 方式传递和接收，在跳转前的页面使用 state 参数。示例代码如下：

```
this.$router.push({name: 'home', state:{uname:'123', pwd:'abc'} })
```

在 home 对应的页面可以使用 history.state 接收数据。

下面通过一个实例讲解编程式导航的应用。

【例 12-4】Vue Router 实战——编程式导航的应用。

本实例在例 12-3 的基础上修改根组件 App.vue 即可（本实例的项目名称为 programming-navigation），其他步骤与例 12-3 相同，具体如下：

1）修改根组件 App.vue

App.vue 的代码如下：

```
<template>
    <h1>编程式导航及嵌套路由</h1>
    <!--编程式导航-->
    <button @click="go1">前进一步</button>
    <button @click="back1">后退一步</button>
    <button @click="goHome">回首页</button>
    <button @click="goProduct">看产品介绍</button>
    <button @click="repAbout">代替关于我们</button>
    <nav>
```

```
        <router-link to="/">首页</router-link> |
        <router-link to="/about">关于我们</router-link> |
        <router-link to="/product">产品介绍</router-link>
    </nav>
    <router-view class="my-view"> </router-view>
</template>
<script>
export default {
    name: 'App',
    methods: {
        go1(){
            this.$router.forward()
        },
        back1(){
            this.$router.back()
        },
        goHome(){
            this.$router.push('/')   //字符串路由 path
        },
        goProduct(){
            this.$router.push({   //对象
                path: '/product'
            })
        },
        repAbout(){
            this.$router.replace({
                name: 'home'   //命名路由
            })
        }
    }
}
</script>
```

2）测试运行

首先在项目 programming-navigation 的 Terminal 终端输入 npm run serve 命令启动服务，然后在浏览器的地址栏中输入"http://localhost:8080/"，运行效果如图 12.19 所示。

图 12.19　例 12-4 的首页面

❹ 命名路由

在链接一个路由或执行跳转时,对路由定义一个名称(name)将显得方便一些。命名路由的示例代码如下:

```
{
    path: '/',
    name: 'home',
    component: HomeView
}
```

如果需要链接到一个命名路由,可以给<router-link>的:to 属性传递一个对象,示例代码如下(注意 to 前面的冒号):

```
<router-link:to="{name: 'home', params: {uname: '123', pwd: 'abc'}}">首页</router-link>
```

上述示例代码的功能与编程式导航 this.$router.push({name: 'home', params:{uname:'123', pwd:'abc'} })的功能相同。

❺ 重定向

用户可通过路由配置完成路由的重定向,例如实现从/first 重定向到/second,路由配置的代码如下:

```
{
    path: '/first',
    redirect: '/second',
    name: 'first',
    component: FirstView
}
```

重定向的目标也可以是一个命名路由,路由配置的代码如下:

```
{
    path: '/first',
    redirect: {name: 'second'},
    name: 'first',
    component: FirstView
}
```

❻ 路由组件 props 传参

在组件中使用$route 将使路由与组件形成高度耦合,从而使组件只能在某些特定的 URL 上使用,限制了组件的灵活性。在配置路由时,使用 props 传参可降低路由与组件的耦合度。路由组件传参的具体示例如下。

1)导航组件

假设导航组件中有如下链接:

```
<router-link to="/">首页</router-link>
```

2)使用 props 传参

可以使用 props 配置 URL "/" 对应的路由,并传递参数给组件 HomeView,具体代码如下:

```
{
    path: '/',
    name: 'home',
    component: HomeView,
    props: {uname: '张三', upwd: '123456'}
}
```

3）通过 props 接收参数

最后在目标视图组件 HomeView 中通过 props 接收参数，具体代码如下：

```
<template>
    <div class="home">
        <h1>{{uname}}</h1>
        <h1>{{upwd}}</h1>
    </div>
</template>
<script>
export default {
    name: 'HomeView',
    props: {
        uname: {type: String, default: 'lisi'},
        upwd: {type: String, default: '000000'}
    }
}
</script>
```

▶ 12.3.4 路由钩子函数

在路由跳转时，用户可能需要一些权限判断或者其他操作，这时需要使用路由的钩子函数。路由钩子函数主要是给使用者在路由发生变化时进行一些特殊处理而定义的函数，又称为路由守卫或导航守卫。

❶ 全局前置钩子函数

在 Vue Router 中，使用 router.beforeEach 注册一个全局前置钩子函数（在路由跳转前执行），注册示例代码如下：

```
const router = new createRouter({ ... })
router.beforeEach((to, from) => {
    // ...
    //返回 false，以取消导航
    return false
})
```

当一个导航触发时，全局前置钩子函数按照创建顺序调用。beforeEach 函数接收两个参数，具体如下：

（1）to: Route：即将进入的目标路由对象。

（2）from: Route：当前导航正要离开的路由。

beforeEach 函数可以返回 false 值或一个路由地址，具体如下。

（1）false：取消当前导航。如果浏览器的 URL 改变（用户手动或者浏览器后退按钮），那么 URL 地址将重置到 from 路由对应的地址。

（2）一个路由地址：跳转到该路由地址，即当前的导航被中断，进行一个新的导航。

例如使用 beforeEach 函数检查用户是否登录，示例代码如下：

```
router.beforeEach(async (to, from) => {
    //在 ES7 标准中新增了 async 和 await 关键字，作为处理异步请求的一种解决方案
    if (
        //检查用户是否登录
        !isAuthenticated &&
        //避免无限重定向
        to.name !== 'Login'
    ) {
        //将用户重定向到登录页面
```

```
            return { name: 'Login' }
    }
})
```

在之前的 Vue Router 版本中，beforeEach 需要使用第三个参数 next，现在是一个可选的参数。next 参数的相关说明具体如下：

（1）next()：进行管道中的下一个钩子函数。如果全部钩子函数执行完毕，则导航的状态就是 confirmed（确认的）。

（2）next(false)：中断当前的导航。如果浏览器的 URL 改变（可能是用户手动或者浏览器后退按钮），那么 URL 地址会重置到 from 路由对应的地址。

（3）next('/')或者 next({ path: '/' })：跳转到一个不同的地址，当前的导航被中断，然后进行一个新的导航，可以向 next 传递任意位置对象，且允许设置诸如 replace: true、name: 'home' 之类的选项以及任何用在 router-link 的 to 属性或 router.push 中的选项。

（4）next(error)：如果传入 next 的参数是一个 Error 实例，则导航被终止，且该错误被传递给 router.onError()注册过的回调。

确保 next()函数在任何给定的前置钩子中被严格调用一次。它可以出现多次，但是只能在所有逻辑路径都不重叠的情况下，否则钩子永远都不会被解析或报错。例如，在用户未能验证身份时重定向到/login 的示例。

```
router.beforeEach((to, from, next) => {
    if (to.name !== 'Login' && !isAuthenticated)
        next({ name: 'Login' })
    else
        next()
})
```

❷ 全局解析钩子函数

在 Vue Router 中，使用 router.beforeResolve 注册一个全局解析钩子函数。与 router.beforeEach 类似，它在每次导航时都会触发，但是要确保在导航被确认之前，同时在所有组件内钩子函数和异步路由组件被解析之后，解析钩子函数被正确调用。例如，确保用户可以访问自定义的路由元信息（meta 属性）requiresCamera 的路由，具体代码如下：

```
router.beforeResolve(async to => {
    if (to.meta.requiresCamera) {
        try {
            await askForCameraPermission()
        } catch (error) {
            if (error instanceof NotAllowedError) {
                //处理错误，然后取消导航
                return false
            } else {
                //意料之外的错误，取消导航并把错误传给全局处理器
                throw error
            }
        }
    }
})
```

router.beforeResolve 是获取数据或执行任何其他操作（例如，用户无法进入页面，希望避免执行的操作）的理想位置。

❸ 全局后置钩子函数

在 Vue Router 中也可以使用 router.afterEach 注册全局后置钩子函数，该钩子函数不接收

next 参数，也不会改变导航本身，在跳转之后判断，对于分析、更改页面标题、声明页面等辅助功能很有用。示例代码如下：

```
router.afterEach((to, from) => {
    // ...
})
```

❹ 某个路由的钩子函数

顾名思义，它是写在某个路由中的函数，本质上跟组件内的函数没有区别。示例代码如下：

```
const routes = [
    {
        path: '/users/:id',
        component: UserDetails,
        beforeEnter: (to, from) => {
            //取消导航
            return false
        },
    },
]
```

路由的 beforeEnter 钩子函数，只在进入路由时触发，不会在 params、query 或 hash 改变时触发。例如，从/users/2 进入/users/3 或者从/users/2#info 进入/users/2#projects。它们只有在从一个不同的路由导航时才会被触发。

用户也可以将一个函数数组传递给路由 beforeEnter 钩子函数，这在为不同的路由重用钩子函数时很有用，示例代码如下：

```
function removeQueryParams(to) {
    if (Object.keys(to.query).length)
        return { path: to.path, query: {}, hash: to.hash }
}
function removeHash(to) {
    if (to.hash) return { path: to.path, query: to.query, hash: '' }
}
const routes = [
    {
        path: '/users/:id',
        component: UserDetails,
        beforeEnter: [removeQueryParams, removeHash],
    },
    {
        path: '/about',
        component: UserDetails,
        beforeEnter: [removeQueryParams],
    },
]
```

❺ 组件内的钩子函数

可以在路由组件内直接定义路由导航钩子函数 beforeRouteEnter、beforeRouteUpdate、beforeRouteLeave，具体示例代码如下：

```
const UserDetails = {
    template: '...',
    beforeRouteEnter(to, from) {
        //在渲染该组件的对应路由被验证前调用
        //不能获取组件实例'this'
        //因为当该钩子函数执行时组件实例还没被创建
```

```
    },
    beforeRouteUpdate(to, from) {
        //在当前路由改变,但是该组件被复用时调用
        //举例来说,对于一个带有动态参数的路径'/users/:id',在'/users/1'和'/users/2'
        //之间跳转的时候,由于渲染同样的'UserDetails'组件,所以组件实例会被复用,此钩
        //子函数在此情况下也被调用。因为在这种情况发生的时候,组件已经加载好了,该钩子
        //函数可以访问组件实例'this'
    },
    beforeRouteLeave(to, from) {
        //在导航离开渲染该组件的对应路由时调用
        //与'beforeRouteUpdate'一样,它可以访问组件实例'this'
    },
}
```

beforeRouteEnter 钩子函数不能访问 this,因为 beforeRouteEnter 在导航确认前被调用,所以即将登场的新组件还没被创建。用户可以通过传一个回调给 next()函数来访问组件实例。在导航被确认的时候执行回调,并且把组件实例作为回调方法的参数,具体示例代码如下:

```
beforeRouteEnter(to, from, next) {
    next(vm => {
        //通过'vm'访问组件实例
    })
}
```

beforeRouteEnter 是支持给 next()函数传递回调的唯一钩子函数。对于 beforeRouteUpdate 和 beforeRouteLeave 来说,this 已经可以使用了,所以不支持传递回调,也没有必要。示例代码如下:

```
beforeRouteUpdate(to, from) {
    //使用'this'
    this.name = to.params.name
}
```

beforeRouteLeave 钩子函数通常用来防止用户在还未保存修改前突然离开,导航可以通过返回 false 来取消离开操作。示例代码如下:

```
beforeRouteLeave(to, from) {
    const answer = window.confirm('真的离开? 你还没保存修改!')
    if (!answer)
        return false
}
```

本小节只是简单地介绍钩子函数的分类与定义,具体应用在 12.3.6 小节中。

▶ 12.3.5 路由元信息

用户有时可能希望将任意信息附加到路由上,例如过渡名称、访问路由权限等。这些工作可以通过接收属性对象的 meta 属性来实现,并且它可以在路由地址和导航守卫(路由钩子函数)中都能被访问到。路由的 meta 属性的配置示例如下:

```
const routes = [
    {
        path: '/posts',
        component: PostsLayout,
        children: [
            {
                path: 'new',
                component: PostsNew,
                //只有经过身份验证的用户才能创建帖子
```

```
                meta: { requiresAuth: true }
            },
            {
                path: ':id',
                component: PostsDetail
                //任何人都可以阅读帖子
                meta: { requiresAuth: false }
            }
        ]
    }
]
```

那么如何访问路由的 meta 属性呢？

首先将 routes 配置中的每个路由对象称为路由记录。路由记录可以是嵌套的，因此当一个路由匹配成功后，它可能匹配多个路由记录。

例如，根据上面的路由配置，/posts/new 这个 URL 将会匹配父路由记录（path: '/posts'）以及子路由记录（path: 'new'）。

一个路由匹配到的所有路由记录被暴露为 $route 对象的 $route.matched 数组，需要遍历这个数组来检查路由记录中的 meta 字段。Vue Router 还提供了一个 $route.meta 方法，它是一个非递归合并所有 meta 字段（从父字段到子字段）的方法。因此用户可以通过 $route.meta 方法简单地获取路由的 meta 属性值，具体示例代码如下：

```
router.beforeEach((to, from) => {
    //不是去检查每条路由记录
    //to.matched.some(record => record.meta.requiresAuth)
    if (to.meta.requiresAuth && !auth.isLoggedIn()) {
        //此路由需要授权，请检查是否已登录
        //如果没有，则重定向到登录页面
        return {
            path: '/login',
            //保存当前路由所在的位置，以便再回来
            query: { redirect: to.fullPath },
        }
    }
})
```

▶ 12.3.6 登录权限验证示例

视频讲解

登录权限验证实例的具体要求如下：

（1）在 App.vue 根组件中通过 <router-link> 访问登录页面组件 Login.vue、主页面组件 Main.vue 以及 Home.vue 组件。

（2）登录成功后才能访问主页面组件 Main.vue 和 Home.vue 组件。

（3）在 main.js 中使用路由钩子函数 beforeEach(to,from) 实现登录权限验证。

【例 12-5】登录权限验证实例。

其具体实现过程如下：

❶ 使用 Vue CLI 搭建基于 Router 功能的项目

参考 12.3.1 节使用 Vue CLI 搭建基于 Router 功能（如图 12.6 所示）的项目 login-validate。

❷ 完善 App.vue

完善项目的根组件 App.vue 的代码，具体如下：

```
<template>
    <nav>
```

```html
        <router-link to="/login">Login</router-link> |
        <router-link to="/main">Main</router-link> |
        <router-link to="/home">Home</router-link>
    </nav>
    <router-view/>
</template>
```

❸ 配置路由

在 src/router 目录的 index.js 文件中配置路由,需要登录验证的路由使用 meta 元信息标注。路由配置的具体代码如下:

```javascript
import { createRouter, createWebHistory } from 'vue-router'
import Login from '../views/LoginView.vue'
import Main from '../views/MainView.vue'
import Home from '../views/HomeView.vue'
const routes = [
    {
        path: '/login',
        name: 'Login',
        component: Login
    },
    {
        path: '/home',
        name: 'Home',
        component: Home,
        meta:{auth:true}
    },
    {
        path: '/main',
        name: 'Main',
        component: Main,
        meta:{auth:true}          //需要验证登录权限
    }
]
const router = createRouter({
    history: createWebHistory(process.env.BASE_URL),
    routes
})
export default router
```

❹ 验证登录权限

在配置文件 main.js 中,使用路由钩子函数 beforeEach(to,from)实现登录权限验证,具体代码如下:

```javascript
import { createApp } from 'vue'
import App from './App.vue'
import router from './router'      //导入 router 目录中的 index.js,路由的创建与配置
                                   //在该文件中
createApp(App).use(router).mount('#app')
// eslint-disable-next-line no-unused-vars
router.beforeEach(to,from)=>{      //提示未使用,ESLint 规则 no-unused vars 关闭为
                                   //eslint-disable-next-line
    //如果路由需要验证
    if(to.meta.auth){
        //对路由进行验证
        if (window.sessionStorage.getItem('user') == null) {
            alert("您没有登录,无权访问!")
            /*未登录则跳转到登录界面,
            query:{ redirect: to.fullPath}表示把当前路由信息传递过去方便登录后跳转回来*/
```

```
            return {
                path: 'login',
                query: {redirect: to.fullPath}
            }
        }
    }
})
```

❺ 新建登录组件 LoginView.vue

在 views 目录中新建登录组件 LoginView.vue，在该组件中使用 window.sessionStorage.setItem() 保存登录状态。LoginView.vue 的代码具体如下：

```
<template>
    <div>
        <h2>登录页面</h2>
        <form>
            用户名:<input type="text" v-model="uname" placeholder="请输入用户名"/>
            <br><br>
            密码：<input type="password" v-model="upwd" placeholder="请输入密码"/>
            <br><br>
            <button type="button" @click="login" :disabled="isDisable">登录</button>
            <button type="reset">重置</button>
        </form>
    </div>
</template>
<script>
export default {
    data () {
        return {
            isDisable:false,
            uname: '',
            upwd: ''
        }
    },
    methods: {
        login () {
            this.isDisable = true
            if (this.uname === 'zhangsan' && this.upwd == '123456') {
                alert('登录成功')
                //将成功登录的用户的信息保存到 session
                window.sessionStorage.setItem('user', this.uname)
                //到达成功之后的页面
                let path = this.$route.query.redirect
                this.$router.replace({path:path==='/'||path===undefined?
                '/main': path})
            }else {
                alert("用户名或密码错误！")
                this.isDisable = false
            }
        }
    }
}
</script>
```

❻ 新建主页面组件 MainView.vue

在 views 目录中新建主页面组件 MainView.vue，MainView.vue 的代码具体如下：

```
<template>
    <div>欢迎{{uname}}登录成功</div>
</template>
```

```
<script>
export default {
    data () {
        return {
            uname : window.sessionStorage.getItem('user')
        }
    }
}
</script>
```

❼ 测试运行

在 Terminal 终端输入 npm run serve 命令启动服务后，在浏览器的地址栏中访问"http://localhost:8080/"即可运行项目 login-validate。在登录界面中输入用户名 zhangsan、密码 123456，即可成功登录。登录成功后打开主页面组件，如图 12.20 所示。

图 12.20　主页面

读者在 login-validate 目录下执行 npm install 命令将自动安装所有依赖，然后执行 npm run serve 命令启动服务，即可运行项目。

12.4　setup 语法糖

在 setup 选项中，属性、方法等必须通过 return 返回暴露出来，然后在 template 中才能使用，这很不方便。Vue.js 3.2 增加了 setup 语法糖，即在<script>标签中添加 setup。

在使用 setup 语法糖时，组件只需要引入不需要注册，属性和方法不需要通过 return 返回，不用再写 setup 函数，也不用写 export default 默认输出。

setup 语法糖是在单文件组件中使用组合式 API 的编译时语法糖，与普通的<script>语法相比，它具有更多优势：

（1）更少的 template 内容，更简洁的代码。

（2）能够使用纯 TypeScript 声明 props 和抛出事件。

（3）更好的运行时性能（其 template 将被编译成与其同一作用域的渲染函数，没有任何的中间代理）。

（4）更好的 IDE 类型推断性能（减少语言服务器从代码中抽离类型的工作）。

许多基于 Vue.js 的开源 UI 组件库的官方示例都是使用 setup 语法糖编写的。为了方便学习 Vue UI 组件库，本节事先学习 setup 语法糖的知识。

▶ 12.4.1　属性与方法的绑定

在使用 setup 语法糖时，属性和方法不需要通过 return 返回暴露出来，可以直接在 template 中使用。下面通过一个实例讲解使用 setup 语法糖时如何进行属性与方法的绑定。因为 setup

语法糖是在单文件组件中使用组合式 API 的编译时语法糖,所以本小节实例在使用 Vue CLI 搭建的 Vue.js 项目的单文件组件中演示。

【例 12-6】使用 setup 语法糖时如何进行属性与方法的绑定。

其具体步骤如下:

❶ 使用 Vue CLI 搭建基于 Router 功能的 Vue.js 项目

参考 12.3.1 节,使用 Vue CLI 搭建基于 Router 功能的 Vue.js 项目 setup-sugar。

❷ 修改根组件 App.vue

在 setup-sugar 的根组件 App.vue 中定义一个非响应式基本数据属性 firstNumber、一个响应式基本数据属性 secondNumber、一个响应式对象(复杂数据)属性 thirdNumber 和一个改变属性的方法 changeNumber,并使用 setup 语法糖进行属性与方法的绑定。App.vue 的代码具体如下:

```
<template>
    <div>firstNumber 是一个基本类型数据: {{firstNumber}}</div>
    <div>secondNumber 是一个响应式且可改变的基本数据对象: {{secondNumber}}</div>
    <div>thirdNumber 是一个响应式且可改变的对象数据: {{thirdNumber}}</div>
    <div><button @click="changeNumber">改变数据</button></div>
</template>
<!--只需添加 setup 属性-->
<script setup>
//注意,和 setup 函数一样,使用 setup 语法糖时没有 this
import { ref,reactive } from 'vue'
//不使用 ref 定义的变量
let firstNumber = 100
//使用 ref 定义的变量,是一个响应式且可改变的基本数据对象(也可以是对象)
let secondNumber = ref(100)
//使用 reactive 定义的变量,是一个响应式且可改变的对象
let thirdNumber = reactive({
    uname: '陈恒',
    age: '58'
})
//定义一个无参数的方法,在使用 setup 语法糖时,方法也被认为属性
let changeNumber = ()=> {
    firstNumber++
    //在模板中使用时,它会自动开箱,无须在模板内额外书写.value
    secondNumber.value++
    thirdNumber.age++
}
</script>
```

❸ 测试与运行

首先在 setup-sugar 项目的 Terminal 终端输入 npm run serve 命令启动服务,然后通过浏览器的地址栏访问"http://localhost:8080/"运行项目 setup-sugar 的根组件 App.vue,页面显示效果如图 12.21 所示。

图 12.21 使用 setup 语法糖进行属性与方法的绑定

12.4.2 路由

在 12.3 节学习路由知识时,大家知道可以使用 this.$router.push 进行编程式导航,但使用 setup 语法糖时并没有 this,如何进行编程式导航呢?具体做法是首先使用 import 语句引入 useRouter,即 import {useRouter} from 'vue-router';然后使用 useRouter 创建路由对象,即 const router = useRouter();最后使用路由对象 router 进行编程式导航。

下面通过一个实例讲解使用 setup 语法糖时如何进行编程式导航。

【例 12-7】使用 setup 语法糖时进行编程式导航。

该实例在例 12-6 的基础上完成,其他具体步骤如下:

❶ 修改根组件 App.vue

在 setup-sugar 的根组件 App.vue 中首先使用 import 语句引入 useRouter,然后使用 useRouter 创建路由对象 router,最后使用路由对象 router 进行编程式导航。App.vue 的代码如下:

```
<template>
    <div><button @click="goToNewPage">去 Vue 主页 home</button></div>
    <router-view/>
</template>
<script setup>
//记住是在 vue-router 中引入 useRouter
import {useRouter} from 'vue-router'
const router = useRouter()
let goToNewPage = ()=>{
    let params = {uname:'123', pwd:'abc'}
    router.push({name: 'home', state: params})
}
</script>
```

❷ 修改路由配置

修改 setup-sugar 的路由配置文件/router/index.js,实现编程式导航。index.js 的代码具体如下:

```
...
const routes = [
    {
        path: '/home',
        name: 'home',
        component: HomeView
    },
...
```

❸ 修改 HomeView.vue

在/views/HomeView.vue 文件的<template>中使用 setup 语法糖简化程序。HomeView.vue 的代码具体如下:

```
<template>
    <div class="home">
        <img alt="Vue logo" src="../assets/logo.png">
        <br>uname: {{historyParams.uname}}
        <br>uname: {{historyParams.pwd}}
        <br>
        <HelloWorld msg="Welcome to Your Vue.js App"/>
    </div>
</template>
<script setup>
//组件只需要引入不需要注册
```

```
import HelloWorld from '@/components/HelloWorld.vue'
//接收传递过来的数据
const historyParams = history.state
</script>
```

❹ 测试运行

首先在 setup-sugar 项目的 Terminal 终端输入 npm run serve 命令启动服务，然后通过浏览器的地址栏访问"http://localhost:8080/"运行项目 setup-sugar 的根组件 App.vue，接着单击"去 Vue 主页 home"按钮通过编程式导航打开 HomeView.vue，HomeView.vue 的显示效果如图 12.22 所示。

图 12.22　HomeView.vue 的显示效果

在例 12-7 中，使用 setup 语法糖时，父组件 HomeView 只需要引入子组件 HelloWorld，并没有注册子组件 HelloWorld。另外，在子组件 HelloWorld 中使用 export default {name: 'HelloWorld', props: { msg: String }}接收父组件 HomeView 传递过来的值。此时无法在子组件 HelloWorld 中使用 setup 语法糖，即无法获取 props、emit 等，幸运的是 setup 语法糖提供了新的 API（defineProps、defineEmits、defineExpose 等）供用户使用。下面讲解 defineProps、defineEmits、defineExpose 等 API 在组件传值中的具体应用。

▶ 12.4.3　组件传值

在子组件中，defineProps 用来接收父组件传来的 props；defineEmits 用来声明触发的事件；defineExpose 用来导出数据，暴露于父组件（在父组件中通过 ref='xxx'的方法来获取子组件实例）。下面通过一个具体实例讲解 defineProps、defineEmits、defineExpose 等 API 在组件传值中的具体应用。

【例 12-8】defineProps、defineEmits、defineExpose 等 API 在组件传值中的具体应用。

该实例在例 12-7 的基础上完成，其他具体步骤如下：

❶ 修改 HomeView.vue

在/views/HomeView.vue 父组件中引入子组件/components/HelloWorld.vue，并传值给子组件，同时接收子组件传递的数据与获得子组件的属性值。HomeView.vue 的代码具体如下：

```
<template>
    <div>
        <!--myAdd 和 myDel 与子组件中使用 defineEmits 声明向父组件抛出的事件名称相同-->
```

```
            <HelloWorld :info="msg" constv="88" @myAdd="myAddAction" @myDel=
"myDelAction" ref="comRef"/>
            <button @click="getSon">获取子组件的属性值</button>
    </div>
</template>
<script setup>
//组件只需要引入不需要注册
import HelloWorld from '@/components/HelloWorld.vue'
import { ref } from 'vue'
const msg = "传给子组件"
//接收子组件传递过来的数据
const myAddAction= (msg) => {
    console.log('单击子组件的新增按钮传值：', msg)
}
//接收子组件传递过来的数据
const myDelAction= (msg) => {
    console.log('单击子组件的删除按钮传值：', msg)
}
const comRef = ref()
//获取子组件的属性值
const getSon = () => {
    console.log('获得子组件中的性别：', comRef.value.prop1)
    console.log('获得子组件中的其他信息：', comRef.value.prop2)
}
</script>
```

❷ 修改 HelloWorld.vue

在/components/HelloWorld.vue 子组件中使用 defineProps 接收父组件 HomeView.vue 的传值，使用 defineEmits 声明向父组件抛出的自定义事件，使用 defineExpose 将数据暴露于父组件。HelloWorld.vue 的代码具体如下：

```
<template>
    <div>
        <h4>接收父组件传值</h4>
        <h4>info: {{ info }}</h4>
        <h4>constv: {{ constv }}</h4>
    </div>
    <div>
        <button @click="add">新增</button>  
        <button @click="del">删除</button>
    </div>
    <div>
        <h4>性别：{{prop1}}</h4>
        <h4>其他信息：{{prop2}}</h4>
    </div>
</template>
<script setup>
import {ref, reactive, defineProps, defineEmits, defineExpose } from 'vue'
const prop1 = ref('男')
const prop2 = reactive({
    uname: '陈恒',
    age: 88
})
//使用 defineExpose 将数据暴露于父组件
defineExpose({
    prop1,
    prop2
})
```

```
//接收父组件传值
defineProps({
    info: {
        type: String,
        default: '-----'
    },
    constv: {
        type: String,
        default: '0'
    }
})
//使用 defineEmits 声明向父组件抛出的自定义事件
const myemits = defineEmits(['myAdd', 'myDel'])
const add = () => {
    //通过抛出 myAdd 事件向父组件传值
    myemits('myAdd', '传向父组件的新增数据')
}
const del = () => {
    myemits('myDel', '传向父组件的删除数据')
}
</script>
<style scoped>
h3 {
    margin: 40px 0 0;
}
</style>
```

❸ 测试运行

首先在 setup-sugar 项目的 Terminal 终端输入 npm run serve 命令启动服务，然后通过浏览器的地址栏访问"http://localhost:8080/"运行项目 setup-sugar 的根组件 App.vue，接着单击"去 Vue 主页 home"按钮打开 HomeView.vue，HomeView.vue 的显示效果如图 12.23 所示。

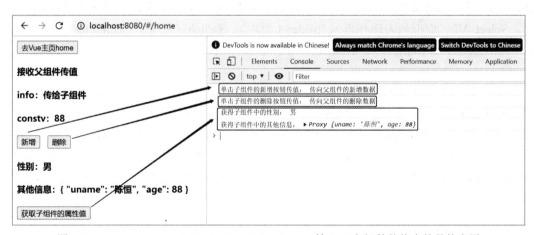

图 12.23　defineProps、defineEmits、defineExpose 等 API 在组件传值中的具体应用

12.5　Element Plus UI 组件库

扫一扫

视频讲解

Element Plus 是一套为开发者、设计师和产品经理准备的基于 Vue 3 的组件库，提供了配套设计资源，简化了常用组件的封装，大大降低了开发难度，帮助使用者的网站快速成型。Element Plus 目前还处于快速开发迭代中，官方文档可以参见官网（https://element-plus.gitee.io/zh-CN/）。

12.5.1 Element Plus 的安装

建议开发者使用包管理器（例如 NPM、Yarn、PNPM）安装 Element Plus，能更好地和 Vite、webpack 等打包工具配合使用，根据实际需要选择一个自己喜欢的包管理器，具体安装命令如下：

```
# NPM
$ npm install element-plus --save
# Yarn
$ yarn add element-plus
# PNPM
$ pnpm install element-plus
```

12.5.2 Element Plus 组件的介绍

Element Plus 组件主要包括 Basic（基础）、Form（表单）、Data（数据展示）、Navigation（导航）、Feedback（反馈）五大类组件，每一类又包含很多组件。下面简单介绍每类组件中所包含的组件。

❶ Basic（基础）

Basic（基础）类组件包括 Button（按钮）、Border（边框）、Color（色彩）、Container（布局容器）、Icon（图标）、Layout（布局）、Link（链接）、Scrollbar（滚动条）、Space（间距）、Typography（排版）等组件。

❷ Form（表单）

Form（表单）类组件包括 Cascader（级联选择器）、Checkbox（多选框）、ColorPicker（颜色选择器）、DatePicker（日期选择器）、DateTimePicker（日期时间选择器）、Form（表单）、Input（输入框）、Input Number（数字输入框）、Radio（单选框）、Rate（评分）、Select（选择器）、Select V2（虚拟列表选择器）、Slider（滑块）、Switch（开关）、TimePicker（时间选择器）、TimeSelect（时间选择）、Transfer（穿梭框）、Upload（上传）等组件。

❸ Data（数据展示）

Data（数据展示）类组件包括 Avatar（头像）、Badge（徽章）、Calendar（日历）、Card（卡片）、Carousel（走马灯）、Collapse（折叠面板）、Descriptions（描述列表）、Empty（空状态）、Image（图片）、Infinite Scroll（无限滚动）、Pagination（分页）、Progress（进度条）、Result（结果）、Skeleton（骨架屏）、Table（表格）、Virtualized Table（虚拟化表格）、Tag（标签）、Timeline（时间线）、Tree（树形控件）、TreeSelect（树形选择）、Tree V2（虚拟化树形控件）等组件。

❹ Navigation（导航）

Navigation（导航）类组件包括 Affix（固钉）、Backtop（回到顶部）、Breadcrumb（面包屑）、Dropdown（下拉菜单）、Menu（菜单）、Page Header（页头）、Steps（步骤条）、Tabs（标签页）等组件。

❺ Feedback（反馈）

Feedback（反馈）类组件包括 Alert（提示）、Dialog（对话框）、Drawer（抽屉）、Loading（加载）、Message（消息提示）、MessageBox（消息弹框）、Notification（通知）、Popconfirm（气泡确认框）、Popover（气泡卡片）、Tooltip（文字提示）等组件。

▶ 12.5.3 Element Plus 组件的应用

Element Plus 提供了一套常用的图标，这些图标还没有默认添加到组件中，需要另外安装才能使用，安装方式与 Element Plus 的安装方式相同。

下面通过实例讲解如何在使用 Vue CLI（Vue 脚手架）搭建的 Vue.js 项目中应用 Element Plus 组件。

【例 12-9】在使用 Vue CLI 搭建的 Vue.js 项目中应用 Element Plus 组件。

其具体步骤如下：

❶ 使用 Vue CLI 搭建 Vue.js 项目

参考 12.3.1 节，使用 Vue CLI 搭建 Vue.js 项目 elementplus-vue。

❷ 安装 Element Plus 和 @element-plus/icons-vue

使用 VS Code 打开项目 elementplus-vue，并进入 Terminal 终端，依次执行 "npm install element-plus --save" 和 "npm install @element-plus/icons-vue" 命令，进行 Element Plus 和 @element-plus/icons-vue 的安装。

❸ 引入 Element Plus 组件并注册图标组件 ElementPlusIconsVue

在 elementplus-vue 的 main.js 文件中完整引入 Element Plus，并注册图标组件 ElementPlusIconsVue，main.js 的代码具体如下：

```js
import { createApp } from 'vue'
import ElementPlus from 'element-plus'
import 'element-plus/dist/index.css'
import * as ElementPlusIconsVue from '@element-plus/icons-vue'
import App from './App.vue'
const app = createApp(App)
//注册所有图标
for (const [key, component] of Object.entries(ElementPlusIconsVue)) {
    app.component(key, component)
}
//使用 ElementPlus
app.use(ElementPlus).mount('#app')
```

❹ 在根组件 App.vue 中使用 Element Plus 组件

在 elementplus-vue 的根组件 App.vue 中使用 Element Plus 组件实现如图 12.24 所示的功能。

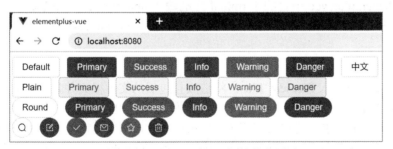

图 12.24 Element Plus 组件的应用

App.vue 的代码具体如下：

```html
<template>
    <el-row class="mb-4">
        <el-button>Default</el-button>
        <el-button type="primary">Primary</el-button>
        <el-button type="success">Success</el-button>
        <el-button type="info">Info</el-button>
```

```html
            <el-button type="warning">Warning</el-button>
            <el-button type="danger">Danger</el-button>
            <el-button>中文</el-button>
        </el-row>
        <el-row class="mb-4">
            <el-button plain>Plain</el-button>
            <el-button type="primary" plain>Primary</el-button>
            <el-button type="success" plain>Success</el-button>
            <el-button type="info" plain>Info</el-button>
            <el-button type="warning" plain>Warning</el-button>
            <el-button type="danger" plain>Danger</el-button>
        </el-row>
        <el-row class="mb-4">
            <el-button round>Round</el-button>
            <el-button type="primary" round>Primary</el-button>
            <el-button type="success" round>Success</el-button>
            <el-button type="info" round>Info</el-button>
            <el-button type="warning" round>Warning</el-button>
            <el-button type="danger" round>Danger</el-button>
        </el-row>
        <el-row>
            <el-button :icon="Search" circle />
            <el-button type="primary" :icon="Edit" circle />
            <el-button type="success" :icon="Check" circle />
            <el-button type="info" :icon="Message" circle />
            <el-button type="warning" :icon="Star" circle />
            <el-button type="danger" :icon="Delete" circle />
        </el-row>
</template>
<script setup>
import {
    Check,
    Delete,
    Edit,
    Message,
    Search,
    Star,
} from '@element-plus/icons-vue'
</script>
```

❺ 测试运行

首先在 elementplus-vue 项目的 Terminal 终端输入 npm run serve 命令启动服务，然后通过浏览器的地址栏访问 "http://localhost:8080/" 运行项目 elementplus-vue。

▶ 12.5.4 按需引入 Element Plus

在例 12-9 中完整引入 Element Plus，将造成打包后的文件很大并且首页加载稍慢的后果。如果用户对打包后的文件大小不是很在乎，那么使用完整引入会更方便。但是，如果用户对打包后的文件大小很在乎，建议按需引入 Element Plus。那么如何按需引入 Element Plus 呢？推荐开发者使用额外的插件 unplugin-vue-components 和 unplugin-auto-import 自动引入要使用的组件。

下面在例 12-9 的基础上讲解如何使用插件 unplugin-vue-components 和 unplugin-auto-import 自动按需引入要使用的 Element Plus 组件，具体步骤如下：

❶ 安装插件

打开项目 elementplus-vue 的 Terminal 终端，执行 "npm install -D unplugin-vue-components unplugin-auto-import" 命令，安装 unplugin-vue-components 和 unplugin-auto-import 插件。

❷ 配置插件

打开项目 elementplus-vue 的配置文件 vue.config.js，并配置 unplugin-vue-components 和 unplugin-auto-import 插件，具体配置代码如下：

```js
const { defineConfig } = require("@vue/cli-service");
const AutoImport = require("unplugin-auto-import/webpack");
const Components = require("unplugin-vue-components/webpack");
const { ElementPlusResolver } = require("unplugin-vue-components/resolvers");
module.exports = defineConfig({
    transpileDependencies: true,
    configureWebpack: {
        plugins: [
            AutoImport({
                resolvers: [ElementPlusResolver()],
            }),
            Components({
                resolvers: [ElementPlusResolver()],
            }),
        ],
    },
});
```

❸ 修改 main.js 文件，去除完整引入

修改项目 elementplus-vue 的 main.js 文件，去除完整引入。main.js 的具体代码如下：

```js
import { createApp } from 'vue'
//import ElementPlus from 'element-plus'   //完整引入
//import 'element-plus/dist/index.css'
//引入图标
import * as ElementPlusIconsVue from '@element-plus/icons-vue'
import App from './App.vue'
const app = createApp(App)
//注册所有图标
for (const [key, component] of Object.entries(ElementPlusIconsVue)) {
    app.component(key, component)
}
//使用 ElementPlus
//app.use(ElementPlus).mount('#app')   //完整引入
app.mount('#app')
```

经过上述 3 个步骤即可按需自动引入 Element Plus 组件，使用起来也极其方便，因此推荐开发者使用此方式按需自动引入 Element Plus 组件。

❹ 测试运行

首先在 elementplus-vue 项目的 Terminal 终端输入 npm run serve 命令启动服务，然后通过浏览器的地址栏访问 "http://localhost:8080/" 运行项目 elementplus-vue，运行效果如图 12.25 所示。

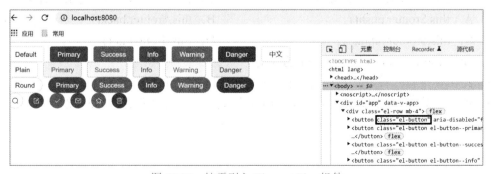

图 12.25　按需引入 Element Plus 组件

从图 12.25 可以看出，在按需自动引入 Element Plus 组件时自动引入了 ElButton 组件，并引入了组件样式，不过当需要使用命令的方式创建 Element Plus 组件时还需要以 import 的方式引入。示例代码如下：

```
<template>
    <el-button v-on:click="gogo">Round</el-button>
</template>
<script setup>
import { ElMessage } from 'element-plus'
const gogo = () => {
    ElMessage.warning('注意注意！')
}
</script>
```

本章小结

本章主要介绍了 Vue CLI、Vue Router 插件以及 Element Plus UI 组件库等知识，希望读者重点学习 Vue CLI 与 Vue Router 插件的用法，为下一章综合项目实战夯实基础。

习题 12

1. 在 Vue Router 中，下列能够设置页面导航的是（　　）。
   A. \<router-link\>   B. \<router-view\>   C. \<router-a\>   D. \<router-nav\>
2. 下列能够显示或渲染匹配到的路由信息的标记是（　　）。
   A. \<router-link\>   B. \<router-view\>   C. \<router-v\>   D. \<router-vue\>
3. 下列能够正确表示跳转到 user/chenheng 的路由是（　　）。
   A. {path: '/user', name: 'user', component: UserView}
   B. {path: '/user/:uname', name: 'user', component: UserView}
   C. {path: '/user/uname', name: 'user', component: UserView}
   D. {path: '/user/name', name: 'user', component: UserView}
4. 定义命名路由使用的属性是（　　）。
   A. component   B. path   C. meta   D. query
5. 定义路由元信息使用的属性是（　　）。
   A. component   B. path   C. meta   D. query
6. 在编程式导航中，能够跳转到新路由并且在历史记录中添加一条新记录的方法是（　　）。
   A. this.$router.push()   B. this.$router.back()
   C. this.$router.replace()   D. this.$router.go()
7. 简述 route、routes 以及 router 的区别。
8. 路由传参有几种方式？如何接收路由传递的参数？请举例说明。
9. 如何安装 Vue CLI？请使用 Vue CLI 的界面引导的方式创建 Vue.js 项目。

第 13 章　电子商务平台的设计与实现（Spring Boot + Vue.js 3 + MyBatis-Plus）

本章为在线电子文档，阅读方式如下。

➢ 移动端阅读方式：通过移动设备扫描二维码，在移动端进行阅读。

➢ 计算机端阅读方式：通过浏览器登录网站 https://read.wqyunpan.com，在"我的图书"页面中进行阅读。

扫一扫

在线文档

附录 A 实验指导

本课程实验教学的目的在于提高学生的实际动手能力，以期更好地掌握 Spring、Spring MVC、Spring Boot、MyBatis、MyBatis-Plus 以及 Vue.js 等框架的相关原理、技术及方法，为后期毕业设计及以后从事 Java 相关开发打下坚实的基础。

与本书对应的实验（含实战）共 11 个（30 学时），分别为 Spring 开发环境的构建、基于注解的依赖注入、Controller 接收请求参数（注册与登录系统）、自定义 Starters、JSR 380 验证（表单验证）、基于 Thymeleaf 模板引擎的 Spring Boot Web 开发、基于 Spring Boot + Thymeleaf + MyBatis-Plus 的注册与登录系统、自定义组件（正整数数字输入框组件）、组件通信（战情互报）、基于 Spring Boot + MyBatis-Plus + Vue.js 的电子商务平台的设计与实现、分组（1～3 人）设计并实现一个基于 Spring Boot + MyBatis-Plus + Vue.js 的企业级应用系统。

本课程实验教学的要求如下。
（1）根据实验指导书中规定的内容在要求的时间内完成实验。
（2）成功部署实验中要求实现的内容，并能演示。
（3）提交实验中规定的开发内容的核心代码。
（4）提交实验报告。

读者扫描下方二维码，可以在线学习。

扫一扫

在线文档